MATRICES
and their
ROOTS

MATRICES and their ROOTS

A Textbook of Matrix Algebra

A R G Heesterman
University of Birmingham

World Scientific
Singapore • New Jersey • London • Hong Kong

Published by

World Scientific Publishing Co. Pte. Ltd.

P O Box 128, Farrer Road, Singapore 9128

USA office: 687 Hartwell Street, Teaneck, NJ 07666

UK office: 73 Lynton Mead, Totteridge, London N20 8DH

MATRICES AND THEIR ROOTS
— A Textbook of Matrix Algebra

ISBN 981-02-0395-0
      981-02-0396-9 (pbk)

Printed in Singapore by Utopia Press.

CONTENTS

INTRODUCTION

This is a textbook of matrix algebra. It addresses itself primarily to two groups of students who need mathematics in an applied context: undergraduates starting at the beginning, and postgraduates who need reference-material, but who, not being mathematics specialists, nevertheless are not best served by an ordinary mathematics textbook, which will generally be at a higher level of abstraction.

The relevant areas of application are, as far as my own background is concerned, economics, econometrics and mathematical programming, but I am well aware of the fact that engineering is the other major area of application.

The book starts at the very beginning, covering the resolution of linear equations, matrix multiplication and addition, block-equations, matrix inversion and rank. It gives full proofs throughout, and is illustrated with a large number of numerical examples, re-inforcing the student's grasp of the topics covered by exercises and corresponding answersheets.

In relation to proofs, I have given some thought to the question of leaving a number of them out, but decided otherwise for two main reasons. One reason is simply that understanding the why helps remembering the what. A theorem which is simply stated, is soon forgotten. The other is, that one has to find, even when teaching students for whom mathematics is a tool for purposes of application, some degree of balance between educating and training, and there is no education of mathematics itself without proofs and some idea how they are built up.

Besides the examples and answersheets in the printed text, numerical illustration is also provided by the companion tutorial program-disk, which will do various calculations, with optional levels of additional information/explanation on the screen. The program 'Illustrate' will run on any IBM-compatible micro-computer.

Chapters I-V are (possibly first year) undergraduate material. For some of the better prepared A-level holders, much of Chapter I-IV in particular is more of a refresher, and provides a common framework of notation and formal theory with the rest of the book.

Indeed, the book begins, and again reinforces something which one would actually hope to be known to a reasonable numerate student: the resolution of a system of linear equations by the elimination method.

There are two main reasons for including, and even repeating this rather elementary material. There are students who actually need a brief on how to solve three unknown variables from a system of three linear equations in a systematic way. For students of average and above-average numeracy, a further educational reason is that familiarity with the particular version of the elimination method which is used in a large part of the book, is essential for the appreciation of some of its implications. Chapter V on rank and Chapter VI on definiteness both contain proofs, which refer to a particular tabulation-arrangement.

While Chapters I to IV are fairly elementary, V begins to introduce a somewhat higher level of abstraction, and proofs by recursive induction. Chapters VI -XI cover material, which is in terms of its substance generally classified as more advanced: definiteness, the characteristic equations of both symmetric and of non-symmetric matrices, similarity and diagonalization, triangularization, the inversion of complex matrices, the Moore-Penrose inverse, principal components, rotation-operators and the properties of non-negative square matrices.

Within these later chapters, there is still some degree of gradual increase in sophistication.

Chapter VI deals with definiteness without using latent roots.

The exposition of latent roots and characteristic vectors is split into two chapters, VII and VIII, one meant for undergraduates and the other not. The seventh chapter contains the basics and does not go much beyond the definition of characteristic vectors and the derivation of the characteristic equation. It does contain an introduction to complex numbers, the main purpose of which is to provide a background for showing that symmetric matrices have only real roots.

The development of theory in the seventh chapter is largely

limited to what is prerequisite for Chapter IX, which deals with the symmetric Eigenvalue problem.

Chapter X (geometrical interpretations) falls itself into a fairly elementary part (starting with elementary coordinate geometry), and a somewhat more advanced part, rotation-operators and n-dimensional geometry. The n-dimensional concept of area is obviously relevant in econometrics, where it arises as the ratio between the density functions in the spaces of observed variables and structural residuals.

The eleventh chapter deals with non-negative square matrices. It briefly deals with the relationship with Input / Output analysis, but only insofar as this concerns the specific mathematical properties of this type of matrices. Non-economists, who are not concerned with that particular application, will obviously skip those few passages. As far as economists, who are indeed concerned with the I/O analysis application are concerned, it should be stressed that this chapter is not meant as an introduction to the basic accounting framework. There are other textbooks which cover that, including chapter III of mine own [22].

There are two sections of the book, which contain a possibly novel contribution to the subject.

In section 7.7 it is shown that, in the special cases of triangular or symmetric matrices, a repeated root is associated with the separate vanishment of each of the principal minors of $[I\lambda-A]$, which are of the appropriate order, not just the sum of a group. For a triangular matrix that is obvious, but for a symmetric matrix that is not so immediately obvious. Yet I have not found that theorem anywhere else.

Section 11.5 identifies the structure of a matrices which are non-negative and in-decomposable, but not also semi-positive.

However, it should be stressed that the main purpose of the book is to provide briefing material for students and research workers.

One other issue which is useful to discuss here, is the not entirely standard approach to defining the inverse. There is, in the literature generally, a development away from starting with determinants, and it probably is now conventional to define the inverse as $\mathbf{A}\,\mathbf{B} = \mathbf{B}\,\mathbf{A} = \mathbf{I}$.

I went one step further, and put the definition of the inverse in the context of systems of equations. The two main advantages of that definition are its more straightforward integration with what I reckon to be the most practical method of its calculation (elementary row operations), and the almost seamless transition from the ordinary inverse to the Moore Penrose generalized inverse.

Here, it should again be made clear that the definition of the generalized inverse as used in this book, which includes the ordinary inverse by straightforward application of the more general definition, does not involve the more usual artefact of taking a limit.

As the book contains both elementary undergraduate material and more complicated material, some briefing as to how it could best be used in undergraduate teaching is in order. This is obviously up to individual lecturers, but I would, following my own practice suggest the following:

Start at the beginning with Chapter I, follow through section by section until section 2.7. For students who are also, or will be doing econometrics, pick up section 2.8 later on, when they have assimilated section 2.7 and generally gained some confidence. For students for whom this section has no practical relevance it could possibly be skipped. This is probably also the right time for the first feedback-class, making sure students know their way with ILLUSTRATE, and generally verifying that they have followed.

In Chapter III, go through sections 3.1 and 3.2, leaving the rest for reference, i.e. explain the terminology, with a reference to Chapter III, if and when one comes to apply it. By that time students should be able to follow such relatively elementary material on their own.

Carry on with chapter IV, initially again section by section. The next stage up in the level of abstraction comes at the start of section 4.7, on block-elimination. As this technique is extensively used in the rest of the book, it is <u>not</u> recommended to skip this section. But it is a suitable point to have a bit of a pause, either simply by timing another class feedback session that way, or by taking up section 2.8.

Section 4.8, on the differentiation of the inverse, can probably be skipped in the context of undergraduate teaching. (If repeated roots of symmetric matrices -the vanishment of each separate $|I\lambda_j - A_{11}|$- in section 7.7 are to be covered, the proof would then

have to be skipped: differentiation of an inverse is essential
there.)

In chapter V, it is again recommended to start with following the
material section for section, but, in part depending on the time
available, to skip some of the proofs. I would, however, urge
that this should not go as far as skipping all proofs at this
level: it is the first time in the book that proofs by recursive
induction are introduced. The other bit of non priority material
in this chapter is section 5.9 on the all-integer elimination
method. This section has been included, largely to emphasize the
equivalence of the inverse, as defined in this book (= as far as
calculation is concerned, calculated by row operations), and the
inverse, defined as the adjoint divided by the determinant.

From Chapter VI onwards, the question of selecting particular
chapters and sections begins to depend increasingly on the areas
of application, and obviously on the available time. For students
of econometrics and/or quadratic optimization, definiteness is
essential, and I would urge Chapter VI to be covered in its
entirety, including the proofs, which are not very complicated.

Chapter VII and VIII are in a sense one chapter. Because
mathematics is built up consecutively, it is inevitable that
earlier sections are on the whole the ones which are the easiest
to follow, and later sections more advanced. Splitting this
amount of material in two was a requirement, if only for reasons
of convenient editing, but if one is looking for a point where to
cut off, it does not have to be at the start of section 8.1. If
an introduction to latent roots is all that is required, 7.1, 7.4
and 7.5. form a coherent whole.

This still gives enough introduction to this material, to
properly follow chapter IX on symmetric eigenvalue problems.
(There are backward references to section 7.6 in chapter IX, but
these are not so fundamental as the very notion that roots also
could be complex.).

Most of Chapter VIII should be considered as reference material
for more advanced students, not material to be taught to
undergraduates as their first introduction to matrix algebra, but
section 8.7 covers idempotent matrices, a topic which is often
applied in undergraduate econometrics courses. Provided the
proofs are skipped, it can be taken on its own.

Chapter IX on symmetric eigenvalue problems, contains its own introduction to latent roots, to the benefit of readers who have not or barely read any of either VII or VIII, but it is necessary to point out the limitations of such a procedure: Chapter IX explains the symmetric eigenvalue problem as a particular case of the general eigenvalue problem, and lists the <u>simplifications</u> which then arise. People who have read only section 9.1 and no other introduction to some of the complications which can arise for <u>non-symmetric</u> matrices, will have little idea, why that are special properties of symmetric matrices only.

The material in chapter IX falls fairly clearly in two parts: sections 9.1 to 9.4 deal with the latent roots of symmetric matrices as such, and their relation to definiteness, whereas section 9.5 really is the first preparation towards dealing with principal components and the Moore-Penrose inverse. Those topics form the rest of Chapter IX. It follows, that if one is looking for a natural cut-off point of a course, the end of section 9.4 is such a point.

Chapter X starts quite elementary, and covers some topics for which it is useful for students to be aware of, but unless one is dealing with people who need a brief in elementary two-dimensional coordinate geometry, (section 10.1), its systematic use as course-material at undergraduate level is not recommended: it climbs quickly, and gives the full hog, including the use of complex numbers in rotation-operators.

Chapter XI, on non-negative square matrices is not recommended for normal undergraduate teaching either: it is relevant only for people who are dealing with matrices of that type.

ILLUSTRATE

The companion tutorial program ILLUSTRATE provides numerical illustration, as well as a limited amount of relevant text-display, which will come up, at certain points as controlled by the program, to which particular text-displays were felt to be useful and relevant.

The display of real numbers is in 'fixed point' format with 2 decimal digits after the decimal point as default, but 1, or 3, 4, or 5 decimals can be specified. If desired, one can <u>key in</u> numbers with more (or fewer) digits after the decimal point, and the number as actually keyed in is used in the calculations. It is also possible to key in as a 'real number' a square root of a number, which may, or may not be preceded by a multiplier. The letter 'r' indicates that. Both parts of such a 'number' must be a valid real number e.g. 0.5r2 is accepted, as is -0.5r2, but 1/2r2 is not.

ILLUSTRATE is <u>not</u> a general calculation-package, on account of the following limitations:

Tableau-matrices are arrays for which the pre-set maximum size is 10 by 10. The final version of the program will be supplied without range-check, and once-in-a-while one may get away with handling a system above that size, but writing outside the pre-set array-sizes usually leads to unpredictable and catastrophic malfunctioning of the program, and in the standard case of input of a tableau-matrix ILLUSTRATE's input-output checking does not permit that anyhow. In practice, the screen-display only meaningfully copes with 3, at best 4 variables, and bigger tableaus, especially if more than 2 decimals are specified, will result either in a jumbled-up screen, or in a refusal to display. (There are some built-in checks against that kind of mis-use.)

There is an obvious trade-off between the order of the specified problem, and the number of decimal digits asked for.

The program should normally run on any IBM compatible micro-computer, but in case of any problems, some more strictly computer-orientated information, as provided in the file ILLUSTRATE.DOC on the diskette, may be useful.

We now list the options offered, and their capabilities, in somewhat more detail, as is possible on the screen-displayed menu.

MULTR   multiply two matrices.

               optional facility:
               Illustrate the calculation of individual elements.

SOLV   solve or part-resolve the system $\mathbf{A}\,\mathbf{x} = \mathbf{B}\,\mathbf{y}$.

               m equations (rows), and n righthand-side columns:
               $\mathbf{B}$ is of order m by n.
               For n=1 this interprets as actually solving, but the
               reduced form is the general case.

               optional facilities:
               Illustration of separate steps,
               calculation of the determinant, following section 5.6

SOLVI   As SOLV using integer numbers only (= section 5.9)

SING   illustrate singularity as row-dependence

BLOCK   relates row-operations to block-pivoting formulae

               This is a modified call to the inversion-procedure
               which is also called under the SOLV option.
               The display of the tableau-matrices associated with
               individual steps is now standard, not optional, and
               there is no option to add a sum-count column.
               Instead, the relevant block-pivoting formulae from
               section 4.7 are screen-displayed at the relevant
               points, and illustrated numerically.

RANK   establishing the rank and vectors $\mathbf{A}\,\mathbf{v} = [0]$

               Subsidiary options as under SOLV, where the determi-
               nant (if calculated) is the determinant of an actu-
               ally inverted block, which may not be the only inver-
               tible block of that order in the matrix.

DIFQ   differentiation of a quadratic function, with, at suitable points, screen-display of the relevant formulae, and additional explanation as related to the figures

CHARD   calculation of the determinant by its definition, by the leading row,
(illustration of the) extraction of the characteristic equation, using a modification of the minors form (section 7.2).

Additional option:
Display of (the calculation of) minors.

DEF   find the type of definiteness of a symmetric matrix

Recapitulates some of the theory, no options.

ROOT   calculation of a real root and vector by repeated multiplication (section 8.11).

This method applies for real roots only.
In the presence of repeated and/or complex roots, it could fail to converge, hence an iteration-count has been built in, causing this option to terminate by program if it fails to converge.

No subsidiary options.

DIVIDE   divide two polynomial functions (section 7.3).

GENER   calculate the Moore-Penrose inverse

subsidiary options:
Calculate by postmultiplication of $\mathbf{A}'$ by the generalized inverse of the product expression $\mathbf{AA}'$, or by pre-multiplication of the generalized inverse of $\mathbf{A}'\mathbf{A}$.

The generalized inverse of the symmetric product expression is calculated by addition of $\mathbf{XX}'$, where $\mathbf{X}$ are the vectors associated with the zero roots.

If the matrix is of full rank, and the 'wrong' (non-invertible) symmetric product expression is chosen, this mis-choice is pointed out.

SOLVECOM   solve a system of complex equations (inc. inversion)

INVERTCOM complex inversion by real numbers (section 8.8)

LOGFILE   This option causes screen-display of some matrices to be written to a log-file, to be specified in response to the relevant prompt. This might possibly be useful for generating examples for teaching purposes.

It cannot be switched off, and the log can only be accessed after EXIT.

DISPLAY   Screen-display an ASCII file. (A prompt for the name of the file will appear.)

Command-line call of Illustrate

Illustrate can also be called with an option as parameter, e.g. ILLUSTRATE ROWR, and the option ROWR will be activated. This form of calling illustrate also by-passes the need to hit RETURN at the start. The DISPLAY option also allows second and following parameters, the name(s) of file(s) to be displayed, in which case all preliminary display is suppressed, and the files are displayed without going through the menu.

Your author has used this option also for general teaching purposes, with a screen-projection device linked to the monitor. It is in that context worth to mention that the WORDSTAR control characters ^X = expanded typeface, ^T = superscript, and ^V = subscript, will trigger the colour-choice as specified in the initialization file, and that the display procedure will stop and wait for a RETURN, if the characterstring 'RETURN to' occurs in any line, which is then not displayed itself.

If wordprocessing to this purpose is done in document mode, the separate program REFORM can be used to convert the file to the ASCII form which ILLUSTRATE requires, removing the word-wrap information. If the wordstar interpretation option is left as in the default initialization file, lines starting with '.', i.e.

dot commands used to control the layout of the file, can then be left in: REFORM will not remove these, but they won't be displayed anyhow, at least not in that mode of calling.

The file ILLUSTRA.DOC on the tutorial-disc contains some more strictly computational information on the use and installation of ILLUSTRATE.

<u>q.e.d.</u>

The two basic building-blocks of the formal structure of mathematics are <u>definitions</u> -they give the terms which set the framework-, and <u>theorems</u>, statements about things which logically follow from the definitions and from other theorems. Theorems require <u>proofs</u> -we need to satisfy ourselves that these statements are indeed logical-. Proofs conventionally end with the abbreviation: q.e.d.

In my own secondary school years, Latin was sufficiently prominent in any surrounding in which there was learning, that you automatically picked up a certain minimum of it, even when, as in my own case, you had no formal instruction in it.

This does not generally hold for the current generation of undergraduate students.

Hence the following explanation of the abbreviation q.e.d.:

The full Latin words are: quod era demonstrandum = which was to be shown. You use it as a standard ending of a formal proof, irrespective of whether you actually repeat what was to be shown, or are otherwise satisfied that the logical argument which proves a theorem, is complete.

Index of Exercises and answersheets.

ACKNOWLEDGEMENTS

The one person to whom I owe an acknowledgement for this book more than to anyone else, is my son, Dr. Peter Heesterman.

He first of all wrote the private printer driver program with its capability to include special characters in ordinary Wordstar-type files, which was used to print this book.

It may be that we did not always smoothly agree on various additional modifications which I desired and initially programmed, but there is no doubt in my mind that his work was decisive.

Secondly, in the process of discussing various modifications, I found myself thoroughly learning Turbopascal, whereas I had been using Algol before. Although ILLUSTRATE has no common code with PRINT and INDEX, I could not have written it without my son's coaching in Turbopascal.

I would also like to thank my wife Wiebina for her patience with the long hours of often late work on the manuscript and various related problems.

Aart Heesterman

CHAPTER I

ROW-OPERATIONS AND TABLEAUS

1.1 The resolution of a system of linear equations

The following is a standardized method of solving a system of n unknown variables and n linear equations.

We apply a series of proportional additions and subtractions of (all coefficients in) one equation, to/from all other equations, multiplied by a multiplier, which is set at a value which is suitable for eliminating reference to a particular variable from all but that one equation. That operation is called an <u>elimination step</u>, hence the name of the method: elimination by row-operations. We illustrate the method of row-operations, by reference to an example-illustration, as follows:

$$0.867\, x_1 \quad - \quad 0.067\, x_2 \qquad\qquad = \quad 240 \qquad (11)$$

$$-0.150\, x_1 \quad + \quad 0.850\, x_2 \qquad\qquad = \quad 210 \qquad (12)$$

$$-0.167\, x_1 \quad - \quad 0.100\, x_2 \quad + \; x_3 \quad = \quad 0 \qquad (13)$$

In order to turn the leading equation (11), into an x_1 equation, we divide all its coefficients by 0.867, to obtain:

$$x_1 \quad - \quad 0.076\, x_2 \qquad\qquad = \quad 277 \qquad (21) = (11)\,/\,0.867$$

To facilitate the elimination of the reference to x_1 from another equation, we first of all multiply (21) by a suitable number, i.e. by minus the coefficient to be eliminated. There are two equations in which a reference to x_1 occurs, and the operation is performed twice, with different multipliers.

We obtain:

$$0.150\ x_1\quad -\ 0.011\ x_2\qquad\qquad =\qquad 42\qquad (22)\ =\ (21)\times 0.15$$

and

$$0.167\ x_1\quad -\ 0.013\ x_2\qquad\qquad =\qquad 46\qquad (23)\ =\ (21)\times 0.167$$

We now re-name (21) as (31), add (22) to (12), and add (23) to (31), to obtain:

$$x_1\quad -\ 0.076\ x_2\qquad\qquad =\qquad 277\qquad (31)\ =\ (21)$$
$$0.839\ x_2\qquad\qquad =\qquad 252\qquad (32)\ =\ (12)+(22)$$
$$-\ 0.113\ x_2\quad +\ x_3\ =\quad 46\qquad (33)\ =\ (13)+(23)$$

Note, that elimination by forming the combination of two equations, is equivalent to substitution. If we write (21) in the explicit form:

$$x_1\ =\ 0.076\ x_2\ +\ 277,$$

and then substitute the righthand side of this expression for x_1 into (12) and (13), we obtain:

$$-0.150\ (0.076\ x_2\ +\ 277)\quad +\ 0.850\ x_2\qquad\qquad =\qquad 210$$

and:

$$-0.167\ (0.076\ x_2\ +\ 277)\quad -\ 0.100\ x_2\quad +\ x_3\ =\quad 0$$

From which:

$$-0.012\ x_2\quad -\ 42\quad +\ 0.850\ x_2\qquad\qquad =\qquad 200$$

and

$$-0.013\ x_2\quad -\ 46\quad -\ 0.100\ x_2\quad +\ x_3\ =\qquad 0$$

From which (32) and (33) are obtained by re-ordering and combining some terms.

Each elimination-step expresses one variable with a unity coefficient in a particular equation, and eliminates reference to that particular variable from all other equations.

We therefore have, so far, made one elimination step, and we now proceed with the second elimination step.

We divide all coefficients in (32) by 0.839, to obtain:

$$x_2 \qquad = \quad 300 \qquad (42) = (32) / 0.839$$

We now once more multiply all the coefficients in (42) by, respectively 0.076 to eliminate reference to x_2 from (31), and by 0.113 to eliminate reference to x_2 from (33). The results are:

$$0.076\ x_2 \qquad = \quad 23 \qquad (41) = (42) \times 0.076$$

and

$$0.113\ x_2 \qquad = \quad 34 \qquad (43) = (42) \times 0.113$$

We now add (41) to (31), to obtain (51), rename (42), which becomes (52), and add (43) to (33), to obtain (53). The successor system is:

$$x_1 \qquad\qquad\qquad = \quad 300 \qquad (51)$$

$$x_2 \qquad\qquad = \quad 300 \qquad (52)$$

$$x_3 \quad = \quad 80 \qquad (53)$$

In general, the resolution of a system of 3 linear equations and 3 variables, would require 3 elimination steps. However, for this particular system, the third elimination step would consist entirely of trivial operations, division by unity, and adding zero times the third equation to the others.

We find the system resolved.

1.2 **The use of a calculation tableau**

In section 1.1 the three equations were written 5 times, giving
rise to 75 times writing 'x_1', 'x_2', and 'x_3'. To dispense with
such frequent re-writing of the variable-names would obviously
economize on effort.

There are actually other advantages to writing tableaus of
coefficients, rather than explicitly written equations as well.
A further device to emphasize any systematic structure, like
certain groups of coefficients being zeros, is to use a single
dash/minus-sign '-' for a zero.

We illustrate that procedure below, as follows:

$$
\begin{array}{cccccc}
x_1 & x_2 & x_3 & = & \text{R.H.S} & \\
\left[\begin{array}{ccc} |0.867| & -0.067 & - \\ -0.150 & 0.850 & - \\ -0.167 & -0.100 & 1 \end{array}\right] & & & & \left[\begin{array}{c} 240 \\ 210 \\ - \end{array}\right] & \begin{array}{c}(11)\\(12)\\(13)\end{array}
\end{array}
$$

$$
\begin{array}{cccccc}
x_1 & x_2 & x_3 & = & \text{R.H.S} & \\
\left[\begin{array}{ccc} 1 & -0.076 & - \\ 0.150 & -0.011 & - \\ 0.167 & -0.013 & - \end{array}\right] & & & & \left[\begin{array}{c} 277 \\ 42 \\ 46 \end{array}\right] & \begin{array}{c}(21)\\(22)\\(23)\end{array}
\end{array}
$$

$$
\begin{array}{cccccc}
x_1 & x_2 & x_3 & = & \text{R.H.S} & \\
\left[\begin{array}{ccc} 1 & -0.076 & - \\ - & |0.839| & - \\ - & -0.113 & 1 \end{array}\right] & & & & \left[\begin{array}{c} 277 \\ 252 \\ 46 \end{array}\right] & \begin{array}{c}(31)\\(32)\\(33)\end{array}
\end{array}
$$

$$
\begin{array}{cccccc}
x_1 & x_2 & x_3 & = & \text{R.H.S} & \\
\left[\begin{array}{ccc} - & 0.076 & - \\ - & 1 & - \\ - & 0.113 & - \end{array}\right] & & & & \left[\begin{array}{c} 23 \\ 300 \\ 34 \end{array}\right] & \begin{array}{c}(41)\\(42)\\(43)\end{array}
\end{array}
$$

$$
\begin{array}{cccc}
x_1 & x_2 & x_3 & = \quad \text{R.H.S} \\
\end{array}
$$

$$
\begin{bmatrix} 1 & - & - \\ - & 1 & - \\ - & - & 1 \end{bmatrix}
\begin{bmatrix} 300 \\ 300 \\ 80 \end{bmatrix}
\qquad
\begin{array}{l} (51) \\ (52) \\ (53) \end{array}
$$

In this example illustration, we put the reference-numbers, which relate the relevant row of figures to the corresponding explicitly written equations in the previous section, beside the tableaus. That is a meaningful procedure, only if there <u>are</u> explicitly written equations in the first place. Writing the equations explicitly as well, may be help to understand what the tableaus mean, but it negates the object of economizing on effort. Once that is dispensed with, there obviously are no reference numbers either.

It is also more or less usual to dispense with the intermediate tableaus, and to write the tabular presentation of the elimination method as follows:

$$
\begin{array}{cccc}
x_1 & x_2 & x_3 & = \quad \text{R.H.S} \\
\end{array}
$$

$$
\begin{bmatrix} |0.867| & -0.067 & - \\ -0.150 & 0.850 & - \\ -0.167 & -0.100 & 1 \end{bmatrix}
\begin{bmatrix} 240 \\ 210 \\ - \end{bmatrix}
$$

$$
\begin{array}{cccc}
x_1 & x_2 & x_3 & = \quad \text{R.H.S} \\
\end{array}
$$

$$
\begin{bmatrix} 1 & -0.076 & - \\ - & |0.839| & - \\ - & -0.113 & 1 \end{bmatrix}
\begin{bmatrix} 277 \\ 252 \\ 46 \end{bmatrix}
$$

$$
\begin{array}{cccc}
x_1 & x_2 & x_3 & = \quad \text{R.H.S} \\
\end{array}
$$

$$
\begin{bmatrix} 1 & - & - \\ - & 1 & - \\ - & - & 1 \end{bmatrix}
\begin{bmatrix} 300 \\ 300 \\ 80 \end{bmatrix}
$$

The marked diagonal elements of the calculation- tableaus, which determine the ratios used in eliminating the reference to a variable from n-1 equations, are known as <u>pivots</u>.

The convention to mark pivots, or pivotal elements, has been developed in linear programming, and is not actually necessary in the tabular presentation of the method of row-operations: We always operate on the diagonal elements of the tableau, even if (as may be necessary in certain cases to be discussed later in this book), this requires reordering of the equations and/or the variables. It was nevertheless felt, that the essential role of these diagonal elements is brought over more forcefully, if their importance is visually enhanced.

For reasons of wordprocessing convenience, a half-open rectangular marking is used in this book, there is no intrinsic difference with the ringmarkings as are conventional in linear programming.

Exercise:

Solve the following system of linear equations. Write the calculations in tabular form.

$$
\begin{aligned}
2x_1 + x_2 - x_3 + x_4 &= 4 \\
x_1 + x_2 + x_3 + x_4 &= 12 \\
-x_1 + 2x_2 - x_3 - 4x_4 &= 3 \\
x_1 - x_2 + x_3 + 2x_4 &= 8
\end{aligned}
$$

(There is an answer-sheet at the end of the chapter.)

1.3 Elimination and back-substitution

The method of solving a system of linear equations by row operations, as outlined in section 1.1, is known (see for example Hohn [25], p. 65), as the Gauss-Jordan method of elimination. The basic idea of elimination by row-operations is due to the German mathematician Carl Friedrich Gauss ([15], Vol VI, pp. 3-24m and VII, pp. 307-308).

Gauss was interested in astronomy, and he needed to solve a system of six linear equations, and he refers to the method which he used in doing so as 'elimination'. It is the opinion of those who could apparently read the source-text (the verbal part of which is for about four fifths in Latin), e.g. Forsythe and Moler [11], pp 27-31, and Koecher [29], pp 92-93, where the author of this book found the bibliographical reference to Gauss, that the original Gaussian method was in fact one which we would now describe as a combination of forward elimination and backward substitution.

At each kth elimination step, reference to x_k is eliminated from equations k+1 to n, not from equations 1 to n-1. The division of the kth equation by the pivotal element is postponed to the phase of backward substitution.

To illustrate the original Gaussian method, we solve the same system once more:

$$0.867\ x_1 \ - \ 0.067\ x_2 \qquad\qquad = \ 240 \qquad\qquad (11)$$

$$-0.150\ x_1 \ + \ 0.850\ x_2 \qquad\qquad = \ 210 \qquad\qquad (12)$$

$$-0.167\ x_1 \ - \ 0.100\ x_2 \ + \ x_3 \ = \ 0 \qquad\qquad (13)$$

or in tabular form:

$$
\begin{array}{cccccc}
x_1 & x_2 & x_3 & = & \text{R.H.S} \\
\left[\begin{array}{ccc} |0.867| & -0.067 & - \\ -0.150 & 0.850 & - \\ -0.167 & -0.100 & 1 \end{array}\right] & & & & \left[\begin{array}{c} 240 \\ 210 \\ - \end{array}\right]
\end{array}
$$

We do not divide the coefficients of the leading equation by 0.867. To eliminate reference to x_1 from the second equation, we add 0.15/0.867 = 0.173 times the leading equation to it. Similarly, we add 0.167/0.867 = 0.193 times the leading equation to the third equation, thereby eliminating the coefficient of minus 0.167 from the third equation. The successor system is:

$$0.867\ x_1 \ - \ 0.067\ x_2 \qquad\qquad = \ 240 \qquad\qquad (11)$$

$$0.839\ x_2 \qquad\qquad = \ 252 \qquad\qquad (32)$$

$$- \ 0.113\ x_2 \ + \ x_3 \ = \ 46 \qquad\qquad (33)$$

or in tabular form:

x_1	x_2	x_3	=	R.H.S
0.867	-0.067	-		240
-	|0.839|	-		252
-	-0.113	1		46

We now add $0.113/0.839 = 0.135$ times the second equation to the third equation, while leaving both the second and the first equations undisturbed. The successor system is:

$$0.867\, x_1 \quad - 0.067\, x_2 \qquad\qquad = 240 \qquad (11)$$

$$0.839\, x_2 \qquad\qquad = 252 \qquad (32)$$

$$x_3 = 80 \qquad (53)$$

or in tabular form:

x_1	x_2	x_3	=	R.H.S
0.867	-0.067	-		240
-	0.839	-		252
-	-	1		80

The complete solution is now obtained by (only now) dividing each kth equation by its pivotal element, and backward substitution. Having solved x_n, the figure is substituted for the variable in

the $(n-1)^{th}$ equation, and we calculate of the value of x_{n-1}, etc.

In our example, this is done as follows: We first calculate x_2 as $252/0.839 = 300$. Substitution of this figure for x_1 into (11) now results in:

$$0.876\, x_1 - 0.067\, x_2 = 0.876\, x_1 - 0.067 \times 300 =$$

$$0.867\, x_1 - 20 = 240.$$

Therefore:

$$0.867 \; x_1 \;\; = \;\; 260 \qquad \rightarrow \qquad x_1 \; = \; 300.$$

There is some element of doubt about the appropriateness of the reference to Jordan. It appears to relate to Wilhelm Jordan (1842-1899), not to the better known French mathematician Camille Jordan [27]. Householder [26], p. 141, and, following him, Stewart [43], p. 131, reckon that this reference to Jordan is inappropriate. This is based on the fact that the method is described in Jordan's Handbuch der Vermessungskunde (Handbook of geodesy), but only in the third edition, which appeared after his death in 1899, and there is an older description of the method. Mehmke [32], refers to Clasen [7], and that puts the date at 1888. Mehmke also reports that Clasens' publication remained 'totally unknown' in Germany ('bei uns 'völlig unbekannt geblieben'). (Mehmke does not refer to Jordan at all, and the author of this book could not get hold of the source text of either Clasen or Jordan).

We shall therefore refer simply to the ('pure') method of elimination by row-operations, as distinct from the original Gaussian method, which is a combination of forward elimination and backward substitution.

The pure elimination method was chosen here, mainly because it extends itself more or less naturally to problems of part-resolution, where it is required to express k variables with the help of k linear equations, in (some combination of) n-k other unknown variables. We shall be concerned with such problems in Chapters IV and V.

In relation to actually solving an ordinary system of n unknown variables from n linear equations, back-substitution requires less effort than elimination of reference to a variable from all previous equations. Accordingly, a combination of elimination (from equations k+1 to n), and back-substitution, is more efficient than a method which works by elimination only.

We shall not be using this particular combination of elimination and back-substitution anymore, using the 'pure' elimination method only.

Answersheet on elimination by row-operations

x_1	x_2	x_3	x_4	= R.H.S
|2|	1	-1	1	4
1	1	1	1	12
-1	2	-1	-4	3
1	-1	1	2	8

x_1	x_2	x_3	x_4	= R.H.S
1	0.5	-0.5	0.5	2
-	|0.5|	1.5	0.5	6
-	2.5	-1.5	-3.5	5
-	-1.5	1.5	1.5	6

x_1	x_2	x_3	x_4	= R.H.S
1	-	-2	-	-8
-	1	3	1	20
-	-	|-9|	-6	-45
-	-	6	3	36

x_1	x_2	x_3	x_4	= R.H.S
1	-	-	1.33	2
-	1	-	-1	5
-	-	1	0.67	5
-	-	-	|-1|	6

x_1	x_2	x_3	x_4	= R.H.S
1	-	-	-	10
-	1	-	-	-1
-	-	1	-	9
-	-	-	1	-6

CHAPTER II

MATRIX NOTATION

2.1 The purpose of matrix notation

Matrix notation provides a compact way of describing certain well-defined numerical operations. It therefore saves writing and reading effort in written communication. This applies both to communication between one human being and another, and to programming computers, where one may refer to an established body of matrix-processing routines. The use of matrix notation has also facilitated the further analysis of numerical problems, such as linear equation-systems or econometric models. Such facilitation really is a corollary of the reduction in effort. Problems which were at one stage too complicated to manage with the available notational devices, have become more manageable.

2.2 Some definitions and conventions

A <u>matrix</u> is a rectangular grouping of numbers, its <u>elements</u>. Those numbers can be either known figures, for example a data-matrix containing statistical information, or they can be variables, e.g. functions of other numbers or variables.

The elements of a matrix are grouped in a number of <u>rows</u>, each row containing the same number of elements, reading from left to right. Alternatively, one can say that the elements of a matrix are grouped into a number of <u>columns</u>, each column containing the same number of elements, reading from top to bottom.

11

Matrices often occur as the numerical part of tableaus, containing either statistical information, or coefficients defining a system of linear equations. But note a difference in detail of the definition:

$$
\begin{array}{cccc}
x_1 & x_2 & x_3 & = \quad R.H.S \\
\end{array}
$$

$$
\begin{bmatrix}
0.867 & -0.067 & - & 240 \\
-0.150 & 0.850 & - & 210 \\
-0.167 & -0.100 & 1 & -
\end{bmatrix}
\quad \text{is a tableau,}
$$

$$
\begin{bmatrix}
0.867 & -0.067 & - & 240 \\
-0.150 & 0.850 & - & 210 \\
-0.167 & -0.100 & 1 & -
\end{bmatrix}
\quad \text{is a matrix.}
$$

The order parameters of a matrix are two non-negative integer numbers, indicating the number of rows and the number of columns in the matrix.

It is convention to put them in a particular order, the number of rows comes first, the number of columns comes second. The tableau-matrix above is therefore of order 3 by 4. Normally, order parameters will not only be non-negative, but also positive non-zero. However, there are cases where the formulation of theorems is simplified, if we admit the trivial case where a matrix has no elements, as one of its order parameters, or both of them, has the non-negative integer value zero. A matrix of which the two order parameters are equal is called a square matrix.

A square matrix, which in addition meets the condition $a_{ij} = a_{ji}$ for all legitimate values of i and j,

$$
\begin{bmatrix}
2 & 4 & -3 \\
4 & 1 & 6 \\
-3 & 6 & 5
\end{bmatrix}
$$

is said to be symmetric.

In verbal text, one may refer to a matrix by an inherently
meaningful word, e.g. tableau-matrix, correlation-matrix, data-
matrix, etc., but in formulae it is conventional to indicate a
matrix with a single capital letter, possibly followed by an
index, if the matrix is one of series of similar matrices.

In printed text, it is usual to emphasize matrix-symbols
further, by using bold capital letters and/or the use of Greek
capital letters. Bold printing will be used in this textbook.

When printing or writing the matrix itself, it is usual to use
brackets. There are textbooks which use large round brackets for
this purpose, but this runs into problems in connection with
certain further developments of matrix notation, most notable
with partitioning, which we shall come to discuss later in this
chapter. In this book, square brackets will be used.

For example, the initial equation system illustrated in sections
1.1 and 1.2 is contained in the tableau-matrix:

$$\mathbf{T} = \begin{bmatrix} 0.867 & -0.067 & - & 240 \\ -0.150 & 0.850 & - & 210 \\ -0.167 & -0.100 & 1 & - \end{bmatrix}$$

and this matrix **T** is said to be of order 3 by 4.

To give a meaningful presentation of the resolution of this
system, one may wish to refer to a series calculation-tableaus,
and the further notational device of an additional index will be
used, e.g.:

$$\mathbf{T}_1 = \begin{bmatrix} 0.867 & -0.067 & - & 240 \\ -0.150 & 0.850 & - & 210 \\ -0.167 & -0.100 & 1 & - \end{bmatrix}$$

$$\mathbf{T}_2 = \begin{bmatrix} 1 & -0.076 & - & 277 \\ - & 0.839 & - & 252 \\ - & -0.113 & 1 & 46 \end{bmatrix}$$

and

$$\mathbf{T}_3 = \begin{bmatrix} 1 & - & - & 300 \\ - & 1 & - & 300 \\ - & - & 1 & 80 \end{bmatrix}$$

To indicate the individual elements of a matrix, we use a lower-case letter, as corresponding to the upper case symbol for the matrix itself, followed by two indices, of which the row-index is the first. Thus, if we start with a tableau-matrix

$$\mathbf{T} = \begin{bmatrix} 0.867 & -0.067 & - & 240 \\ -0.150 & 0.850 & - & 210 \\ -0.167 & -0.100 & 1 & - \end{bmatrix}$$

the first pivotal element used in solving the corresponding equations-system by row-operations, is $t_{1,1} = 0.867$, and some other elements are $t_{1,2} = -0.067$, and $t_{3,2} = -0.1$.

Note that this could give rise to ambiguity, if the matrix itself also carries an index on account of belonging to some sort of group or series. It is therefore desirable to avoid indexation of matrices, if one also needs to refer to their individual elements. The alternative is to use some kind of ad hoc notation e.g. $\mathbf{T}_{i,j}(k)$ being the ith row, j^{th} column element of $\mathbf{T}_k$. There is no standard notational convention for this situation.

A matrix of which one of the order parameters is 1 = one, is also called a <u>vector</u>. To indicate an element of a vector, we would again use the corresponding unemphasized lower-case letter, and we would need only one index. Vectors can be <u>columns</u> or <u>rows</u>, but it is conventional to assume that a vector is a column, or else to explicitly indicate the row-presentation.

This leads us to the term <u>transposition</u>. The <u>transpose</u> of an m by n matrix $\mathbf{A}$ is is an n by m matrix (indicated as $\mathbf{A}'$) and the rows of $\mathbf{A}'$ contain the same elements as the columns of $\mathbf{A}$, and the columns of $\mathbf{A}'$ contain the elements of the rows of A:

i.e for $\mathbf{A} = \begin{bmatrix} 11 & 12 & 13 & 14 \\ 21 & 22 & 23 & 24 \\ 31 & 32 & 33 & 34 \end{bmatrix}$

we have: $\mathbf{A}' = \begin{bmatrix} 11 & 21 & 31 \\ 12 & 22 & 32 \\ 13 & 23 & 33 \\ 14 & 24 & 34 \end{bmatrix}$

Vectors are conventionally indicated by lower-case letters, also with some special way of emphasizing. In type-written or manuscript text, that is usually underlining, but heavy (bold) print or the use of Greek letters is usual in printed books, bold print will be used in this one. Row-vectors are then indicated as the transposes of the corresponding column-vectors.

For example, a tableau-matrix which specifies a system of n linear equations is often given in two parts, a square matrix which specifies the coefficients of the variables, and a separate vector of righthand-side coefficients, e.g.

$$\mathbf{A} = \begin{bmatrix} 0.867 & -0.067 & - \\ -0.150 & 0.850 & - \\ -0.167 & -0.100 & 1 \end{bmatrix}$$

and

$$\mathbf{b} = \begin{bmatrix} 240 \\ 210 \\ - \end{bmatrix}$$

If we then wish to present the three righthand-side coefficients in one line, we may write: $\mathbf{b}' = [\ 240,\ 210,\ -\ ]$.

Note that in this case the matrix-notation dovetails into specifying a system of linear equations by means of Σ notation:

$$\sum_{j=1}^{n} a_{ij}\, x_j = b_i \quad (i = 1, 2, \ldots\ldots n)$$

correctly relates to the elements of $\mathbf{A}$ and $\mathbf{b}$ as illustrated above, but it would also be familiar to someone who had never seen any matrix notation.

Obviously, vectors can be columns or rows, extracted from a matrix. In that case, we need an index to indicate the position of the column or row within the matrix.

Thus t_2 will relate to the tableau-matrix

$$\mathbf{T} = \begin{bmatrix} 0.867 & -0.067 & - & 240 \\ -0.150 & 0.850 & - & 210 \\ -0.167 & -0.100 & 1 & - \end{bmatrix}$$

as $\mathbf{t}_2 = \begin{bmatrix} -0.067 \\ 0.850 \\ -0.100 \end{bmatrix}$

There is, unfortunately, some element of ambiguity in relation to
rows. As far as standard notational conventions is concerned, $t'3$
could be either the third row of $\mathbf{T}$, e.g.:

$$t'_3 = [\ -0.167 \quad -0.100 \quad 1 \quad -\ ],$$

or it could be the transpose of t_3, i.e. of the third column of
T. We can just about make the distinction, by strictly sticking
to the transposition-convention, i.e. the transpose is obtained
by adding the transposition-sign at the end.

That makes t_3' into the transposed third column of T:

$$\mathbf{t}_3' = [\ - \quad - \quad 1\ ]$$

whereas t'_3 is the third row. This is not, however, a sufficiently clear visual distinction, to avoid the need for an
explicit statement of the nature of such operands, when they
are used.

2.3 Addition and subtraction of matrices and vectors

The arithmetic operations of addition and subtraction permit
immediate generalization from ordinary numbers to matrices and
vectors. The one additional complication is, that we must assume
the order parameters to fit.

Thus, the statement: $$C = A + B$$

pre-supposes that **A** and **B** are of the same order. If that condition is not satisfied, such a statement has no meaning, just is not legitimate. Once we know that **A** and **B** are of the same order (let us say, both of order m by n), it conveys the information:

$$c_{ij} = a_{ij} + b_{ij}$$

$$(i = 1, 2, \ldots\ldots n, \ j = 1, 2, \ldots\ldots n)$$

The generalization of these principles to vectors, and to subtraction, will be obvious.

2.4 **Matrix-multiplication**

The product-expression **A** x **B**, (or **AB**), is legitimate, only if the number of columns of **A** matches the number of rows of **B**.

If that condition is satisfied, e.g. if **A** is of order m by n, and B of order n by k, the statement:

$$C = A\ B$$

implies that **C** is of order m by k, and gives the numerical information:

$$c_{ij} = \sum_{h=1}^{n} a_{ih}\, b_{hj}$$

$$(i = 1, 2, \ldots\ldots m, \quad j = 1, 2, \ldots\ldots k) \hspace{3cm} (2.4.1)$$

Example:

Evaluate $\quad \mathbf{C} = \mathbf{AB} = \begin{bmatrix} 1 & 2 \\ -1 & 1 \end{bmatrix} \begin{bmatrix} 1 & 2 & -2 \\ 1 & 2 & 3 \end{bmatrix}$

$$c_{1,1} = a_{1,1} \times b_{1,1} + a_{1,2} \times b_{2,1} = 1 \times 1 + 2 \times 1 = 3$$

$$c_{1,2} = a_{1,1} \times b_{1,2} + a_{1,2} \times b_{2,2} = 1 \times 2 + 2 \times 2 = 6$$

$$c_{1,3} = a_{1,1} \times b_{1,3} + a_{1,2} \times b_{2,3} = 1 \times (-2) + 2 \times 3 = 4$$
$$c_{2,1} = a_{2,1} \times b_{1,1} + a_{2,2} \times b_{2,1} = -1 \times 1 + 1 \times 1 = 0$$
$$c_{2,2} = a_{2,1} \times b_{1,2} + a_{2,2} \times b_{2,2} = -1 \times 2 + 1 \times 2 = 0$$
$$c_{2,3} = a_{2,1} \times b_{1,3} + a_{2,2} \times b_{2,3} = -1 \times (-2) + 1 \times 3 = 5$$

Therefore:

$$\mathbf{C} = \mathbf{A}\,\mathbf{B} = \begin{bmatrix} 1 & 2 \\ -1 & 1 \end{bmatrix} \begin{bmatrix} 1 & 2 & -2 \\ 1 & 2 & 3 \end{bmatrix} = \begin{bmatrix} 3 & 6 & 4 \\ - & - & 5 \end{bmatrix}$$

Note, that whereas for ordinary numbers or functions, we have:

$$a \times b = b \times a$$

this does not apply to matrices, even if both expressions exist. For $\mathbf{A}$ and $\mathbf{B}$ as in the example above, $\mathbf{BA}$ does not even constitute a legitimate expression, as the number of columns of $\mathbf{B}$, i.e. 3, does not match the number of rows of $\mathbf{A}$, which is 2. Even if both expressions are legitimate, they can be of different order, and even if they are of the same order, they do not generally have the same content.

Examples:

$$\begin{bmatrix} 1 & 2 \end{bmatrix} \begin{bmatrix} 3 \\ 4 \end{bmatrix} = \begin{bmatrix} 11 \end{bmatrix}, \qquad \text{order 1 by 1,}$$

$$\begin{bmatrix} 3 \\ 4 \end{bmatrix} \begin{bmatrix} 1 & 2 \end{bmatrix} = \begin{bmatrix} 3 & 6 \\ 4 & 8 \end{bmatrix}, \qquad \text{order 2 by 2}$$

$$\begin{bmatrix} 1 & 2 \\ 2 & 3 \end{bmatrix} \begin{bmatrix} 5 & -1 \\ -1 & 2 \end{bmatrix} = \begin{bmatrix} 3 & 3 \\ 7 & 4 \end{bmatrix}$$

$$\begin{bmatrix} 5 & -1 \\ -1 & 2 \end{bmatrix} \begin{bmatrix} 1 & 2 \\ 2 & 3 \end{bmatrix} = \begin{bmatrix} 3 & 7 \\ 3 & 4 \end{bmatrix}$$

Even in the case of the product of two symmetric matrices of equal order, the products, though both legitimate and of the same order, are not always equal !

In this context we should mention the terms <u>pre-multiplication</u> and <u>post-multiplication</u>, i.e. $\mathbf{A}$ pre-multiplied by $\mathbf{B}$ is $\mathbf{BA}$, and $\mathbf{A}$

post-multiplied by **B** is **AB**.

The generalization of these definitions to the product of a matrix and a vector, and to the product of two vectors is straightforward: a vector is a matrix, of which one of the order-parameters is 1.

Some obvious restrictions on the legitimacy of the product of a matrix and a vector arise from the fact that one of the order-parameters of the vector is 1 by definition.

A matrix can be post-multiplied by a column-vector, provided the number of columns of the matrix matches the number of elements of the column:

$$\begin{bmatrix} 1 & 2 & -3 \\ -2 & 5 & -1 \end{bmatrix} \begin{bmatrix} 2 \\ 1 \\ -1 \end{bmatrix} = \begin{bmatrix} 7 \\ 2 \end{bmatrix}$$

As a column is a matrix containing <u>by definition</u> only one column, we do not consider the statement:

$$c = b\ A$$

as legitimate, even if it might be the case that the number of rows contained in **A**, was only one.

A matrix can be pre-multiplied by a row-vector, provided the number of elements in the row matches the number of rows in the matrix:

$$\begin{bmatrix} 1 & 2 \end{bmatrix} \begin{bmatrix} 1 & 2 & 3 \\ 4 & 5 & 6 \end{bmatrix} = \begin{bmatrix} 9 & 12 & 15 \end{bmatrix}$$

When we wish to evaluate the product of two matrices, it is often convenient, to do that either per column, post-multiplying the lefthand-side matrix by the separate columns of the righthand-side matrix, e.g.:

$$\begin{bmatrix} 1 & 2 & -3 \\ -2 & 5 & -1 \end{bmatrix} \begin{bmatrix} 2 & 5 \\ 1 & -3 \\ -1 & 2 \end{bmatrix} = \begin{bmatrix} 7 & -7 \\ 2 & -27 \end{bmatrix}$$

is obtained as the combination of:

$$\begin{bmatrix} 1 & 2 & -3 \\ -2 & 5 & -1 \end{bmatrix} \begin{bmatrix} 2 \\ 1 \\ -1 \end{bmatrix} = \begin{bmatrix} 7 \\ 2 \end{bmatrix}$$

and

$$\begin{bmatrix} 1 & 2 & -3 \\ -2 & 5 & -1 \end{bmatrix} \begin{bmatrix} 5 \\ -3 \\ 2 \end{bmatrix} = \begin{bmatrix} -7 \\ -27 \end{bmatrix}$$

Then again, one may find it convenient to operate on the righthand-side matrix as a whole, which then needs to be pre-multiplied by each row of the lefthand-side matrix:

$$\begin{bmatrix} 1 & 2 & -3 \\ -2 & 5 & -1 \end{bmatrix} \begin{bmatrix} 2 & 5 \\ 1 & -3 \\ -1 & 2 \end{bmatrix} = \begin{bmatrix} 7 & -7 \\ 2 & -27 \end{bmatrix}$$

is obtained as the combination of:

$$\begin{bmatrix} 1 & 2 & -3 \end{bmatrix} \begin{bmatrix} 2 & 5 \\ 1 & -3 \\ -1 & 2 \end{bmatrix} = \begin{bmatrix} 7 & -7 \end{bmatrix}$$

and:

$$\begin{bmatrix} -2 & 5 & -1 \end{bmatrix} \begin{bmatrix} 2 & 5 \\ 1 & -3 \\ -1 & 2 \end{bmatrix} = \begin{bmatrix} 2 & -27 \end{bmatrix}$$

There are two types of products of two vectors the <u>inner</u> product, which is a row post-multiplied by a column of equal order. In that case, the order of the result is of order 1 by 1, i.e. a single number.

Example:

$$[1 \quad 2 \quad 3] \begin{bmatrix} 4 \\ 5 \\ 6 \end{bmatrix} = [4 + 10 + 18] = [32]$$

We can in fact use this definition of the inner product of two vectors, to streamline the general definition of matrix-multiplication:

$$c_{i,j} = a'_i \, b_j \qquad (2.4.2)$$

where a'_i is the ith row of **A**, and b_j is the jth column of **B**, is equivalent to the general definition of the product **AB**, as given in (2.4.1): application of that definition to the particular case of the inner product of a'_i and b_j confirms:

$$a'_i \, b_j = \sum_{h=1}^{n} a_{i,h} \, b_{h,j} = c_{i,j} \qquad (2.4.3)$$

as may be illustrated in relation to c_{12} as calculated above:

$$c_{12} = a'_1 \, b_2 = [1 \quad 2] \begin{bmatrix} 2 \\ 2 \end{bmatrix} =$$

$$1 \times 2 + 2 \times 2 = 2 + 4 = 6.$$

The other legitimate product expression of two vectors is the <u>outer product</u>, i.e. the column is post-multiplied by the row. In that case there is no restriction on the orders, as the matching order parameter is the definitional 1=one.

Example:

$$\begin{bmatrix} 4 \\ 5 \end{bmatrix} [1 \quad 2 \quad 3] = \begin{bmatrix} 4 & 8 & 12 \\ 5 & 10 & 15 \end{bmatrix}$$

Matrices (as well as vectors), can also be multiplied by simple numbers. In that case, there are no order-restrictions. Each separate element of the matrix (or the vector) is multiplied by the number. If such a number (or variable) figures as a symbolic constant or a variable, it is conventional to use a Greek letter.

Example:

$$\alpha \times \begin{bmatrix} x_{1,1} & x_{1,2} & x_{1,3} \\ x_{2,1} & x_{2,2} & x_{2,3} \end{bmatrix} = \begin{bmatrix} x_{1,1} & x_{1,2} & x_{1,3} \\ x_{2,1} & x_{1,2} & x_{2,3} \end{bmatrix} \times \alpha =$$

$$\begin{bmatrix} \alpha x_{1,1} & \alpha x_{1,2} & \alpha x_{1,3} \\ \alpha x_{2,1} & \alpha x_{2,2} & \alpha x_{2,3} \end{bmatrix} = \begin{bmatrix} x_{1,1}\alpha & x_{1,2}\alpha & x_{1,3}\alpha \\ x_{2,1}\alpha & x_{2,2}\alpha & x_{2,3}\alpha \end{bmatrix}$$

However, in the case of a vector it is good practice, to put the order of the operands in such a way, that the expression is also legitimate, if the single number is treated as a vector of order 1=one.

Thus,

$$\lambda \mathbf{x} = \begin{bmatrix} \lambda x_1 \\ \vdots \\ \lambda x_n \end{bmatrix}$$

is legitimate, but the expression $\mathbf{x}\lambda$ is preferable.

Similarly, the expression $\mathbf{w}'\lambda$ is not actually illegal, but $\lambda \mathbf{w}'$ is preferable.

Matrix multiplication does not only apply to known numbers or constants. It also applies to variables.

An obvious example is the replacement of the notation:

$$\sum_{j=1}^{n} a_{ij} x_j = b_j \qquad (i = 1, 2, \ldots\ldots\ldots n)$$

to indicate an ordinary system of n linear equations, by the more compact expression:

$$\mathbf{Ax} = \mathbf{b}$$

We finish this section by one last comment on streamlining notation:

We have, so far, separated the two indices of an element of a matrix, if they are numbers, by a comma ','. In examples, where no indices > 9 occur, and in relation to indices which need to be a pair anyhow, this is not really needed. In the rest of this book, it should be understood, that a_{12}, when being an element of the matrix **A**, is equivalent to $a_{1,2}$, not the 12th element of A.

Exercise:

For $\mathbf{A} = \begin{bmatrix} 2 & 3 \\ 4 & 5 \end{bmatrix}$, $\mathbf{B} = \begin{bmatrix} -1 & 2 & -2 \\ 5 & 3 & 2 \\ -2 & - & 8 \end{bmatrix}$

$\mathbf{c} = \begin{bmatrix} 2 \\ 4 \end{bmatrix}$, $\mathbf{d} = \begin{bmatrix} 2 \\ -1 \\ 4 \end{bmatrix}$,

$$\mathbf{v'} = [1 \quad -1], \quad \mathbf{z'} = [2 \quad -2 \quad 3],$$

you are asked:

(1) State which pairs of products are illegal, and why.

(2) Evaluate the legitimate expressions, insofar as **A**, **B**, **c**, **v'** or **z'** figure as the lefthand-factor.

(3) State the orders of any remaining legitimate expressions.

There is an answersheet at the end of the chapter.

2.5 Substitutions, recursive products and sums

When two expressions are identical, one is allowed to interchange them, to <u>substitute</u> the one for the other. This postulate is vital to all algebra. It applies to matrix algebra as well.

In the proof of the theorem stated below, we substitute the relevant Σ expression for the equivalent definition of matrix multiplication, in certain matrix expressions.

Theorem:

  If **AB** and **BC** both exist,

then :

  $[A\ B]\ C\ =\ A\ [B\ C]$ (2.5.1)

Proof:

We first deal with the special case where **A** is of order 1 by m, and **C** of order n by 1 (and therefore **B** of order m by n). We shall refer to this case as 'case 1'.

In that case we evaluate $[a'B]c$ as:

$$[a'B]\ c\ =\ \sum_{j=1}^{n} \left(\sum_{i=1}^{m} a_i b_{ij} \right) c_j\ =\ \sum_{j=1}^{n} \sum_{i=1}^{m} a_i\ b_{ij}\ c_j \qquad (2.5.2)$$

$$\left(\quad \text{where} \quad \left(\sum_{i=1}^{m} a_i b_{ij} \right) \quad \text{is the } j^{\text{th}} \text{ element of } a'B \quad \right)$$

Therefore, on re-ordering the $\Sigma\ \Sigma$ notation:

$$[a'B]\ c\ =\ \sum_{i=1}^{m} a_i \sum_{j=1}^{n} b_{ij}\ c_j$$

$$=\ \sum_{i=1}^{m} a_i \left(\sum_{j=1}^{n} b_{ij}\ c_j \right)\ =\ a'\ [B\ c] \qquad (2.5.3)$$

$$\left(\quad \text{where} \quad \left(\sum_{j=1}^{n} b_{ij}\ c_j \right) \quad \text{is the } i^{\text{th}} \text{ element of } Bc \quad \right)$$

q.e.d. for case 1.

Now, collecting q separate applications of (2.5.3) for q different rows a'_h into a q by n matrix **A**, we find:

  $[AB]\ c\ =\ A\ [B\ c]$ (2.5.4) and (2.5.1) follows by application of (2.5.4), for each separate column of **C**.

q.e.d.

Once we know that the expressions [**AB**]**C** and **A**[**BC**] are identical, there is no point in writing the '[' brackets at all, we might as well write **ABC**.

Recursive application of the same result, now leads to a more general statement concerning a recursive product expression:

$$\mathbf{A} \, [\mathbf{B} \, \mathbf{C} \, \mathbf{D} \, \ldots \ldots \, \mathbf{Z}] = [\mathbf{A} \, \ldots \ldots \ldots \, \mathbf{Y}] \, \mathbf{Z} = \mathbf{A} \, [\mathbf{B} \, \mathbf{C} \, \ldots \ldots \, \mathbf{Y}] \, \mathbf{Z}$$

and again, there is no point in writing the brackets at all.

The analogous property of the sum of some matrices:

$$\mathbf{A} + [\mathbf{B} + \mathbf{C}] = [\mathbf{A} + \mathbf{B}] + \mathbf{C} \ (= \mathbf{A} + \mathbf{B} + \mathbf{C} \ )$$

will be obvious.

Exercise:

Find the numerical content of:

$$\mathbf{A} \, \mathbf{B} \, \mathbf{C} \ = [1 \ 2 \ 3] \begin{bmatrix} 1 & 3 \\ -1 & -2 \\ 2 & 5 \end{bmatrix} \begin{bmatrix} 1 & 1 \\ 1 & 2 \end{bmatrix}$$

Check by independently calculating [**A B**] **C** and **A** [**B C**], finding the result the same.

2.6 The transposition of recursive products

The appropriate expression for the transpose of two matrices, and by implication, for the transpose of the product of a matrix and a vector, or of two vectors, follows immediately from the definition of matrix multiplication:

If **A** is of order m by n, and **B** is of order n by k, then:

  C = **A B**

gives the information:

$$c_{ij} = \sum_{h=1}^{n} a_{ih}\, b_{hj} \quad (i = 1, 2, \ldots\ldots m, \quad j = 1, 2, \ldots\ldots k)$$

Therefore, reversing the order of the factors in each term:

$$c_{ij} = \sum_{h=1}^{n} b_{hj}\, a_{ih} \quad (j = 1, 2, \ldots\ldots k, \quad i = 1, 2, \ldots\ldots m)$$

which is the definition of the matrix-product:

  C' = **B' A'**

Note the way in which this result works through recursively:

  [**A B C** .... **Z**]' = **Z'** [**A B C** .... **Y**]'

  = **Z' Y'** [**A B C** ...... **V**]' = **Z' Y' V'** .... **C' B' A'**

i.e. we need to interchange the order of the operands throughout.

2.7 The differentiation of some matrix expressions

When functions, variables, or algebraic relations are expressed in the form of matrix notation, it is often useful to express differential variations of such expressions in matrix notation as well.

Rules for the differentiation of a number of not too complicated matrix expressions are obtained by applying the first principles of calculus to the appropriate matrix expressions and their definitions.

The differentiation of sums and products of matrices is discussed here as follows:

For the sum of two matrices **A** and **B** we have:

$$d[\mathbf{A} + \mathbf{B}] = d\mathbf{A} + d\mathbf{B} \qquad (2.7.1)$$

The generalization of this formula to differences, summations of more than one term, and to sums and differences of vectors, will be obvious.

Product expressions become more complicated, and we shall need to start from the first principles of calculus.

From:

$$\mathbf{P} = \mathbf{A}\,\mathbf{B} \qquad (2.7.2)$$

we first of all obtain an expression for a finite change, for which we shall be using the symbol Δ.

$$
\begin{aligned}
\mathbf{P} + \Delta\mathbf{P} &= [\mathbf{A} + \Delta\mathbf{A}]\,[\mathbf{B} + \Delta\mathbf{B}] \\[4pt]
&= \mathbf{A}\,[\mathbf{B} + \Delta\mathbf{B}] + \Delta\mathbf{A}\,[\mathbf{B} + \Delta\mathbf{B}] \\[4pt]
&= \mathbf{A}\,\mathbf{B} + \mathbf{A}\,\Delta\mathbf{B} + \Delta\mathbf{A}\,\mathbf{B} + \Delta\mathbf{A}\,\Delta\mathbf{B} \qquad (2.7.3)
\end{aligned}
$$

Subtraction of the product expression **P** at its original level, as expressed by (2.7.2) from (2.7.3), now yields the following formula for a change in the product expression.

$$\Delta\mathbf{P} = \mathbf{A}\,\Delta\mathbf{B} + \Delta\mathbf{A}\,\mathbf{B} + \Delta\mathbf{A}\,\Delta\mathbf{B} \qquad (2.7.4)$$

For small changes in **A** and **B**, the third term on the righthand side of (2.7.4), which consists entirely of products of changes, becomes insignificant in comparison with the two others, and a first-order linear approximation of (2.7.4) is:

$$\Delta\mathbf{P} \simeq \mathbf{A}\,\Delta\mathbf{B} + \Delta\mathbf{A}\,\mathbf{B} \qquad (2.7.5)$$

As usual in calculus, we are allowed to take this approximation as exact, for differential variations approaching zero:

$$d\mathbf{P} = \mathbf{A}\,d\mathbf{B} + d\mathbf{A}\,\mathbf{B} \qquad (2.7.6)$$

Note that we <u>cannot</u> extend (2.7.6) to

$$\partial P/\partial A = B$$

as we would do in the case of:

$$p = a\,b \qquad \rightarrow \qquad \partial p/\partial a = b$$

The reason, that the matrix generalization of this particular
exercise in partial differentiation is incorrect, is primarily
the inconsistency in the order parameters.

A (partial) derivative of a matrix expression <u>is</u> the collection,
in the form of a suitable matrix expression, of all the (partial)
derivatives of the separate elements of the parent-expression.
The first thing we need to do, therefore is to establish how many
derivatives there are.

If A is of order m by n, and B of order n by k, the product
expression $P = AB$ is of order m by k. If the $m \times k$ elements of P
are all differentiated with respect to the $n \times k$ elements of B,
there are going to be $(m \times k) \times (n \times k) = m \times n \times k^2$ partial
derivatives.

Differentiation of compound matrix expressions, with respect to
<u>matrices</u> is not very widely practiced. Much more common is the
differentiation of vectorial, and of scalar expressions, in which
matrices occur as arrangements of known coefficients.

In particular, for the symmetric expression:

$$\emptyset = x'Ax \qquad\qquad\qquad (2.7.7)$$

(which is known as a <u>quadratic form</u>), the differential
expression is:

$$d\emptyset = x'd(Ax) + dx'Ax$$

$$ = x'Adx \quad + dx'Ax \quad = 2\,x'Adx \qquad\qquad (2.7.8)$$

For reasons to be discussed in more detail in Chapter VI, it is
conventional to require the matrix which defines a quadratic form
to be symmetric ($a_{ij} = a_{ji}$), hence the identity between the two
terms on the righthand side of (2.7.8).

As in ordinary differentiation, the (vector of) partial derivatives of a non-linear function, only attains a specific numerical value, given some specific values of the variables.

Example:

$$A = \begin{bmatrix} 1 & 2 \\ 2 & 1 \end{bmatrix} \qquad\qquad x = \begin{bmatrix} 1 \\ 2 \end{bmatrix}$$

$$\emptyset = [x_1 \quad x_2] \begin{bmatrix} 1 & 2 \\ 2 & 1 \end{bmatrix} \begin{bmatrix} x_1 \\ x_2 \end{bmatrix} = x_1{}^2 + 4x_1 x_2 + x_2{}^2$$

$$d\emptyset = 2[1 \quad 2] \begin{bmatrix} 1 & 2 \\ 2 & 1 \end{bmatrix} \begin{bmatrix} dx_1 \\ dx_2 \end{bmatrix} = 2\,[5 \quad 4] \begin{bmatrix} dx_1 \\ dx_2 \end{bmatrix} = 10\,dx_1 + 8\,dx_2$$

2.8 Some detail-guidelines for vectorial differentiation

We now limit ourselves mainly to the development of an expression of partial derivatives of a scalar expression. If a scalar function is differentiated relative to a vector, the result is a vector of partial derivatives.

We will illustrate the procedure of developing such a vectorial expression, in relation to the sum of squares of sample residuals, as it arises in the context of Ordinary Least Squares regression

The following stages of manipulation arise:

First stage:

We need a suitable expression of the dependent variable. For straightforward differentiation, this expression must refer explicitly to the independent argument-vector, relative to which we wish to differentiate.

For the OLS example, we develop the following relations:

$$\hat{u} = y - X\,\hat{a} \tag{2.8.1}$$

Here $\hat{\mathbf{u}}$ is the n by 1 vector of sample-residuals, n being the number of sample-observations. The vector $\mathbf{y}$ is the similar vector of observed variables to be explained. The k by 1 vector $\hat{\mathbf{a}}$ is the vectorial collection of k regression coefficients. The matrix $\mathbf{X}$ is the n by k matrix of the sample-observations of the explanatory variables.

In variation of the usual matrix algebra convention, we refer to the sum of squares of the residuals as S, and express S as:

$$S = \hat{\mathbf{u}}'\hat{\mathbf{u}} = [\mathbf{y} - \mathbf{X}\hat{\mathbf{a}}]'\,[\mathbf{y} - \mathbf{X}\hat{\mathbf{a}}] =$$

$$[\mathbf{y}' - \hat{\mathbf{a}}'\mathbf{X}']\,[\mathbf{y} - \mathbf{X}\hat{\mathbf{a}}] = \mathbf{y}'[\mathbf{y} - \mathbf{X}'\hat{\mathbf{a}}] - \hat{\mathbf{a}}'\mathbf{X}'[\mathbf{y} - \mathbf{X}\hat{\mathbf{a}}] =$$

$$\mathbf{y}'\mathbf{y} - \mathbf{y}'\mathbf{X}\hat{\mathbf{a}} - \hat{\mathbf{a}}'\mathbf{X}'\mathbf{y} + \hat{\mathbf{a}}'\mathbf{X}'\mathbf{X}\hat{\mathbf{a}} =$$

$$\mathbf{y}'\mathbf{y} - 2\mathbf{y}'\mathbf{X}\hat{\mathbf{a}} + \hat{\mathbf{a}}'\mathbf{X}'\mathbf{X}\hat{\mathbf{a}} \tag{2.8.2}$$

Note the combination of two terms in the penultimate member of (2.8.2). These are each others transpose, and, as they are both of order 1 by 1, that makes them equal (identical):

$$[\mathbf{y}'\mathbf{X}\hat{\mathbf{a}}]' = \hat{\mathbf{a}}'\mathbf{X}'\mathbf{y} = \mathbf{y}'\mathbf{X}\hat{\mathbf{a}}$$

Second stage:

Develop an expression for a (finite) change.

We refer to the new level of S as S*, and to increments with the prefix Δ. We develop the following relation:

$$S^* = S + \Delta S = \mathbf{y}'\mathbf{y} - 2\mathbf{y}'\mathbf{X}[\hat{\mathbf{a}} + \Delta\hat{\mathbf{a}}] + [\hat{\mathbf{a}} + \Delta\hat{\mathbf{a}}]'\,\mathbf{X}'\mathbf{X}\,[\hat{\mathbf{a}} + \Delta\hat{\mathbf{a}}] =$$

$$\mathbf{y}'\mathbf{y} - 2\mathbf{y}'\mathbf{X}[\hat{\mathbf{a}} + \Delta\hat{\mathbf{a}}] + [\hat{\mathbf{a}}' + \Delta\hat{\mathbf{a}}']\,\mathbf{X}'\mathbf{X}\,[\hat{\mathbf{a}} + \Delta\hat{\mathbf{a}}] =$$

$$\mathbf{y}'\mathbf{y} - 2\mathbf{y}'\mathbf{X}[\hat{\mathbf{a}} + \Delta\hat{\mathbf{a}}] + \hat{\mathbf{a}}'\,\mathbf{X}'\mathbf{X}\,[\hat{\mathbf{a}} + \Delta\hat{\mathbf{a}}] + \Delta\hat{\mathbf{a}}'\,\mathbf{X}'\mathbf{X}\,[\hat{\mathbf{a}} + \Delta\hat{\mathbf{a}}] =$$

$$\mathbf{y}'\mathbf{y} - 2\mathbf{y}'\mathbf{X}\hat{\mathbf{a}} - 2\mathbf{y}'\mathbf{X}\,\Delta\hat{\mathbf{a}} + \hat{\mathbf{a}}'\mathbf{X}'\mathbf{X}\hat{\mathbf{a}} + \hat{\mathbf{a}}'\mathbf{X}'\mathbf{X}\,\Delta\hat{\mathbf{a}}$$

$$+ \Delta\hat{\mathbf{a}}'\mathbf{X}'\mathbf{X}\hat{\mathbf{a}} + \Delta\hat{\mathbf{a}}'\mathbf{X}'\mathbf{X}\,\Delta\hat{\mathbf{a}} =$$

$$\mathbf{y}'\mathbf{y} - 2\mathbf{y}'\mathbf{X}\hat{\mathbf{a}} - 2\mathbf{y}'\mathbf{X}\,\Delta\hat{\mathbf{a}} + \hat{\mathbf{a}}'\mathbf{X}'\mathbf{X}\hat{\mathbf{a}}$$

$$+ 2\hat{\mathbf{a}}'\mathbf{X}'\mathbf{X}\,\Delta\hat{\mathbf{a}} + \Delta\hat{\mathbf{a}}'\mathbf{X}'\mathbf{X}\,\Delta\hat{\mathbf{a}} \tag{2.8.3}$$

Note again the combination of two terms in the penultimate member of (2.8.3), which are each others transpose (and therefore, being of order 1 by 1, identical), into a single term $2\hat{a}'X'X\,\Delta\hat{a}$.

An expression for the increment of S is now obtained, by subtracting (2.8.2) from (2.8.3):

$$\Delta S = -2y'X\,\Delta\hat{a} + 2\hat{a}'X'X\,\Delta\hat{a} + \Delta\hat{a}'X'X\,\Delta\hat{a} \qquad (2.8.4)$$

Third stage:

As usual in differential calculus, we assume that the change in the (vector of) independent argument variables is only very small, approaching zero, and terms which involve a product of two such small changes (or the square of one), are dropped.

Therefore:

$$d S = -2\,y'X\,d\hat{a} + 2\,\hat{a}'X'X\,d\hat{a} \qquad (2.8.5)$$

Fourth stage:

Present the result in a standardized form. Normally, the easiest transition from the third to the fourth stage is:

$$f = f(x) \;\rightarrow\; df = v'dx \;\rightarrow\; \partial f/\partial x = v$$

This follows the usual convention of presenting vectors in column form.

For example, for $f(x) = a'x$, we would have: $df = a'dx$, $\partial f/\partial x = a$.

In the example we are dealing with, the fourth stage is therefore initiated with:

$$dS = [-2y'X + 2\hat{a}'X'X]\,d\hat{a} \qquad (2.8.6)$$

In this example, this part of the fourth stage is rather trivial, it simply amounts to grouping the two terms together.

There are, however, functions and ways of developing them, where the third stage can end with the vector of independent increments somewhere in the middle of at least some of the terms, and a slightly more substantial operation, such as transposition of one or more product-expressions, may be needed.

We now need to conform with the standard matrix notation convention, that the 'normal' way of presenting a vector is the column form. Therefore, on transposing the vector within [ ] in (2.8.6), which is a row-vector, we obtain:

$$\partial S / \partial \hat{a} = -2 X' y + 2 X'X \hat{a} \qquad (2.8.7)$$

Exercises on matrix differentiation:

Verify the order of the operands, and the accuracy of the following formulae, including any restrictions on the matrices and vectors involved, implied by the result:

1:

$$\frac{\partial x'[b\ b']x}{\partial b} = 2(x'b)x = 2xx'b \ , \quad n \text{ by } 1.$$

2:

$$\frac{\partial x'\hat{B}x}{\partial b} = \hat{X}\,x, \quad n \text{ by } 1$$

(Refer to section 3.1 concerning the notation used for diagonal matrices)

3:

$$\frac{\partial^2 x'Ax}{\partial^2 x} = 2 A, \quad n \text{ by } n$$

4:

$$\frac{\partial x'a\,b'x}{\partial x} \quad = \quad [a\,b' + b\,a']\,x\,, \qquad n \text{ by } 1.$$

Answersheet on matrix-multiplication

Av', **Az'**, **Bv'**, and **Bz'** are illegal on account of the attempt to post-multiply a matrix by a row.

cA, **cB**, **dA**, and **dB** are illegal, on account of the attempt to pre-multiply a matrix by a column.

Ac, **Bc**, **BA**, **z'A**, and **z'c** are illegal, on account of the attempt to match 3 columns (or elements of a row-vector), to only 2 rows.

v'B and **v'd** are illegal, on account of the attempt to match 2 elements of a row (=2 columns), to 3 rows.

The legitimate expressions which are asked for are:

$$AB = \begin{bmatrix} 1 & 2 & 3 \\ 3 & 4 & 5 \end{bmatrix} \begin{bmatrix} -1 & 2 & -2 \\ 5 & 3 & 2 \\ -2 & - & 8 \end{bmatrix} = \begin{bmatrix} 3 & 8 & 26 \\ 7 & 18 & 42 \end{bmatrix}$$

$$Ad = \begin{bmatrix} 1 & 2 & 3 \\ 3 & 4 & 5 \end{bmatrix} \begin{bmatrix} 2 \\ -1 \\ 4 \end{bmatrix} = \begin{bmatrix} 12 \\ 22 \end{bmatrix}$$

$$Bd = \begin{bmatrix} -1 & 2 & -2 \\ 5 & 3 & 2 \\ -2 & - & 8 \end{bmatrix} \begin{bmatrix} 2 \\ -1 \\ 4 \end{bmatrix} = \begin{bmatrix} -12 \\ 15 \\ 28 \end{bmatrix}$$

$$\mathbf{c}\,\mathbf{v'} = \begin{bmatrix} 2 \\ 4 \end{bmatrix} \begin{bmatrix} 1 & -1 \end{bmatrix} = \begin{bmatrix} 2 & -2 \\ 4 & -4 \end{bmatrix}$$

$$\mathbf{c}\,\mathbf{z'} = \begin{bmatrix} 2 \\ 4 \end{bmatrix} \begin{bmatrix} 2 & -2 & 3 \end{bmatrix} = \begin{bmatrix} 4 & -4 & 6 \\ 8 & -8 & 12 \end{bmatrix}$$

$$\mathbf{v'}\,\mathbf{A} = \begin{bmatrix} 1 & -1 \end{bmatrix} \begin{bmatrix} 1 & 2 & 3 \\ 3 & 4 & 5 \end{bmatrix} = \begin{bmatrix} -2 & -2 & -2 \end{bmatrix}$$

$$\mathbf{v'}\,\mathbf{c} = \begin{bmatrix} 1 & -1 \end{bmatrix} \begin{bmatrix} 2 \\ 4 \end{bmatrix} = \begin{bmatrix} -2 \end{bmatrix}$$

$$\mathbf{z'}\,\mathbf{B} = \begin{bmatrix} 2 & -2 & 3 \end{bmatrix} \begin{bmatrix} -1 & 2 & -2 \\ 5 & 3 & 2 \\ -2 & - & 8 \end{bmatrix} = \begin{bmatrix} -18 & -2 & 16 \end{bmatrix}$$

and

$$\mathbf{z'}\mathbf{d} = \begin{bmatrix} 2 & -2 & 3 \end{bmatrix} \begin{bmatrix} 2 \\ -1 \\ 4 \end{bmatrix} = 18.$$

The expressions $\mathbf{d}\,\mathbf{v'}$ (order 2 by 2), and $\mathbf{d}\,\mathbf{z'}$ (order 2 by 3) are legitimate, but their calculations were not asked.

Answersheet on matrix differentiation

<u>1</u>:

$$\emptyset(\mathbf{b}) = \mathbf{x'}[\mathbf{b}\,\mathbf{b'}]\mathbf{x} = \mathbf{x'}\mathbf{b}\,\mathbf{b'}\mathbf{x}$$

is a 1 by 1 expression.

$\mathbf{x'}$ is of order 1 by n. The outer product $[\mathbf{b}\,\mathbf{b'}] = \mathbf{b}\,\mathbf{b'}$ is of order n by n, and $\mathbf{x}$ is of order n by 1. Differentiation of a scalar expression, relative to an n by 1 column vector, gives rise to an n by 1 column of partial derivatives, confirming the stated order.

We now develop an expression for the changed value of ϕ, as follows:

$$f(\mathbf{b} + \Delta\mathbf{b}) = \mathbf{x}'[\mathbf{b} + \Delta\mathbf{b}][\mathbf{b}' + \Delta\mathbf{b}']\mathbf{x}$$

$$= \mathbf{x}'\mathbf{b}[\mathbf{b}' + \Delta\mathbf{b}']\mathbf{x} + \mathbf{x}'\Delta\mathbf{b}[\mathbf{b}' + \Delta\mathbf{b}']\mathbf{x}$$

$$= \mathbf{x}'\mathbf{b}\,\mathbf{b}'\mathbf{x} + \mathbf{x}'\mathbf{b}\,\Delta\mathbf{b}'\mathbf{x} + \mathbf{x}'\Delta\mathbf{b}\,\mathbf{b}'\mathbf{x} + \mathbf{x}'\Delta\mathbf{b}\,\Delta\mathbf{b}'\mathbf{x}$$

Therefore:

$$\Delta\phi = \phi(\mathbf{b} + \Delta\mathbf{b}) - \phi(\mathbf{b}) = \mathbf{x}'\mathbf{b}\,\Delta\mathbf{b}'\mathbf{x} + \mathbf{x}'\Delta\mathbf{b}\,\mathbf{b}'\mathbf{x} + \mathbf{x}'\Delta\mathbf{b}\,\Delta\mathbf{b}'\mathbf{x}$$

The expressions $\mathbf{x}'\mathbf{b}$ and $\Delta\mathbf{b}'\mathbf{x}$, resp $\mathbf{x}'\Delta\mathbf{b}$ and $\mathbf{b}'\mathbf{x}$, are each others transpose. As an expression of order order 1 by 1 always is its own transpose, we have:

$$\mathbf{x}'\mathbf{b} = \mathbf{b}'\mathbf{x}, \text{ and } d\mathbf{b}'\mathbf{x} = \mathbf{x}'d\mathbf{b}.$$

This consideration simplifies $\Delta\phi$ to:

$$\Delta\phi = 2\mathbf{x}'\mathbf{b}\,\Delta\mathbf{b}'\mathbf{x} + \Delta\mathbf{b}'\mathbf{x}\,\mathbf{x}'\mathbf{b}$$

$$= 2\mathbf{b}'\mathbf{x}\,\mathbf{x}'\mathbf{b} + \Delta\mathbf{b}'\mathbf{x}\,\mathbf{x}'\Delta\mathbf{b}$$

For differential changes approaching zero, we may neglect the last term on the righthand side, as it refers to products of changes.

Hence:

$$d\phi = 2\mathbf{b}'\mathbf{x}\,\mathbf{x}'d\mathbf{b}$$

The vector of first-order partial derivatives, presented in row-form is now obtained by leaving the vector $d\mathbf{b}$ out. However, to conform with the standard convention of presenting vectors in column form, we need to transpose, and confirm the stated relation:

$$\frac{\partial\mathbf{x}'[\mathbf{b}\,\mathbf{b}']\mathbf{x}}{\partial\mathbf{b}} = 2\mathbf{x}\,\mathbf{x}'\mathbf{b} = 2(\mathbf{x}'\mathbf{b})\mathbf{x}$$

Note, that the factor $(\mathbf{x}'\mathbf{b})$ is of order 1 by 1, and has been treated as a scalar variable:

b x is not a valid expression, therefore $2\mathbf{x'b}$ **x** is not either.

<u>2</u>:

$$\phi(\mathbf{b}) = \mathbf{x'\hat{B}x} = \mathbf{x'[\hat{b}]x}$$

$$\phi(\mathbf{b} + \Delta\mathbf{b}) = \mathbf{x'[\hat{b}} + \Delta\hat{\mathbf{b}}]\mathbf{x}$$

Therefore:

$$\Delta\phi = \phi(\mathbf{b} + \Delta\mathbf{b}) - \phi(\mathbf{b}) = \mathbf{x'[\Delta\hat{b}]x}$$

The product of a diagonal matrix, post-multiplied by a column vector, permits interchanging:

$$\begin{bmatrix} \Delta b_1 & . & . & . \\ . & \Delta b_2 & . & . \end{bmatrix} \begin{bmatrix} x_1 \\ x_2 \end{bmatrix} = \begin{bmatrix} \Delta b_1 x_1 \\ \Delta b_2 x_2 \end{bmatrix} = \begin{bmatrix} x_1 & . & . & . \\ . & x_2 & . & . \end{bmatrix} \begin{bmatrix} \Delta b_1 \\ \Delta b_2 \end{bmatrix}$$

Therefore:

$$\Delta\phi = \mathbf{x'[\hat{x}]\Delta b}$$

The row-presentation of $\mathrm{O}[\mathbf{x'\hat{B}x}]/\mathrm{O}\mathbf{b}$ therefore is:

$$\mathrm{O}[\mathbf{x'\hat{B}x}]/\mathrm{O}\mathbf{b'} = \mathbf{x'[\hat{x}]} = \mathbf{x'\hat{X}}$$

To obtain the conventional column-presentation, we need to transpose, and confirm the formula as given.

<u>3</u>:

The legitimacy of the expression **x'Ax** already implies that **A** is square. The expression **x'Ax** is of order 1 by 1, and its transpose is **x'A'x**.

Therefore, given that the transpose of a 1 by 1 expression is the expression itself:

$$\mathbf{x'Ax} = \mathbf{x'A'x} = \mathbf{x'}[0.5\ \mathbf{A} + 0.5\ \mathbf{A'}]\mathbf{x}$$

It follows that all (first and second order) derivatives of **x'Ax** are those of **x'**[0.5 **A** + 0.5 **A'**]**x**, irrespective of the symmetry of **A**.

As $\phi(\mathbf{x})$ is of order 1 by 1, there is an n by 1 vector of first-order derivatives, and an n by n Hessian matrix of second order derivatives, confirming the stated order.

We first develop the vector of first-order derivatives:

From $\phi(\mathbf{x}) = \mathbf{x'Ax}$ we infer:

$$\phi(\mathbf{x} + \Delta\mathbf{x}) = [\mathbf{x'} + \Delta\mathbf{x'}]\,\mathbf{A}\,[\mathbf{x} + \Delta\mathbf{x}]$$

$$= \mathbf{x'}\,\mathbf{A}\,[\mathbf{x} + \Delta\mathbf{x}] + \Delta\mathbf{x'}\,\mathbf{A}\,[\mathbf{x} + \Delta\mathbf{x}]$$

$$= \mathbf{x'Ax} + \mathbf{x'A}\,\Delta\mathbf{x} + \Delta\mathbf{x'Ax} + \Delta\mathbf{x'A}\,\Delta\mathbf{x}$$

Again, the first two terms in the last member above are each other's transpose, and, being of order 1 by 1, they are identical.

Therefore:

$$\Delta\phi = \mathbf{x'}[\mathbf{A} + \mathbf{A'}]\Delta\mathbf{x} + \Delta\mathbf{x'A}\,\Delta\mathbf{x}$$

From which, on suppressing the second-order term for variations approaching zero:

$$d\phi = \mathbf{x'}[\mathbf{A} + \mathbf{A'}]d\mathbf{x}$$

This result is always symmetric, and becomes identical to the formula as given, if $\mathbf{A}$ was symmetric. Hence we must assume that (as is usual with quadratic forms $\mathbf{x'Ax}$), $\mathbf{A}$ is a symmetric matrix.

<u>4</u>

For the expression $\mathbf{x'a\,b'x}$ to be legitimate, we need to assume, that $\mathbf{x}$, $\mathbf{a}$, and $\mathbf{b}$, are all of the same order n by 1.

This makes the outer product $\mathbf{a\,b'}$ into a (square) n by n matrix. The derivation of the vector of first-order derivatives of the quadratic form $\mathbf{x'}[\mathbf{a\,b'}]\mathbf{x}$ is then the same as given for $\mathbf{x'Ax}$ above.

The one presentational difference is that, unless $\mathbf{b}$ is proportional to $\mathbf{a}$, $[\mathbf{a\,b'}]$ is not symmetric, and

$$\partial\phi/\partial\mathbf{x} = [\mathbf{A} + \mathbf{A'}]\mathbf{x}\ \text{will apply, rather than } 2\,\mathbf{Ax}.$$

CHAPTER III

SOME STRUCTURED MATRICES AND OPERATORS

3.1 Some matrices of special structure

In this section we first of all introduce the names of some
matrices (and vectors) of special structure, as we shall find
cause to refer to such matrices and vectors.

We recall first of all (from section 2.2), the terms <u>square</u>
matrix (= number of columns equals number of rows), and, <u>within</u>
the class of square matrices, the <u>symmetric</u> matrix ($a_{ij} = a_{ji}$).

We now introduce some more types of special matrices.

A symmetric matrix **A**, which meets the condition:

$a_{ij} = 0$ for all admissible $i \neq j$,

is said to be a <u>diagonal</u> matrix.

Example:

$$\mathbf{A} \quad = \quad \begin{bmatrix} 2 & - & - \\ - & - & - \\ - & - & - \end{bmatrix}$$

A matrix, of which all the elements are zeros, is a <u>null matrix</u>.

Note, that the requirement that the off-diagonal elements of a diagonal matrix must be zeros, does not imply that the diagonal elements are non-zero. A symmetric null matrix is therefore a special case of a diagonal matrix.

Following Stone [44], we shall use the hat '$\hat{}$', to indicate a diagonal matrix.

The notation to be used in this book is, however, somewhat different from the one used by Stone. According to Stone's notation, $\hat{p}$ is a diagonal matrix. The notation to be followed in this book consistently uses upper case letters or brackets [ ] for matrices. We shall do so, either in relation to the vector of the diagonal elements, e.g.

$$\mathbf{a} = \begin{bmatrix} 1 \\ 2 \\ 3 \end{bmatrix} \qquad\qquad [\hat{\mathbf{a}}] = \begin{bmatrix} 1 & - & - \\ - & 2 & - \\ - & - & 3 \end{bmatrix}$$

or simply put the hat on the matrix itself, i.e. the symbol $\hat{\mathbf{A}}$ will indicate a diagonal matrix.

A symmetric matrix, of which all the non-zero elements occur on or on one side of the diagonal (or main diagonal, we do not really use any other type of diagonal), is called a triangular matrix.

We distinguish between upper triangular matrices ($i > j$ implies $a_{ij} = 0$), and lower triangular matrices ($j > i$ implies $a_{ij} = 0$).

Note, that a diagonal matrix, and a symmetric null matrix, are both upper- and lower-triangular.

$$\mathbf{A} = \begin{bmatrix} 2 & - & - \\ - & - & - \\ - & - & - \end{bmatrix} \quad\text{and}\quad \mathbf{B} = \begin{bmatrix} 1 & - & - \\ 2 & 3 & - \\ 4 & 5 & 6 \end{bmatrix}$$

are both lower-triangular matrices

A diagonal matrix of, of which all the diagonal elements are

1=one, is called a <u>unit matrix</u> or <u>identity matrix</u>. The identity
matrix is conventionally indicated by the symbol I, or sometimes
I plus an integer number in index-position, indicating its
order. I_n is an identity matrix of order n.

Identity matrices have the property that any legitimate product-
expressions in which they occur, are unaffected by their
presence:

$$
\begin{bmatrix} 1 & - & - \\ - & 1 & - \\ - & - & 1 \end{bmatrix}
\begin{bmatrix} x_1 \\ x_2 \\ x_3 \end{bmatrix}
=
\begin{bmatrix} x_1 \\ x_2 \\ x_3 \end{bmatrix}
$$

3.2 Partitioning and blocks

A number of rows of a matrix may be grouped into a <u>block-row</u>.
Then again, a number of columns of a matrix may be grouped in a
<u>block-column</u>. The intersection of a block-row and a block-column
is called a <u>block</u>. The operation of defining block-rows and/or
block-columns, and by implication blocks, is called <u>partitioning</u>.

Normally, partitioning will arise, i.e. be found useful, if the
matrix has a particular structure as regards to its zero and non-
zero elements, giving rise to a partitioning in which some of the
blocks are null blocks.

Example:

$$
\mathbf{A} = \begin{bmatrix} 1 & 2 & - & - \\ 3 & 4 & - & - \\ - & - & 5 & 6 \\ - & - & 7 & 8 \end{bmatrix}
$$

We conventionally indicate partitioning with some further lines,
e.g. the partitioning which **A** as indicated above suggests more or
less naturally, is indicated as:

$$
\mathbf{A} = \begin{bmatrix} 1 & 2 & - & - \\ 3 & 4 & - & - \\ - & - & 5 & 6 \\ - & - & 7 & 8 \end{bmatrix}
$$

A has now been partitioned into two block-columns, which are indicated with the same letter as the matrix, i.e. in this case:

$$
\mathbf{A_1} = \begin{bmatrix} 1 & 2 \\ 3 & 4 \\ - & - \\ - & - \end{bmatrix} \qquad\qquad \mathbf{A_2} = \begin{bmatrix} - & - \\ - & - \\ 5 & 6 \\ 7 & 8 \end{bmatrix}
$$

It is necessary to state explicitly, that $\mathbf{A_1}$ and $\mathbf{A_2}$ are block-columns of $\mathbf{A}$, as otherwise there could be confusion with some sort of series-index, such as we saw for example in Chapter II for the tableau-matrices $\mathbf{T_1}$, $\mathbf{T_2}$, etc.

The same matrix $\mathbf{A}$ also partitions naturally into two block-rows:

$$
\mathbf{A'_1} = \begin{bmatrix} 1 & 2 & - & - \\ 3 & 4 & - & - \end{bmatrix}
$$

and

$$
\mathbf{A'_2} = \begin{bmatrix} - & - & 5 & 6 \\ - & - & 7 & 8 \end{bmatrix}
$$

As with rows of matrices, we shall be following the convention of putting the (') transposition sign before the index if we refer to a (block-)row of a matrix, and the other way round, if we refer to a transposed block-column:

$$
\mathbf{A_2'} = \begin{bmatrix} - & - & 5 & 7 \\ - & - & 6 & 8 \end{bmatrix}
$$

is the transpose of $\mathbf{A_2}$, the second block-column of A.

The intersection of a block-column and a block-row is called a
block, and is indicated by the same capital letter as the matrix,
followed by two indices, the first of which indicates to which
block-row the block belongs, the second indicates the block-
column:

$$A_{11} = \begin{bmatrix} 1 & 2 \\ 3 & 4 \end{bmatrix} \qquad A_{12} = \begin{bmatrix} - & - \\ - & - \end{bmatrix}$$

$$A_{21} = \begin{bmatrix} - & - \\ - & - \end{bmatrix} \qquad A_{22} = \begin{bmatrix} 5 & 6 \\ 7 & 8 \end{bmatrix}$$

Blockstructure does not always relate to zero and non-zero
blocks. It can also be useful, if different parts of a tableau of
numbers, even if all non-zero, have a different relevance, and/or
require different operations to be performed on them.

The following is a demonstration-example of an inter-industry
analysis tableau or input-output table, drawn from my book
Forecasting Models for National Economic Planning [23]:

	s_1	s_2	d_1	d_2	d_3	d_4	d_5
s_1	40	20	50	20	130	50	-10
s_2	45	45	115	23	50	20	2
f_1	50	30	40	10	-	20	-10
f_2	100	120	-	80	-	-	-
f_3	35	55	-	-	-	-	-

This type of tableau gives for each sector of production the
names of the sectors being indicated here for only two as s_1 and
s_2, along the row, the destinations where its output went, i.e.
either some other sector of production as semi-finished product
(including supplies to other firms in the same industry, which
are entered in the diagonal cells), or to the final sales
destinations, d_1 = domestic sales to households, d_2 = domestic
sales to the government, d_3 = export, d_4 = domestic sales of
fixed capital equipment to industry, and d_5 = supplies of
inventories going into stocks, where the latter item can be
negative.

The bottom 4 rows of the tableau give the similar account of the income-value arising to the suppliers of production-factors and other beneficiaries: f_1 = imports, f_2 = taxes on the sold products, f_3 = employment, f_4 = residual income to industry.

In this example, the partitioning purely relates to the different meaning of the figures in the four quadrants.

3.3 Some operators and their use

An <u>operator</u> is a matrix (or a vector), which serves the purpose of putting a certain operation into effect, rather than providing itself meaningful numerical information. Some commonly used operators are:

The summation-vector:

We reserve the symbol **s** (or, to indicate its order, **s**, followed by an integer number in index-position, e.g. $\mathbf{s}_n$), for a vector consisting of nothing but unity elements, e.g.:

$$\mathbf{s}_3 = \begin{bmatrix} 1 \\ 1 \\ 1 \end{bmatrix}$$

For example, the input-output table matrix **T** given in the previous section, contains just the raw 'data', but input-output tables are normally presented with sumcount columns and rows (as well as some sub-total columns and rows).

$$\mathbf{T} = \begin{bmatrix} 40 & 20 & 50 & 20 & 130 & 50 & -10 \\ 45 & 45 & 115 & 23 & 50 & 20 & 2 \\ 50 & 30 & 40 & 10 & - & 20 & -10 \\ 100 & 120 & - & 80 & - & - & - \\ 35 & 55 & - & - & - & - & - \end{bmatrix} \qquad \mathbf{Ts} = \begin{bmatrix} 300 \\ 300 \\ 140 \\ 60 \\ 90 \end{bmatrix}$$

and the fact that **s** is a vector consisting of seven unity elements, serving to add up the elements in each row of **T** into an entry into the column of sum-totals, should be understood, rather than strictly needing to be stated explicitly.

Likewise, $\mathbf{s'T}$ = [300   300   205   133   180   90   -18], where $\mathbf{s'}$ is not the transpose of $\mathbf{s}$ as above, but a row of unity elements, of an order which makes the product $\mathbf{s'T}$ legitimate, i.e. 5. In case of any doubt on that point, one may wish to add an index to the symbol $\mathbf{s}$, (or $\mathbf{s'}$, as the case may be), which is in that case its order.

A vector consisting of a single unity element and otherwise zeros, is known as a <u>unit vector</u>. It normally occurs in a context where its order is known already, and the single integer number in index-position which is normally associated with a unit vector, indicates the position of the unity element.

It is at this point necessary to mention a potential confusion in terminology. There is a different more general definition of the term 'unit' vector, where a vector is called a unit vector if it meets the condition that the inner product of the vector and itself equals 1, i.e. the vector $\mathbf{x}$ is said to be a unit vector if it meets the condition $\mathbf{x'x}$ = 1. (See for example Pullman [37], p.199.) This is related to the geometrical concept of the length of the vector. We shall come to discuss this issue in Chapter X, where we shall refer to such a vector as a vector of length 1, but the main thrust of this book is on more specifically algebraic problems, and the definition used here is well established in a branch of applied mathematics, linear programming (Dantzig [8], p.179). Given the essential overlap between linear programming and the resolution of linear equations, we need the same term, with the same meaning and rationale, and it makes no sense to use a different term for it in different textbooks.

Unit vectors serve the purpose of isolating columns (or rows) out of a matrix:

$$\begin{bmatrix} 11 & 12 & 13 & 14 \\ 21 & 22 & 23 & 24 \\ 31 & 32 & 33 & 34 \end{bmatrix} \begin{bmatrix} - \\ - \\ 1 \\ - \end{bmatrix} = \begin{bmatrix} 13 \\ 23 \\ 33 \end{bmatrix}$$

$$[- \quad - \quad 1] \begin{bmatrix} 11 & 12 & 13 & 14 \\ 21 & 22 & 23 & 24 \\ 31 & 32 & 33 & 34 \end{bmatrix} = [31 \quad 32 \quad 33 \quad 34]$$

Some slightly more complicated operators are the <u>permutation operator</u>, and the <u>aggregation matrix</u>.

A <u>permutation operator</u> is a square matrix of order n by n, containing n non-zero elements, all of them unity, which are arranged in such a way, that all its columns, and all its rows, are unit-vectors.

As the name indicates, a permutation-operator serves the purpose of permuting, or re-ordering a matrix. Post-multiplication by a permutation-operator interchanges columns, pre-multiplication by a permutation-operator interchanges rows.

We shall be following the convention, of indicating the re-ordered matrix with the same letter as the original matrix, but with an asterisk added to it:

$$\mathbf{A} = \begin{bmatrix} 11 & 12 & 13 & 14 \\ 21 & 22 & 23 & 24 \\ 31 & 32 & 33 & 34 \end{bmatrix}$$

$$\mathbf{A^*} = \begin{bmatrix} 11 & 12 & 13 & 14 \\ 21 & 22 & 23 & 24 \\ 31 & 32 & 33 & 34 \end{bmatrix} \begin{bmatrix} 1 & - & - & - \\ - & - & - & 1 \\ - & - & 1 & - \\ - & 1 & - & - \end{bmatrix} = \begin{bmatrix} 11 & 14 & 13 & 12 \\ 21 & 24 & 23 & 22 \\ 31 & 34 & 33 & 32 \end{bmatrix}$$

An <u>aggregation matrix</u> also has the property that all its elements are either 0 = zero, or 1 = one. Its purpose is best discussed by distinguishing column-aggregation matrices, whose purpose is the aggregation of several columns of a matrix into one column of its transformation, and row-aggregation matrices, whose purpose is the aggregation of several rows of the original matrix into a single row of of the transformed matrix. In fact, there is no clear distinction between those two types of aggregation matrices, except that the one operation is performed by post-multiplication, and the other by pre-multiplication.

$$
\begin{bmatrix} 11 & 12 & 13 & 14 \\ 21 & 22 & 23 & 24 \\ 31 & 32 & 33 & 34 \end{bmatrix}
\begin{bmatrix} 1 & - \\ 1 & - \\ - & 1 \\ - & 1 \end{bmatrix}
=
\begin{bmatrix} 23 & 27 \\ 43 & 47 \\ 63 & 67 \end{bmatrix}
$$

and

$$
\begin{bmatrix} 1 & 1 & - \\ - & - & 1 \end{bmatrix}
\begin{bmatrix} 11 & 12 & 13 & 14 \\ 21 & 22 & 23 & 24 \\ 31 & 32 & 33 & 34 \end{bmatrix}
=
\begin{bmatrix} 32 & 34 & 36 & 36 \\ 31 & 32 & 33 & 34 \end{bmatrix}
$$

In this example, the aggregation matrices served to aggregate,
i.e. to add up only. Aggregation can also occur in combination
with also copying the original information, and one may, or may
not, still want to use the term aggregation operator.

Consider for example, the input-output tableau matrix as
illustrated in the previous section. In presenting such tableaus,
it is in fact usual to include sub-total columns for the supplies
delivered to industry and to final demand separately, as well as
a column of total supplies. We again use an asterisks to indicate
a transformed matrix.

We illustrate the operation first for a column-aggregation
and copying operator:

$$
C = \left[\begin{array}{cc|cccccc|c}
1 & - & 1 & - & - & - & - & - & 1 \\
- & 1 & 1 & - & - & - & - & - & 1 \\ \hline
- & - & - & 1 & - & - & - & 1 & 1 \\
- & - & - & - & 1 & - & - & 1 & 1 \\
- & - & - & - & - & 1 & - & 1 & 1 \\
- & - & - & - & - & - & 1 & - & 1 & 1 \\
- & - & - & - & - & - & - & 1 & 1 & 1
\end{array}\right]
$$

For the tableau matrix illustrate in the previous section,

$$
T = \begin{bmatrix}
40 & 20 & 50 & 20 & 130 & 50 & -10 \\
45 & 45 & 115 & 23 & 50 & 20 & 2 \\
50 & 30 & 40 & 10 & - & 20 & -10 \\
100 & 120 & - & 80 & - & - & - \\
35 & 55 & - & - & - & - & -
\end{bmatrix}
$$

we then have:

	s_1	s_2	Σ	d_1	d_2	d_3	d_4	d_5	Σ	Σ
$T\,C =$	40	20	60	50	20	130	50	-10	240	300
	45	45	90	115	23	50	20	2	210	300
	50	30	40	40	10	-	20	-10	60	140
	30	30	60	-	-	-	-	-	-	60
	100	120	220	-	80	-	-	-	80	300
	35	55	90	-	-	-	-	-	-	90

To obtain sumcount rows as well, we now need to pre-multiply by a row-aggregation and copying operator matrix, which is:

$$
R = \begin{bmatrix}
1 & - & - & - & - & - \\
- & 1 & - & - & - & - \\
1 & 1 & - & - & - & - \\
- & - & 1 & - & - & - \\
- & - & - & 1 & - & - \\
- & - & - & - & 1 & - \\
- & - & - & - & - & 1 \\
- & - & 1 & 1 & 1 & 1 \\
1 & 1 & 1 & 1 & 1 & 1
\end{bmatrix}
$$

The tabulation with subtotal rows and subtotal columns, as well as overall total row and column, is now obtained as:

$\mathbf{T}^* = \mathbf{R} \mathbf{T} \mathbf{C} =$

	s_1	s_2	Σ	d_1	d_2	d_4	d_4	d_5	Σ	Σ
s_1	40	20	60	50	20	130	50	−10	240	300
s_2	45	45	90	115	23	50	20	2	210	300
Σ	85	65	150	165	43	180	70	−8	450	600
f_1	50	30	40	40	10	–	20	−10	60	140
f_2	30	30	60	–	–	–	–	–	–	60
f_3	100	120	220	–	80	–	–	–	80	300
f_4	35	55	90	–	–	–	–	–	–	90
Σ	215	235	450	40	90	–	20	−10	140	590
Σ	300	300		205	133	180	90	−18	590	1180

3.4 Matrix multiplication per outer product

In the previous chapter, we illustrated how the product of two matrices is evaluated, immediately for each separate element, i.e. we calculate the element of the product-expression in position 1,1 first, than deal with the element in position 1,2 (or 2,1), etc. That way we evaluate the product $\mathbf{C} = \mathbf{AB}$ per group of terms, as figuring in:

$$c_{ij} = \sum_{h=1}^{n} a_{ih} b_{hj} \qquad\qquad (3.4.1)$$

successively for all i, and all j.

This corresponds to evaluating each c_{ij} as an inner product:

$$c_{ij} = \mathbf{a}'_i \mathbf{b}_j \quad (i = 1, 2, \ldots\ldots m, \; j = 1, 2, \ldots\ldots\ldots k)$$

But we may also separately state all the products $a_{ih} b_{hj}$, group them in a matrix, and then add n such matrices together.

In that case $\mathbf{C}$ is expressed as a summation of n outer products:

$$C = \sum_{h=1}^{n} a_h \, b'_h \qquad\qquad (3.4.2)$$

Example (from section 2.4):

$$\mathbf{A}\,\mathbf{B} = \begin{bmatrix} 1 & 2 & -3 \\ -2 & 5 & -1 \end{bmatrix} \begin{bmatrix} 2 & 5 \\ 1 & -3 \\ -1 & 2 \end{bmatrix} = \begin{bmatrix} 7 & -7 \\ 2 & -27 \end{bmatrix}$$

The product matrix can also be obtained as the summation of 3 outer products, as follows:

$$h = 1: \quad a_1\, b'_1 = \begin{bmatrix} 1 \\ -2 \end{bmatrix} \begin{bmatrix} 2 & 5 \end{bmatrix} = \begin{bmatrix} 2 & 5 \\ -4 & -10 \end{bmatrix}$$

$$h = 2: \quad a_2\, b'_2 = \begin{bmatrix} 2 \\ 5 \end{bmatrix} \begin{bmatrix} 1 & -3 \end{bmatrix} = \begin{bmatrix} 2 & -6 \\ 5 & -15 \end{bmatrix}$$

$$h = 3: \quad a_3\, b'_3 = \begin{bmatrix} -3 \\ -1 \end{bmatrix} \begin{bmatrix} -1 & 2 \end{bmatrix} = \begin{bmatrix} 3 & -6 \\ 1 & -2 \end{bmatrix}$$

Therefore, adding the three terms together:

$$C = \begin{bmatrix} 7 & -7 \\ 2 & -27 \end{bmatrix}$$

In the standard case of multiplying two matrices, this order of performing the calculations is probably not the most practical one. The situation becomes different if we are dealing with a recursive product of three matrices, of which the middle one is a diagonal matrix.

The product $\mathbf{A}\,\hat{\mathbf{D}}$ is obtained, by multiplication of each jth column of $\mathbf{A}$ by d_{jj}.

Therefore, if $\hat{\mathbf{D}}$ is of order n by n, application of (3.4.2) leads to:

$$\mathbf{A}\ \hat{\mathbf{D}}\ \mathbf{B}\ =\ \sum_{j=1}^{n} \mathbf{a}_j\, d_{jj}\, \mathbf{b}'_j\ =\ \sum_{j=1}^{n} d_{jj}\, \mathbf{a}_j\, \mathbf{b}'_j \qquad (3.4.3)$$

Example:

$$\mathbf{A}\ \hat{\mathbf{D}}\ \mathbf{B}\ =\ \begin{bmatrix} 2 & 3 \\ -1 & 2 \\ 1 & 7 \end{bmatrix} \begin{bmatrix} -2 & - \\ - & 3 \end{bmatrix} \begin{bmatrix} 1 & -5 & 2 \\ 4 & 3 & -5 \end{bmatrix}$$

$$d_{11}\ \mathbf{a}_1\ \mathbf{b}'_1\ =\ -2 \begin{bmatrix} 2 & -10 & 4 \\ -1 & 5 & -2 \\ 1 & -5 & 2 \end{bmatrix} = \begin{bmatrix} -4 & 20 & -8 \\ 2 & -10 & 4 \\ -2 & 10 & -4 \end{bmatrix}$$

$$d_{22}\ \mathbf{a}_2\ \mathbf{b}'_2\ =\ 3 \begin{bmatrix} 12 & 9 & -15 \\ 8 & 6 & -10 \\ 28 & 21 & -35 \end{bmatrix} = \begin{bmatrix} 36 & 27 & -45 \\ 24 & 18 & -30 \\ 84 & 63 & -105 \end{bmatrix}$$

The recursive product therefore is:

$$\mathbf{A}\ \hat{\mathbf{D}}\ \mathbf{B}\ =\ \begin{bmatrix} 32 & 47 & -53 \\ 26 & 8 & -26 \\ 82 & 73 & -109 \end{bmatrix}$$

Exercise:

For:

$$\mathbf{A} = \begin{bmatrix} 1 & 2 & 3 \\ -2 & 3 & -1 \end{bmatrix} \qquad \hat{\mathbf{D}} = \begin{bmatrix} 2 & - & - \\ - & -1 & - \\ - & - & 4 \end{bmatrix} \qquad \mathbf{B} = \begin{bmatrix} 2 & 4 \\ -3 & 5 \\ 1 & -1 \end{bmatrix}$$

evaluate $\mathbf{A}\ \hat{\mathbf{D}}\ \mathbf{B}$, conform (3.4.3), and cross-check by evaluating $[\mathbf{A}\ \hat{\mathbf{D}}]\mathbf{B}$, as well as $\mathbf{A}[\hat{\mathbf{D}}\ \mathbf{B}]$.

CHAPTER IV

BLOCK-EQUATIONS AND INVERSION

4.1 Solvable and non-solvable systems

We begin this chapter with some recapitulation from the two previous chapters.

The matrix notation formulation of an 'ordinary' system of linear equations is:

$$\mathbf{A}\,\mathbf{x} \;=\; \mathbf{b} \tag{4.1.1}$$

Here $\mathbf{A}$ is a matrix, containing known coefficients, $\mathbf{b}$ is a known vector, $\mathbf{x}$ is the vectorial collection of the unknown variables. The <u>solution</u> of such a system consists of evaluating $\mathbf{x}$ in such a way that all the equations are met, i.e. that $\mathbf{Ax}$ actually is equal to $\mathbf{b}$. For this to be possible for only one vector $\mathbf{x}$, we normally need to assume that the number of equations is the same as the number of unknown variables, i.e. that $\mathbf{A}$ is a square matrix. If $\mathbf{A}$ is of order n by n, than both $\mathbf{b}$ and $\mathbf{x}$ must by implication be assumed to be of order n (by 1). If the system (4.1.1) has been solved, the evaluated vector $\mathbf{x}$ is called the <u>solution vector</u>.

The condition, that $\mathbf{A}$ is square, is not sufficient to ensure that one and only one vector $\mathbf{x}$ can be solved from (4.1.1).

We consider the following contra-examples:

51

$$x_1 + x_2 = 3$$
$$2x_1 + 2x_2 = 6$$

and

$$x_1 + x_2 = 3$$
$$2x_2 + 2x_2 = 15$$

The lefthand side system is satisfied by $x_1 = 0$, $x_2 = 3$. And also by $x_1 = 2$, $x_2 = 2$, as well as by $x_1 = -500$, $x_2 = 503$. Any vector **x**, which satisfies $x_1 + x_2 = 3$, also satisfies $2x_1 + 2x_2 = 6$.

The second system has no solution at all.

The two systems have in common, a special condition of the matrix

$$\mathbf{A} = \begin{bmatrix} 1 & 1 \\ 2 & 2 \end{bmatrix}$$

The second row of this matrix is proportional to the first one.

A square matrix of which the addition or subtraction of (suitable multiples of) some of the rows, yields another row as their weighted sum, is known as a <u>singular</u> matrix. Singularity need not always be so evident at glance, as in this example.

The matrix
$$\begin{bmatrix} - & 5 & 3 \\ 2 & -3 & -7 \\ 3 & 13 & - \end{bmatrix}$$

is singular as well. (The third row is obtained, as the sum of 3 and a half times the leading row, plus 1 and a half times the second row.)

We therefore define singularity as:

$$\mathbf{w}' \, \mathbf{A} = [0]' , \text{ for some } \mathbf{w}' \neq [0]'$$

For $\mathbf{A} = \begin{bmatrix} 1 & 1 \\ 2 & 2 \end{bmatrix}$, the vector $\mathbf{w}' = [2 \quad -1]$ will suffice,

for $\mathbf{A} = \begin{bmatrix} - & 5 & 3 \\ 2 & -3 & -7 \\ 3 & 13 & - \end{bmatrix}$, the vector $\mathbf{w}' = [3.5 \quad 1.5 \quad -1]$ will suffice to confirm singularity.

If **A** is singular, there also is a (at least one) vector $\mathbf{x} \neq [0]$, for which $\mathbf{Ax} = [0]$ applies, but no proof of that statement is offered at this stage. (We will come back to this point in Chapter V.)

Some of the more obvious cases of singularity are: proportionality of two rows (see above), proportionality of two columns, a zero row, and a zero column.

If a matrix contains a zero row, it will meet the definition of singularity, with **w'** being a unit vector:

$$[- \quad - \quad 1] \begin{bmatrix} -1 & 3 & -5 \\ 2 & 6 & 6 \\ - & - & - \end{bmatrix} = [- \quad - \quad -]$$

while the corresponding vector **x**, satisfying $\mathbf{Ax} = [0]$, will not normally have such a structure.

If **A** is singular, the corresponding equations-system can be <u>dependent</u> or <u>contradictory</u>. This will depend on the righthand side vector **b**.

If the ith row of **A** can be expressed as a combination of suitable multiples of other rows (or none at all, if it is itself a zero row), and the similar combination of the elements of **b** gives b_i as result, then the ith equation is met by any vector **x** which meets the others. If that combination of the other elements

of **b** does not equal b_i, then the combination of the other equations will contradict the ith equation. (In the example at the beginning of this section:

$$x_1 + x_2 = 3 \qquad \text{implies} \qquad 2x_1 + 2x_2 = 6,$$

but contradicts $\qquad\qquad\qquad 2x_1 + 2x_2 = 15.)$

4.2 **The elimination process**

We state, at this stage without proof, that a linear system of the description $\mathbf{Ax} = \mathbf{b}$, can be solved, i.e. we can calculate $\mathbf{x}$, given $\mathbf{A}$ and $\mathbf{b}$, provided $\mathbf{A}$ is square and not singular. The method of doing this, was introduced already in section 1.1 and another example of it, and some more computational details are given below.

In the context of matrix algebra, it is useful to restrict the method to a minimum of manipulation, and write the calculations in a standardized form in a tableau. This is not only because that saves effort, but also because it will be easier to explore the algebraic implications of the calculations to be performed.

The method of row operations consists of a series of elimination steps. These elimination steps are made in the order of the indices of the variables. At each kth step, the kth variable is expressed with a unity coefficient on the lefthand side of a particular equation.

From that point onwards, that particular equation is then the x_k equation. For the time being, we will assume, that we can always use the kth equation to express x_k. In fact, there are systems where a re-ordering of the equations may be needed, but we postpone a discussion of that complication until section 4.5. Reference to x_k is then eliminated from the other equations, by adding, or subtracting, the x_k equation to/from it, with a suitable multiplier.

Example:

$$2x_1 + 4x_2 + 6x_3 = 14$$
$$x_1 + x_2 + 2x_3 = 5$$
$$3x_1 + 3x_2 + 3x_3 = 12$$

Below, the lefthand-side of the page gives the tabular form of the resolution of this system, the meaning is written on the righthand-side:

$$x_1 \quad x_2 \quad x_3 \quad =$$

$$T_1 : \begin{bmatrix} |2| & 4 & 6 & 14 \\ 1 & 1 & 2 & 5 \\ 3 & 3 & 3 & 12 \end{bmatrix} \qquad \begin{array}{rcrcrcr} 2x_1 & + & 4x_2 & + & 6x_3 & = & 14 \\ x_1 & + & x_2 & + & 2x_3 & = & 5 \\ 3x_1 & + & 3x_2 & + & 3x_3 & = & 12 \end{array}$$

To transform the leading equation into an x_1 equation, we divide the leading equation by 2, and then once more multiply it by, respectively, -1 and -3, in preparation for eliminating reference to x_1 from the two other equations. We obtain:

$$x_1 \quad x_2 \quad x_3 \quad =$$

$$S_1 : \begin{bmatrix} 1 & 2 & 3 & 7 \\ -1 & -2 & -3 & -7 \\ -3 & -6 & -9 & -21 \end{bmatrix} \qquad \begin{array}{rcrcrcr} x_1 & + & 2x_2 & + & 3x_3 & = & 7 \\ -x_1 & - & 2x_2 & - & 3x_3 & = & -7 \\ -3x_1 & - & 6x_2 & - & 9x_3 & = & -21 \end{array}$$

To eliminate reference to x_1 from the second and third equations, we now add the last two rows of S_1, to the last two rows of T_1, and combine them with the already transformed leading row, as present in S_1. We obtain the successor tableau matrix:

$$x_1 \quad x_2 \quad x_3 \quad =$$

$$T_2 : \begin{bmatrix} 1 & 2 & 3 & 7 \\ & |-1| & -1 & -2 \\ & -3 & -6 & -9 \end{bmatrix} \qquad \begin{array}{rcrcrcr} x_1 & + & 2x_2 & + & 3x_3 & = & 7 \\ & - & x_2 & - & x_3 & = & -2 \\ & - & 3x_2 & - & 6x_3 & = & -9 \end{array}$$

The second elimination step is now initiated. To transform the second row into an x_2 row, we divide it by -1, and again multiply the transformed row by -2, and by 3, respectively. We now obtain the S_2 tableau matrix, which is:

$$x_1 \quad x_2 \quad x_3 \quad =$$

$$S_2: \begin{bmatrix} -2 & -2 & \bigm| & -4 \\ 1 & 1 & \bigm| & 2 \\ 3 & 3 & \bigm| & 6 \end{bmatrix}$$

$$\begin{aligned} -2x_2 - 2x_3 &= -4 \\ x_2 + x_3 &= 2 \\ 3x_2 + 3x_3 &= 6 \end{aligned}$$

Now adding the first row of S_2 to the first row of T_2, and the third row of S_2 to the third row of T_2, while keeping the second row of S_2 in unamended form, we obtain our next tableau, which is:

$$x_1 \quad x_2 \quad x_3 \quad =$$

$$T_3: \begin{bmatrix} 1 & & 1 & \bigm| & 3 \\ & 1 & 1 & \bigm| & 2 \\ & & |-3| & \bigm| & -3 \end{bmatrix}$$

$$\begin{aligned} x_1 \qquad + x_3 &= 3 \\ x_2 + x_3 &= 2 \\ -3x_3 &= -3 \end{aligned}$$

In order to express the third equation as an x_3 equation, we now divide the third row by -3. This transformed x_3 row is allocated

position 3 in S_3. The two other rows of S_3 will serve to eliminate reference to x_3 from the two leading equations. To that purpose, the transformed x_3 equation is multiplied by -1 i.e.

changed in sign into the first and second rows of S_3. The S_3 tableau is:

$$x_1 \quad x_2 \quad x_3 \quad =$$

$$S_3: \begin{bmatrix} & & -1 & \bigm| & -1 \\ & & -1 & \bigm| & -1 \\ & & 1 & \bigm| & 1 \end{bmatrix}$$

$$\begin{aligned} -x_3 &= -1 \\ -x_3 &= -1 \\ x_3 &= 1 \end{aligned}$$

Our final tableau is now obtained, by adding row 1 of S_3 to row 1

of T_3 into row 1 of T_4, adding row 2 of S_3 to row 2 of T_3 into row 2 of T_4, and copying row 3 of S_3 into row 3 of T_4. That final tableau is:

$$
\mathbf{T_3}: \quad
\begin{array}{ccc|c}
x_1 & x_2 & x_3 & = \\
1 & & & 2 \\
& 1 & & 1 \\
& & 1 & 1
\end{array}
\qquad
\begin{array}{rcl}
x_1 & = & 2 \\
x_2 & = & 1 \\
x_3 & = & 1
\end{array}
$$

The system has now been solved, as $x_1 = 2$, $x_2 = 1$, $x_3 = 1$.

When performing these calculations on paper, it is useful to add a further column to the tableau. This additional column contains simply the sums of the coefficients in each row.

Because one always adds/subtracts multiples of the kth row to/from other rows, (and does the same with the entry in the summation column), the sumcount property is maintained throughout the calculations, unless a mistake is made. In that case, the sumcount column is often helpful to trace the mistake.

We give the first two tableaus in tabular form once more, this time with the sumcount column, and a deliberately introduced 'mistake'.

$$
\mathbf{T_1}: \quad
\begin{array}{ccc|c c}
x_1 & x_2 & x_3 & = & \Sigma \\
|2| & 4 & 6 & 14 & 26 \\
1 & 1 & 2 & 5 & 9 \\
3 & 3 & 3 & 12 & 21
\end{array}
$$

$$
\mathbf{S_1}: \quad
\begin{array}{ccc|c c}
x_1 & x_2 & x_3 & = & \Sigma \\
1 & 2 & 3 & 7 & 13 \\
-1 & -2 & -3 & -7 & -13 \\
-3 & -6 & -9 & -21 & -39
\end{array}
$$

$$
\mathbf{T_2}: \quad
\begin{array}{ccc|c c}
x_1 & x_2 & x_3 & = & \Sigma \\
1 & 2 & 3 & 7 & 13 \\
& -1 & 1 & -2 & -4 \\
& -3 & -6 & -9 & -18
\end{array}
$$

In the second equation in the T_2 tableau, the minus sign in x_3 column has been omitted. The -4 in the sumcount column has been calculated in the usual way, i.e. the 9 which was there in the T_1 tableau, plus the -13 which figures in the same position in the

S_1 tableau. That is what the sum-count of the row <u>should</u> be. As the sumcount now does not fit, the first thing to be done now, is to find the mistake, i.e. to put the minus sign.

In the example above, we have included the 'S' tableaus, which contain the appropriate multiples of the pivotal row. There is no need for that (and no objection against, other than that writing more tableaus is more work), but more or less the usual for of presenting these calculations in tabular form is:

$$T_1: \quad \begin{array}{ccc|cc} x_1 & x_2 & x_3 & = & \Sigma \\ \boxed{2} & 4 & 6 & 14 & 26 \\ 1 & 1 & 2 & 5 & 9 \\ 3 & 3 & 3 & 12 & 21 \end{array}$$

$$T_2: \quad \begin{array}{ccc|cc} x_1 & x_2 & x_3 & = & \Sigma \\ 1 & 2 & 3 & 7 & 13 \\ & \boxed{-1} & -1 & -2 & -4 \\ & -3 & -6 & -9 & -18 \end{array}$$

$$T_3: \quad \begin{array}{ccc|cc} x_1 & x_2 & x_3 & = & \Sigma \\ 1 & & 1 & 3 & 5 \\ & 1 & 1 & 2 & 4 \\ & & \boxed{-3} & -3 & -6 \end{array}$$

$$T_3: \quad \begin{array}{ccc|cc} x_1 & x_2 & x_3 & = & \Sigma \\ 1 & & & 2 & 3 \\ & 1 & & 1 & 2 \\ & & 1 & 1 & 3 \end{array}$$

4.3 Unknown righthand-side variables

Consider the system:

$$0.867\ y_1 - 0.067\ y_2$$
$$= 0.244\ x_1 + 0.150\ x_2 + 0.722\ x_3 + 0.556\ x_4$$

$$-\ 0.150\ y_1 + 0.850\ y_2$$
$$= 0.561\ x_1 + 0.173\ x_2 + 0.278\ x_3 + 0.222\ x_4$$

Or, using the tabular form:

$$
\begin{array}{cc}
y_1 \qquad\quad y_2 \\
\begin{bmatrix} 0.867 & -0.066 \\ -0.150 & 0.850 \end{bmatrix}
\end{array}
=
\begin{array}{cccc}
x_1 \quad\ x_2 \quad\ x_3 \quad\ x_4 \\
\begin{bmatrix} 0.244 & 0.150 & 0.722 & 0.556 \\ 0.561 & 0.173 & 0.278 & 0.222 \end{bmatrix}
\end{array}
$$

The more general form of this system in matrix notation is:

$$\mathbf{A\,y} = \mathbf{B\,x} \qquad\qquad\qquad (4.3.1)$$

e.g. in this example:

$$
\begin{bmatrix} 0.867 & -0.066 \\ -0.150 & 0.850 \end{bmatrix}
\mathbf{y}
=
\begin{bmatrix} 0.244 & 0.150 & 0.722 & 0.556 \\ 0.561 & 0.173 & 0.278 & 0.222 \end{bmatrix}
\mathbf{x}
$$

Here, both $\mathbf{y}$ and $\mathbf{x}$ are unknown vectors. It is therefore impossible to 'solve' the system in the sense of establishing the

values of all the variables. However, we <u>can</u> express y_1 and y_2 into (the so far unknown values of) x_1, x_2, x_3, and x_4.

The computational procedure is a straightforward extension of the method of elimination by row-operations, and we restrict ourselves to summarizing the computations in tabular form:

$$
\begin{array}{cccccccc}
y_1 & y_2 & = & x_1 & x_2 & x_3 & x_4 & \Sigma \\
\left[\begin{array}{cc} |0.867| & -0.066 \\ -0.150 & 0.850 \end{array}\right] & & & \left[\begin{array}{cccc} 0.244 & 0.150 & 0.722 & 0.556 \\ 0.561 & 0.173 & 0.278 & 0.222 \end{array}\right] & & & & \begin{array}{c} 2.437 \\ 1.934 \end{array}
\end{array}
$$

$$
\begin{array}{cccccccc}
y_1 & y_2 & = & x_1 & x_2 & x_3 & x_4 & \Sigma \\
\left[\begin{array}{cc} 1 & -0.076 \\ - & |0.839| \end{array}\right] & & & \left[\begin{array}{cccc} 0.281 & 0.173 & 0.833 & 0.641 \\ 0.603 & 0.199 & 0.403 & 0.318 \end{array}\right] & & & & \begin{array}{c} 2.852 \\ 2.362 \end{array}
\end{array}
$$

$$
\begin{array}{cccccccc}
y_1 & y_2 & = & x_1 & x_2 & x_3 & x_4 & \Sigma \\
\left[\begin{array}{cc} 1 & - \\ - & 1 \end{array}\right] & & & \left[\begin{array}{cccc} 0.336 & 0.191 & 0.869 & 0.670 \\ 0.719 & 0.237 & 0.480 & 0.379 \end{array}\right] & & & & \begin{array}{c} 3.067 \\ 2.815 \end{array}
\end{array}
$$

We are at this stage assuming that the system $\mathbf{Ay} = \mathbf{b}$ is solvable by row-operations, and therefore, that the corresponding computations, performed on the $\mathbf{Ay} = \mathbf{Bx}$ system, will lead to the type of tableau matrix illustrated above.

The resulting system:

$$\mathbf{y} = \mathbf{R}\,\mathbf{x} \qquad\qquad (4.3.2)$$

is known as a reduced form.

The term reduced form originally relates to econometrics, i.e. the statistical estimation of economic relationships. (See Klein and Goldberger [28], p. 14, following the ideas set out by Haavelmo [18], pp. 26-28).

Note, however, a difference in emphasis: We assume that $\mathbf{A}$ and $\mathbf{B}$ are known, but the problem which these econometricians were concerned with, was the estimation of $\mathbf{A}$ and $\mathbf{B}$.

We shall refer to the matrix $\mathbf{R}$ as the reduced form matrix.

In this example, we find: $\mathbf{R} = \begin{bmatrix} 0.336 & 0.191 & 0.869 & 0.670 \\ 0.719 & 0.237 & 0.480 & 0.379 \end{bmatrix}$

Exercise:

For $\begin{bmatrix} 2 & -1 \\ -1 & 1 \end{bmatrix} \mathbf{y} = \begin{bmatrix} 1 & 2 \\ 3 & 4 \end{bmatrix} \mathbf{x}$

calculate the matrix $\mathbf{R}$, as corresponding to $\mathbf{y} = \mathbf{Rx}$
(There is an answersheet at the end of the chapter).

4.4 The inverse

Consider the following equations-system:

$$
\begin{aligned}
x_1 + 2x_2 + 3x_3 &= y_1 \\
x_1 + x_2 + 2x_3 &= \quad\quad y_2 \\
x_1 + x_2 + x_3 &= \quad\quad\quad y_3
\end{aligned} \Bigg\}
$$

Or, more generally:

$$\mathbf{A\,x} = \mathbf{y} \tag{4.4.1}$$

where $\mathbf{A}$ is a square and non-singular matrix, and $\mathbf{x}$ and $\mathbf{y}$ are unknown column-vectors of the same order as the matrix $\mathbf{A}$.

Like the system $\mathbf{Ay} = \mathbf{Bx}$ as discussed in section 4.3., this system cannot be 'solved' in the sense of evaluating $\mathbf{x}$. Both $\mathbf{x}$ and $\mathbf{y}$ are unknown vectors. The usual elimination process leads, however, to a matrix expression of $\mathbf{x}$ in $\mathbf{y}$, of the type:

$$\mathbf{x} = \mathbf{B\,y} \tag{4.4.2}$$

For the example at hand, this elimination is performed as follows:

x_1	x_2	x_3	$=$	Y_1	Y_2	Y_3	Σ
$\boxed{1}$	2	3		1	-	-	7
1	1	2		-	1	-	5
1	1	1		-	-	1	4

x_1	x_2	x_3	$=$	Y_1	Y_2	Y_3	Σ
1	2	3		1	-	-	7
-	$\boxed{-1}$	-1		-1	1	-	-2
-	-1	-2		-1	-	1	-3

x_1	x_2	x_3	$=$	Y_1	Y_2	Y_3	Σ
1	-	1		-1	2	-	3
-	1	1		1	-1	-	2
-	-	$\boxed{-1}$		-	-1	1	-1

x_1	x_2	x_3	$=$	Y_1	Y_2	Y_3	Σ
1	-	-		-1	1	1	2
-	1	-		1	-2	1	1
-	-	1		-	1	-1	1

We have therefore found:

$$\mathbf{x} = \begin{bmatrix} -1 & 1 & 1 \\ 1 & -2 & 1 \\ - & 1 & -1 \end{bmatrix} \mathbf{y}$$

or equivalently:

$$\begin{bmatrix} -1 & 1 & 1 \\ 1 & -2 & 1 \\ - & 1 & -1 \end{bmatrix} \mathbf{y} = \mathbf{x}$$

In the latter form, we have a new system:

$$\mathbf{B}\,\mathbf{y} \;=\; \mathbf{x} \tag{4.4.3}$$

We may perform the same operation again, to obtain an expression of $\mathbf{y}$ in $\mathbf{x}$. But we know that expression already, since $\mathbf{Ax} = \mathbf{y}$, and therefore $\mathbf{y} = \mathbf{Ax}$ was given in the first place. The relationship between $\mathbf{A}$ and $\mathbf{B}$, as present in (4.4.1) and (4.4.3) is named an inverse one.

The matrix $\mathbf{B}$ is the <u>inverse</u> of $\mathbf{A}$.

This relationship is written in symbolic form as $\mathbf{B} = \mathbf{A}^{-1}$.

A formal definition is as follows:

The n by n matrix $\mathbf{B}$ is the inverse of the n by n matrix $\mathbf{A}$, if we can find, for any n by 1 column vector $\mathbf{y}$, a corresponding $\mathbf{x}$, where $\mathbf{x}$ and $\mathbf{y}$ satisfy both (4.4.1) and (4.4.3).

This is not the only possible definition of the inverse. Another possible definition is:

The n by n matrix $\mathbf{B}$ is the inverse of the n by n matrix $\mathbf{A}$, if and only if $\mathbf{AB} = \mathbf{I}$ applies. Yet another definition (which, at one stage was the usual one), will be discussed in the next chapter.

The choice of a particular property as a definition, while then showing other properties as theorems, is to some extent one of convenience: It is a question of how readily other properties follow, and how well the definition relates to what the corresponding word conveys in ordinary language.

Note, that our definition of the inverse requires that $\mathbf{A}\,\mathbf{x} = \mathbf{y}$ applies for <u>all</u> $\mathbf{y}$ and a corresponding $\mathbf{x}$ and that for all corresponding $\mathbf{x}$, we must have $\mathbf{B}\,\mathbf{y} = \mathbf{x}$.

For $\mathbf{B}$ to be the inverse of a given matrix $\mathbf{A}$, the calculation of $\mathbf{y}$ as $\mathbf{y} = \mathbf{Ax}$ conform (4.4.1), and then to find that, for that one pair of vectors $\mathbf{x}$ and $\mathbf{y}$, (4.4.3) is also satisfied, is not sufficient. We must first of all be in a position to calculate a vector $\mathbf{x}$, by resolving (4.4.1), for an arbitrary vector $\mathbf{y}$. If that cannot be done, there is no inverse matrix, and if we find $\mathbf{A}\,\mathbf{x} = \mathbf{y}$ and $\mathbf{B}\,\mathbf{x} = \mathbf{y}$, for some but not all $\mathbf{y}$, then that particular matrix $\mathbf{B}$ is not the inverse of $\mathbf{A}$.

Thus, for $\mathbf{A} = \begin{bmatrix} 1 & 2 & 3 \\ 1 & 1 & 2 \\ 1 & 1 & 1 \end{bmatrix}$; $\mathbf{y} = \mathbf{s} = \begin{bmatrix} 1 \\ 1 \\ 1 \end{bmatrix}$

we have for $\mathbf{x} = \mathbf{e}_1$

$$\begin{bmatrix} 1 & 2 & 3 \\ 1 & 1 & 2 \\ 1 & 1 & 1 \end{bmatrix} \begin{bmatrix} 1 \\ - \\ - \end{bmatrix} = \begin{bmatrix} 1 \\ 1 \\ 1 \end{bmatrix}, \qquad \begin{bmatrix} -1 & 2 & - \\ 1 & -1 & - \\ - & 2 & -2 \end{bmatrix} \begin{bmatrix} 1 \\ 1 \\ 1 \end{bmatrix} = \begin{bmatrix} 1 \\ - \\ - \end{bmatrix}$$

$\mathbf{A}\,\mathbf{x} = \mathbf{y}$ as well as $\mathbf{B}\,\mathbf{y} = \mathbf{x}$

but for $\mathbf{x} = \mathbf{s}$ we have:

$$\begin{bmatrix} 1 & 2 & 3 \\ 1 & 1 & 2 \\ 1 & 1 & 1 \end{bmatrix} \begin{bmatrix} 1 \\ 1 \\ 1 \end{bmatrix} = \begin{bmatrix} 6 \\ 4 \\ 3 \end{bmatrix}, \qquad \begin{bmatrix} -1 & 2 & - \\ 1 & -1 & - \\ - & 2 & -2 \end{bmatrix} \begin{bmatrix} 6 \\ 4 \\ 3 \end{bmatrix} = \begin{bmatrix} 2 \\ 2 \\ 2 \end{bmatrix} \neq \begin{bmatrix} 1 \\ 1 \\ 1 \end{bmatrix}$$

$\mathbf{A}\,\mathbf{x} = \mathbf{y}$; $\mathbf{B}\,\mathbf{y} \neq \mathbf{x}$

Therefore $\mathbf{B} = \begin{bmatrix} -1 & 1 & 1 \\ 1 & 2 & 1 \\ - & 1 & -1 \end{bmatrix}$

is _not_ the inverse of $\mathbf{A}$, despite the fact that for _one_ vector $\mathbf{y}$ we _found_ a vector $\mathbf{x}$, for which both $\mathbf{A}\,\mathbf{x} = \mathbf{y}$ and $\overline{\mathbf{B}\,\mathbf{y}} = \mathbf{x}$ are satisfied.

We now come to discuss some properties of inverses, and their relation to (square and) non-singular matrices.

First of all, since (4.4.1) and (4.4.3) need to apply for _all_

vectors $\mathbf{y}$, we may evaluate n vectors $\mathbf{x}$ for $\mathbf{y} = \mathbf{e}_1$, $\mathbf{y} = \mathbf{e}_2$, etc.

Application of (4.4.3), and (4.4.1), for these vectors $\mathbf{y}$ leads to:

$B\ y_1 = B\ e_1 = b_1 = x_1 \rightarrow$ by (4.4.1): $A\ x_1 = A\ b_1 = y_1 = e_1$

$B\ y_2 = B\ e_2 = b_2 = x_2 \rightarrow$ by (4.4.1): $A\ x_2 = A\ b_2 = y_2 = e_2$

. . . .

$B\ y_n = B\ e_n = b_n = x_n \rightarrow$ by (4.4.1): $A\ x_n = A\ b_n = y_n = e_n$

Therefore, now using matrix notation:

$B\ I = B\ I = B = X$ implies by (4.4.1): $A\ X = A\ B = I$

or recapitulating:

If B is the inverse of A:

$$A\ B = A\ A^{-1} = I \qquad (4.4.4)$$

is true.

Furthermore, given any $B = A^{-1}$, a vector y, for which x and y will satisfy both $A\ x = y$ and $B\ y = x$, can be calculated for any given x, by means of (4.4.1).

Therefore A meets the definition of B^{-1}:

$$B^{-1} = [A^{-1}]^{-1} = A \qquad (4.4.5)$$

Application of (4.4.4), to $B = A^{-1}$ now results in:

$$B\ A = B\ B^{-1} = A^{-1}\ A = I \qquad (4.4.6)$$

Thus, while $AB = BA$ is <u>not</u> generally true, this <u>is</u> true for a matrix and its inverse:

$$A\ A^{-1} = A^{-1}\ A = I \qquad (4.4.7)$$

These properties provide an obvious practical test in checking whether a calculated inverse has been calculated correctly, and actually is the inverse.

Example:

$$\mathbf{A} = \begin{bmatrix} 1 & 2 & 3 \\ 1 & 1 & 2 \\ 1 & 1 & 1 \end{bmatrix}, \qquad \mathbf{A}^{-1} = \begin{bmatrix} -1 & 1 & 1 \\ 1 & -2 & 1 \\ - & 1 & -1 \end{bmatrix}$$

(calculated above):

$$\begin{bmatrix} 1 & 2 & 3 \\ 1 & 1 & 2 \\ 1 & 1 & 1 \end{bmatrix} \begin{bmatrix} -1 & 1 & 1 \\ 1 & -2 & 1 \\ - & 1 & -1 \end{bmatrix} = \begin{bmatrix} 1 & - & - \\ - & 1 & - \\ - & - & 1 \end{bmatrix}$$

as well as:

$$\begin{bmatrix} -1 & 1 & 1 \\ 1 & -2 & 1 \\ - & 1 & -1 \end{bmatrix} \begin{bmatrix} 1 & 2 & 3 \\ 1 & 1 & 2 \\ 1 & 1 & 1 \end{bmatrix} = \begin{bmatrix} 1 & - & - \\ - & 1 & - \\ - & - & 1 \end{bmatrix}$$

But if by error, the minus sign is omitted from the top lefthand element of the 'inverse', we have:

$$\begin{bmatrix} 1 & 1 & 1 \\ 1 & -2 & 1 \\ - & 1 & -1 \end{bmatrix} \begin{bmatrix} 1 & 2 & 3 \\ 1 & 1 & 2 \\ 1 & 1 & 1 \end{bmatrix} = \begin{bmatrix} 3 & 4 & 6 \\ - & 1 & - \\ - & - & 1 \end{bmatrix} \neq \begin{bmatrix} 1 & - & - \\ - & 1 & - \\ - & - & 1 \end{bmatrix}$$

If the product-relation (4.4.6) is not met by two matrices of appropriate order, they are not each others inverse. Conversely, relation (4.4.6) is also sufficient to prove that two n by n matrices $\mathbf{A}$ and $\mathbf{B}$ are each others inverse.

Theorem:

If the n by n matrices **A** and **B** conform to:

A B = **I**

then **B** is the inverse of **A**.

Proof:

Pre-multiply one of the two inverse relations,

$$\mathbf{B}\,\mathbf{y} = \mathbf{x} \qquad\qquad (4.4.3)$$

by **A**, and obtain:

$$\mathbf{y} = \mathbf{A}\,\mathbf{x} \qquad\qquad (4.4.8)$$

or equivalently:

$$\mathbf{A}\,\mathbf{x} = \mathbf{y} \qquad\qquad (4.4.1)$$

Therefore, given any **y**, a vector **x**, for which **x** and **y** meet the inverse relations, may be calculated by (4.4.3). This is the definition of **B** being inverse of **A**.

q.e.d.

Corollary:

If **B A** = **I** holds, then **A** is the inverse of **B**.

It is at this stage useful to mention, that some textbooks, e.g. Wilde [49], and Schreier & Sperner [41], p.250, clearly drawing on an established tradition, distinguish the terms 'right-inverse', defining that term as **AB** = **I**, and 'left-inverse', i.e. **B** is the left-inverse of **A**, if **BA** = **I** is true.

However, the operational relevance of that distinction is somewhat limited, because the one implies the other.

The following property is stated here without proof at this stage:

Every (square and) non-singular matrix has an inverse. (A proof of this statement will be supplied in the next chapter).

Furthermore, from (4.4.7) we obtain, on pre-multiplication by any **w**':

$$\mathbf{w}' \ \mathbf{A}^{-1} \ \mathbf{A} \ = \ \mathbf{w}' \qquad\qquad (4.4.9)$$

The lefthand side expression in (4.4.9) can be non-zero, only if

$$\mathbf{w}'\mathbf{A}^{-1} \ \neq \ [0] \text{ applies.}$$

Hence **w**'$\neq[0]$' implies **w**'$\mathbf{A}^{-1}\neq[0]$',

i.e.:

all inverses are non-singular.

Therefore, if (as will be shown in the next chapter) all non-singular matrices have an inverse, we have the following corollary:

A singular matrix does not have an inverse.

We have, so far, spoken of 'the' inverse, without actually proving that such a matrix is unique. In fact, an inverse matrix is a unique function of the inverted matrix, and we have the following

Theorem:

If the n by n matrices **B** and **C** are both inverses of the n by n matrix **A**, then

$$\mathbf{B} \ = \ \mathbf{C}$$

is true.

Proof:

Apply (4.4.7) for both inverses, to obtain:

$$\mathbf{A} \ \mathbf{B} \ = \ \mathbf{I} \qquad \text{and} \qquad \mathbf{A} \ \mathbf{C} \ = \ \mathbf{I}$$

Therefore:

$$\mathbf{A}\,\mathbf{B} - \mathbf{A}\,\mathbf{C} = \mathbf{I} - \mathbf{I} = [0] \qquad (4.4.10)$$

Since $\mathbf{A}$ has an inverse, $\mathbf{A}$ is non-singular, and

$$\mathbf{z}' = \mathbf{w}'\mathbf{A} \neq [0]' \text{ will apply for all } \mathbf{w}' \neq [0]',$$

and can be required to apply for any $\mathbf{z}' \neq [0]'$, by setting $\mathbf{w}'$ at:

$$\mathbf{w}' = \mathbf{z}'\mathbf{A}^{-1} \qquad (4.4.11)$$

(This is true, regardless of whether $\mathbf{A}^{-1}$ is $\mathbf{B}$, $\mathbf{C}$, or possibly a third inverse of $\mathbf{A}$.)

Therefore, on pre-multiplication of (4.4.10) by any

$$\mathbf{w}' = \mathbf{z}'\mathbf{A}^{-1} \neq [0]' :$$

$$\mathbf{w}'\mathbf{A}\mathbf{B} - \mathbf{w}'\mathbf{A}\mathbf{C} = \mathbf{z}'\mathbf{B} - \mathbf{z}'\mathbf{C} = \mathbf{z}'[\mathbf{B}-\mathbf{C}] = \mathbf{w}' - \mathbf{w}' = [0]' \quad (4.4.12)$$

This can only apply for all $\mathbf{z}' \neq [0]'$, if $[\mathbf{B}-\mathbf{C}]$ is a null matrix.

q.e.d.

Exercise:

Verify (by means of inversion by row-operations, as well as by pre- and post-multiplication, that the inverse of

$$\mathbf{A} = \begin{bmatrix} 1 & 1 & 1 \\ 2 & 1 & 1 \\ 3 & 2 & 1 \end{bmatrix} \quad \text{is:} \quad \mathbf{A}^{-1} = \begin{bmatrix} -1 & 1 & - \\ 1 & -2 & 1 \\ 1 & 1 & -1 \end{bmatrix}$$

4.5 **Row-permutation during inversion**

There are matrices (and equations-systems), which are invertible, (or solvable), but where the method of row operations as outlined so far, runs into a complication, because a zero element appears on the diagonal of the calculation tableau.

Example:

$$\mathbf{A} = \begin{bmatrix} - & 1 & 2 \\ 1 & -1 & - \\ 2 & 1 & -2 \end{bmatrix}$$

This matrix <u>has</u> an inverse. This may be seen by transforming the $\mathbf{Ax} = \mathbf{y}$ system, as follows:

$$
\begin{array}{ccc}
x_1 & x_2 & x_3 \\
\end{array}
\;=\;
\begin{array}{ccc}
y_1 & y_2 & y_3 \\
\end{array}
$$

$$
\begin{bmatrix} - & 1 & 2 \\ 1 & -1 & - \\ 2 & 1 & -2 \end{bmatrix}
\begin{bmatrix} 1 & - & - \\ - & 1 & - \\ - & - & 1 \end{bmatrix}
\qquad
\begin{array}{rcl}
x_2 +x_3 &=& y_1 \\
x_1 -x_2 &=& y_2 \\
2x_1 +x_2 -2x_3 &=& y_3
\end{array}
$$

Interchange the first and the second equations (and correspondingly, the rows of the calculation tableau, to obtain:

$$
\begin{array}{ccc}
x_1 & x_2 & x_3 \\
\end{array}
\;=\;
\begin{array}{ccc}
y_1 & y_2 & y_3 \\
\end{array}
$$

$$
\begin{bmatrix} |1| & -1 & - \\ - & 1 & 2 \\ 2 & 1 & -2 \end{bmatrix}
\begin{bmatrix} - & 1 & - \\ 1 & - & - \\ - & - & 1 \end{bmatrix}
\qquad
\begin{array}{rcl}
x_1 - x_2 &=& y_2 \\
x_2 +2x_3 &=& y_1 \\
2x_1 +x_2 -2x_3 &=& y_3
\end{array}
$$

Now proceed in the usual way, to obtain the successor tableau:

$$
\begin{array}{ccc}
x_1 & x_2 & x_3 \\
\end{array}
\;=\;
\begin{array}{ccc}
y_1 & y_2 & y_3 \\
\end{array}
$$

$$
\begin{bmatrix} 1 & -1 & - \\ - & |1| & 2 \\ - & 3 & -2 \end{bmatrix}
\begin{bmatrix} - & 1 & - \\ 1 & - & - \\ - & -2 & 1 \end{bmatrix}
\qquad
\begin{array}{rcl}
x_1 - x_2 &=& y_2 \\
x_2 +2x_3 &=& y_1 \\
3x_2 -2x_3 &=& -2y_2 +y_3
\end{array}
$$

The next elimination step can now be made in the usual way.

$$
\begin{array}{ccc}
x_1 & x_2 & x_3
\end{array} \quad = \quad
\begin{array}{ccc}
y_1 & y_2 & y_3
\end{array}
$$

$$
\begin{bmatrix}
1 & - & 2 \\
- & 1 & 2 \\
- & - & |-8|
\end{bmatrix}
\begin{bmatrix}
1 & 1 & - \\
1 & - & - \\
-3 & -2 & 1
\end{bmatrix}
\qquad
\begin{aligned}
x_1 & & +2x_3 & = & y_1 & +y_2 \\
& x_2 & +2x_3 & = & y_1 & \\
& & -8x_3 & = & -3y_1 & -2y_2 & +y_3
\end{aligned}
$$

The last inversion step is also made in the usual way, pivoting on the diagonal. We obtain the final calculation tableau:

$$
\begin{array}{ccc}
x_1 & x_2 & x_3
\end{array} \quad = \quad
\begin{array}{ccc}
y_1 & y_2 & y_3
\end{array}
$$

$$
\begin{bmatrix}
1 & - & - \\
- & 1 & - \\
- & - & 1
\end{bmatrix}
\begin{bmatrix}
0.25 & 0.50 & 0.25 \\
0.25 & -0.50 & 0.25 \\
0.37 & 0.25 & -0.12
\end{bmatrix}
\qquad
\begin{aligned}
x_1 & = 0.25y_1 +0.50y_2 +0.25y_3 \\
x_2 & = 0.25y_1 -0.50y_2 +0.25y_3 \\
x_3 & = 0.37y_1 +0.25y_2 -0.12y_3
\end{aligned}
$$

We therefore have established the inverse relationship:

$$
\begin{bmatrix}
- & 1 & 2 \\
1 & -1 & - \\
2 & 1 & -2
\end{bmatrix}^{-1}
=
\begin{bmatrix}
0.25 & 0.50 & 0.25 \\
0.25 & -0.50 & 0.25 \\
0.37 & 0.25 & -0.12
\end{bmatrix}
$$

Searching for a non-zero element in the pivotal column of the calculation tableau, on <u>or below</u> the diagonal, and, if necessary interchanging the rth row in which such an element is found, with the kth row, is an essential feature of the algorithm of elimination by row-operations.

4.6 Inversion of recursive products and transposes

The inversion of a recursive product of several matrices gives rise to a remarkable change in their ordering.

If **A** and **B** are square and non-singular matrices of the same order, we have the following formula:

$$
[\mathbf{A}\,\mathbf{B}]^{-1} \quad = \quad \mathbf{B}^{-1}\,\mathbf{A}^{-1}
\tag{4.6.1}
$$

i.e. we write then inverses of the individual matrices, but in reverse order.

A proof of (4.6.1) follows either from pre-multiplication, i.e.:

$$[\mathbf{B}^{-1}\mathbf{A}^{-1}]\,[\mathbf{AB}] = \mathbf{B}^{-1}\mathbf{A}^{-1}\mathbf{A}\,\mathbf{B} = \mathbf{B}^{-1}[\mathbf{A}^{-1}\mathbf{A}]\,\mathbf{B} = \mathbf{B}^{-1}\mathbf{B} = \mathbf{I} \qquad (4.6.2)$$

or by post-multiplication, i.e.:

$$[\mathbf{AB}]\,[\mathbf{B}^{-1}\mathbf{A}^{-1}] = \mathbf{A}\,[\mathbf{B}\,\mathbf{B}^{-1}]\,\mathbf{A}^{-1} = \mathbf{A}\,\mathbf{A}^{-1} = \mathbf{I} \qquad (4.6.3)$$

If three or more square and invertible matrices of the same order are multiplied, the inverse of their recursive product has the similar property:

$$[\mathbf{A}\;\mathbf{B}\;\mathbf{C}\;\mathbf{D}]^{-1} = \mathbf{D}^{-1}[\mathbf{A}\;\mathbf{B}\;\mathbf{C}]^{-1} = \mathbf{D}^{-1}\;\mathbf{C}^{-1}\;[\mathbf{A}\;\mathbf{B}]^{-1} =$$

$$\mathbf{D}^{-1}\;\mathbf{C}^{-1}\;\mathbf{B}^{-1}\;\mathbf{A}^{-1} \qquad (4.6.4)$$

Example:

Calculate the inverse of:

$$\mathbf{A}\,\mathbf{B} \;=\; \begin{bmatrix} 1 & - & - \\ 2 & 1 & - \\ 3 & 4 & 1 \end{bmatrix} \begin{bmatrix} 1 & 2 & 3 \\ - & 1 & 4 \\ - & - & 1 \end{bmatrix} = \begin{bmatrix} 1 & 2 & 3 \\ 2 & 5 & 10 \\ 3 & 10 & 26 \end{bmatrix}$$

Manual calculation of the inverse of a 3 by 3 matrix which is full of non-zero elements would take some time, and could easily lead to calculation errors. The triangular matrices are rather easier to invert, especially because the computation is simplified by the fact that the diagonal elements are all unity.

$$\begin{bmatrix} 1 & - & - \\ 2 & 1 & - \\ 3 & 4 & 1 \end{bmatrix}^{-1} = \begin{bmatrix} 1 & - & - \\ -2 & 1 & - \\ 5 & -4 & 1 \end{bmatrix}$$

and

$$\begin{bmatrix} 1 & 2 & 3 \\ - & 1 & 4 \\ - & - & 1 \end{bmatrix}^{-1} = \begin{bmatrix} 1 & -2 & 5 \\ - & 1 & -4 \\ - & - & 1 \end{bmatrix}$$

The inverse for which we are looking therefore is:

$$
\left[\begin{bmatrix} 1 & - & - \\ 2 & 1 & - \\ 3 & 4 & 1 \end{bmatrix} \begin{bmatrix} 1 & 2 & 3 \\ - & 1 & 4 \\ - & - & 1 \end{bmatrix}\right]^{-1} =
$$

$$
\begin{bmatrix} 1 & -2 & 5 \\ - & 1 & -4 \\ - & - & 1 \end{bmatrix} \begin{bmatrix} 1 & - & - \\ -2 & 1 & - \\ 5 & -4 & 1 \end{bmatrix} = \begin{bmatrix} 30 & -22 & 5 \\ -22 & 17 & -4 \\ 5 & -4 & 1 \end{bmatrix}
$$

The correctness of this result is shown by verification of the identity matrix product relationship:

$$
\begin{bmatrix} 1 & 2 & 3 \\ 2 & 5 & 10 \\ 3 & 10 & 26 \end{bmatrix} \begin{bmatrix} 30 & -22 & 5 \\ -22 & 17 & -4 \\ 5 & -4 & 1 \end{bmatrix} = \begin{bmatrix} 1 & - & - \\ - & 1 & - \\ - & - & 1 \end{bmatrix}
$$

This example also suggest another relationship.

If $\mathbf{A}^{-1}$ exists, e.g.:

$$
\begin{bmatrix} 1 & - & - \\ 2 & 1 & - \\ 3 & 4 & 1 \end{bmatrix}^{-1} = \begin{bmatrix} 1 & - & - \\ -2 & 1 & - \\ 5 & -4 & 1 \end{bmatrix}
$$

then $[\mathbf{A}']^{-1}$ also exists, and does not require independent calculation:

$[\mathbf{A}']^{-1}$ can be written out, by transposing $\mathbf{A}^{-1}$.

$$
\begin{bmatrix} 1 & 2 & 3 \\ - & 1 & 4 \\ - & - & 1 \end{bmatrix}^{-1} = \begin{bmatrix} 1 & -2 & 5 \\ - & 1 & -4 \\ - & - & 1 \end{bmatrix}
$$

This is indeed a generally valid property, as may be shown by verification of the unity matrix product property. We transpose two members of (4.4.7):

$$
\mathbf{A}\ \mathbf{A}^{-1} = \mathbf{I}
$$

te obtain:

$$
[\mathbf{A}\ \mathbf{A}^{-1}]' = [\mathbf{A}^{-1}]'\ \mathbf{A}' = \mathbf{I}' = \mathbf{I} \tag{4.6.5}
$$

4.7 Block-Elimination

So far, we have applied the method of elimination by row operations, to coefficients and equations in numbers. A very similar technique can also be applied to matrices (blocks), and block-equations.

We will illustrate this application of the elimination technique to symbolically presented matrix formulae, in relation to the following partitioned system:

$$\left. \begin{array}{l} A_{11}x_1 + A_{12}x_2 = B'_1y \\ A_{21}x_1 + A_{22}x_2 = B'_2y \end{array} \right\} \qquad (4.7.1)$$

With a partitioned system, it may, or may not, be known that the structure of the system is such that certain blocks are invertible. In other cases, it may be necessary to search for invertible blocks, as with ordinary inversion (see myself [22]). Our interest here is on the development of formulae, and we will assume the existence of an invertible block A_{11}, possibly after re-ordering the rows and/or the columns of A.

A partial resolution of x_1, in terms of unspecified values of x_2 and y is obtained, by pre-multiplication of the first block-equation in (4.7.1) by the inverse of A_{11}.

$$x_1 + A_{11}^{-1}A_{12}x_2 = A_{11}^{-1}B'_1y \qquad (4.7.2)$$

Re-ordering the terms in (4.7.2) yields a reduced form expression for x_1 in terms of x_2 and y:

$$x_1 = A_{11}^{-1}B'_1y - A_{11}^{-1}A_{12}x_2 \qquad (4.7.3)$$

Substitution of this reduced form expression for x_1 into the second block-equation contained in (4.7.1) yields:

$$[A_{22} - A_{21}A_{11}^{-1}A_{12}]\,x_2 = [B'_2 - A_{21}A_{11}^{-1}B'_1]\,y \qquad (4.7.4)$$

We now combine (4.7.3) and (4.7.4), to obtain:

$$\left. \begin{array}{l} x_1 + A_{11}^{-1}A_{12}x_2 = A_{11}^{-1}B'_1y \\ [A_{22} - A_{21}A_{11}^{-1}A_{12}]x_2 = [B'_2 - A_{21}A_{11}^{-1}B'_1]y \end{array} \right\} \qquad (4.7.5)$$

If we now compare (4.7.1) with (4.7.5), we note a remarkable similarity between the elimination process, as applicable to a system of equations in numbers, and a system of block-equations and partitioned vectors.

The similar operation, put in tabular form, with the corresponding rules for <u>block-pivoting</u>, is illustrated below.

We first of all give the block-column headings. Without block-row partitioning, this implies a one blockrow tableau matrix:

$$\mathbf{x}_1 \quad \mathbf{x}_2 \quad = \quad \mathbf{y}$$

$$[\; \mathbf{A}_1 \;|\; \mathbf{A}_2 \;]\; [\; \mathbf{B} \;] \qquad \mathbf{A}_1\mathbf{x}_1 + \mathbf{A}_2\mathbf{x}_2 \;=\; \mathbf{B}\mathbf{y}$$

As in ordinary pivoting, we go from left to right along the (block)-columns. Above, it was <u>assumed</u> that $\mathbf{A}_{11}$ is non-singular.

There are, however, systems where it can be necessary to re-order the block-rows, and/or to re-define the partitioning, until a (block)-pivot is found. Once that is the case (if indeed the system is non-singular), we have a partitioning conform (4.7.1), and the block-inversion step can be made.

The equivalent of the $\mathbf{T}_1$ tableau is:

$$\mathbf{x}1 \qquad \mathbf{x}_2 \qquad = \qquad \mathbf{y}$$

$$\left[\begin{array}{c|c} \underline{|\mathbf{A}_{11}|} & \mathbf{A}_{12} \\ \hline \mathbf{A}_{21} & \mathbf{A}_{22} \end{array}\right] \left[\begin{array}{c} \mathbf{B'}_1 \\ \hline \mathbf{B'}_2 \end{array}\right] \qquad \begin{array}{l} \mathbf{A}_{11}\mathbf{x}_1 + \mathbf{A}_{12}\mathbf{x}_2 = \mathbf{B'}_1\mathbf{y} \\[6pt] \mathbf{A}_{21}\mathbf{x}_1 + \mathbf{A}_{22}\mathbf{x}_2 = \mathbf{B'}_2\mathbf{y} \end{array}$$

Now <u>pre</u>-multiply the pivotal block-row by the inverse of the block-pivot. As in ordinary pivoting, the substitution and elimination are equivalent, and the equivalent of the $\mathbf{S}_1$ tableau is obtained by pre-multiplying the transformed leading block-row once more by $-\mathbf{A}_{21}$:

$$\mathbf{x}1 \qquad\qquad \mathbf{x}_2 \qquad\qquad = \qquad\qquad \mathbf{y}$$

$$\left[\begin{array}{c|c} \mathbf{I} & \mathbf{A}_{11}^{-1}\mathbf{A}_{12} \\ \hline -\mathbf{A}_{21} & -\mathbf{A}_{21}\mathbf{A}_{11}^{-1}\mathbf{A}_{12} \end{array}\right] \left[\begin{array}{c} \mathbf{A}_{11}^{-1}\mathbf{B'}_1 \\ \hline -\mathbf{A}_{21}\mathbf{A}_{11}^{-1}\mathbf{B'}_1 \end{array}\right]$$

On addition of the second block-row, as obtained by this pre-multiplication, to the second block-row of the original tableau, we obtain the T_2 tableau:

$$
\begin{array}{ccc}
x_1 & x_2 & = \qquad\qquad y
\end{array}
$$

$$
\left[
\begin{array}{c|c}
I & A_{11}{}^{-1}A_{12} \\ \hline
 & A_{22} - A_{21}A_{11}{}^{-1}A_{12}
\end{array}
\right]
\left[
\begin{array}{c}
A_{11}{}^{-1}B'_1 \\ \hline
B'_2 - A_{21}A_{11}{}^{-1}B'_1
\end{array}
\right]
$$

The similar technique is applied below, to the $Ax = y$ system, giving, besides the symbolic form of the tableau (for a 1 by 1 block-pivot), also the corresponding ordinary pivoting operation.

$$
A =
\left[
\begin{array}{c|c}
A_{11} & A_{12} \\ \hline
A_{21} & A_{22}
\end{array}
\right]
=
\left[
\begin{array}{c|cc}
2 & 4 & 6 \\ \hline
1 & 1 & 2 \\
3 & 3 & 3
\end{array}
\right]
$$

As the leading top lefthand block A_{11} is of order 1 by 1, the first block-inversion step is also an ordinary inversion step. For each inversion-step, we give both the symbolic form and the numerical form, with the corresponding (block)equations written explicitly as well.

The symbolic T_1 tableau:

$$
\begin{array}{cccc}
x_1 & x_2 & = & y_1 \qquad y_1
\end{array}
$$

$$
\left[
\begin{array}{c|c}
\boxed{A_{11}} & A_{12} \\ \hline
A_{21} & A_{22}
\end{array}
\right]
\left[
\begin{array}{c|c}
I & \\ \hline
 & I
\end{array}
\right]
\qquad
\begin{array}{ccc}
A_{11}x_1 + A_{12}x_2 & = & y_1 \\
A_{21}x_1 + A_{22}x_2 & = & y_2
\end{array}
$$

Ditto, in numbers:

$$
\begin{array}{ccc}
x_1 & x_2 & = \quad y_1 \quad y_2
\end{array}
$$

$$
\left[
\begin{array}{c|cc}
\boxed{2} & 4 & 6 \\ \hline
1 & 1 & 2 \\
3 & 3 & 3
\end{array}
\right]
\left[
\begin{array}{c|cc}
1 & & \\ \hline
 & 1 & \\
 & & 1
\end{array}
\right]
\qquad
\begin{array}{rl}
2x_1 + 4x_2 + 6x_3 & = y_1 \\
x_1 + x_2 + 2x_3 & = \quad y_2 \\
3x_1 + 3x_2 + 3x_3 & = \qquad y_3
\end{array}
$$

The first (block) inversion step is now prepared.

To obtain the leading block-row of the S_1 tableau, we <u>pre</u>-multiply the leading block-row of the T_1 tableau by the inverse of the block-pivot A_{11}. In preparation of eliminating reference to x_1 from the second block-equation, the leading block-row of S_1 is once more pre-multiplied by $-A_{21}$ (in general, by minus the intersection of the current pivotal block-column), to become, for the time being, the second block-row of the S_1 tableau.

$$
\begin{array}{cc}
\mathbf{x}_1 & \mathbf{x}_2
\end{array}
\qquad = \qquad
\begin{array}{cc}
\mathbf{y}_1 & \mathbf{y}_2
\end{array}
$$

$$
\left[\begin{array}{c|c}
I & A_{11}{}^{-1}A_{12} \\
\hline
-A_{21} & -A_{21}A_{11}{}^{-1}A_{12}
\end{array}\right]
\left[\begin{array}{c|c}
A_{11}{}^{-1} & \\
\hline
-A_{21}A_{11}{}^{-1} &
\end{array}\right]
$$

Explicit presentation of the formulae:

$$x_1 \qquad\qquad + A_{11}{}^{-1}A_{12}x_2 \;=\; A_{11}{}^{-1}y_1$$

$$-A_{21}x_1 \qquad - A_{21}A_{11}{}^{-1}A_{12}x_2 \;=\; -A_{21}A_{11}{}^{-1}y_1$$

Ditto, in numbers:

$$
\begin{array}{cc}
\mathbf{x}_1 & \mathbf{x}_2
\end{array}
\; = \;
\begin{array}{cc}
\mathbf{y}_1 & \mathbf{y}_2
\end{array}
$$

$$
\left[\begin{array}{c|cc}
1 & 2 & 3 \\
\hline
-1 & -2 & -3 \\
-3 & -6 & -9
\end{array}\right]
\left[\begin{array}{c|c}
0.5 & \\
\hline
-0.5 & \\
-10.5 &
\end{array}\right]
\qquad
\begin{aligned}
x_1 & +2x_2 +3x_3 = 0.5\,y_1 \\
-x_1 & -2x_2 -3x_3 = -0.5\,y_1 \\
-3x_1 & -6x_2 -9x_3 = -10.5\,y_1
\end{aligned}
$$

The actual elimination-step is now made. We store the leading (block-)row if the S_1 tableau in the T_2 tableau, and the sum of the second blockrows of the T_1 and S_1 tableaus in the second blockrow of the T_2 tableau, which is:

$$
\begin{array}{cc}
\mathbf{x}1 & \mathbf{x}_2
\end{array}
\qquad = \qquad
\begin{array}{cc}
\mathbf{y}_1 & \mathbf{y}_2
\end{array}
$$

$$
\left[\begin{array}{c|c}
I & A_{11}{}^{-1}A_{12} \\
\hline
A_{22} - A_{21}A_{11}{}^{-1}A_{12} &
\end{array}\right]
\left[\begin{array}{c|c}
A_{11}{}^{-1} & \\
\hline
-A_{21}A_{11}{}^{-1} & I
\end{array}\right]
$$

Explicit presentation of the formulae:

$$x_1 \qquad\qquad + A_{11}{}^{-1}A_{12}x_2 \;=\; A_{11}{}^{-1}y_1$$

$$[A_{22} - A_{21}A_{11}{}^{-1}A_{12}]x_2 \;=\; -A_{21}A_{11}{}^{-1}y_1 + y_2$$

Ditto, in numbers:

$$\mathbf{x_1} \qquad \mathbf{x_2} \;=\; \mathbf{y_1} \qquad \mathbf{y_2}$$

$$
\left[\begin{array}{c|cc} 1 & 2 & 3 \\ \hline & -1 & -1 \\ & -3 & -6 \end{array}\right]
\left[\begin{array}{c|cc} 0.5 & \\ \hline -0.5 & 1 \\ -10.5 & & 1 \end{array}\right]
\qquad
\begin{array}{rll}
x_1 & +2x_2 +3x_3 & = \quad 0.5\; y_1 \\
& -x_2 \;-x_3 & = \;-0.5\; y_1 \qquad +y_2 \\
& -3x_2 \;-6x_3 & = -10.5\; y_1 \qquad\qquad +y_3
\end{array}
$$

We have now made one complete (block-)inversion step.

Note, that all the matrix-expressions which were developed above as formulae, are present in the calculation-tableau in numbers, as may be illustrated as follows:

$$\mathbf{A_{11}}^{-1} = [2]^{-1} = [0.5].$$

$$\mathbf{A_{11}}^{-1}\mathbf{A_{12}} = [2]^{-1}[4 \;\; 6] = [0.5][4 \;\; 6] = [2 \;\; 3].$$

$$-\mathbf{A_{21}}\mathbf{A_{11}}^{-1} = -\begin{bmatrix} 1 \\ 3 \end{bmatrix}[2]^{-1} = -\begin{bmatrix} 1 \\ 3 \end{bmatrix}[0.5] = \begin{bmatrix} -0.5 \\ -1.5 \end{bmatrix}$$

$$\mathbf{A_{22}} - \mathbf{A_{21}}\mathbf{A_{11}}^{-1}\mathbf{A_{12}} = \begin{bmatrix} 1 & 2 \\ 3 & 3 \end{bmatrix} - \begin{bmatrix} 1 \\ 3 \end{bmatrix}[2]^{-1}[1 \;\; 2] \;=$$

$$\begin{bmatrix} 1 & 2 \\ 3 & 3 \end{bmatrix} - \begin{bmatrix} 1 \\ 3 \end{bmatrix}[0.5]\,[4 \;\; 6] \;=$$

$$\begin{bmatrix} 1 & 2 \\ 3 & 3 \end{bmatrix} - \begin{bmatrix} 2 & 3 \\ 6 & 9 \end{bmatrix} = \begin{bmatrix} -1 & -1 \\ -3 & -6 \end{bmatrix}$$

Exercise:

Block-invert (i.e. find the formula for the inverse of):

$$\mathbf{A} = \left[\begin{array}{c|c} \mathbf{A_{11}} & \mathbf{A_{12}} \\ \hline \mathbf{A_{21}} & \end{array}\right]$$

where it is known that the three non-zero blocks of **A** are all square and invertible and therefore by implication all of the same order. (There is an answersheet at the end of the chapter.)

4.8 The differentiation of an inverse

As in ordinary calculus, a relation needs to hold before and after a change in the lefthand-side. From the identity matrix product relationship $A A^{-1} = I$, we obtain, for a change in A and a resulting change in A^{-1}, which needs to leave the product the same: $d[A A^{-1}] = [0]$, and therefore:

$$\Delta[A A^{-1}] = A \Delta A^{-1} + \Delta A A^{-1} + [\Delta A]^2 = [0] \qquad (4.8.1)$$

for a finite change, and, for a differential change approaching the null matrix:

$$d[A A^{-1}] = A dA^{-1} + dA A^{-1} = [0] \qquad (4.8.2)$$

(See also sections 2.7. and 2.8, on matrix differentiation generally)

Pre-multiplication of (4.8.2) by A^{-1} gives rise to:

$$dA^{-1} + A^{-1} dA A^{-1} = [0] \qquad (4.8.3)$$

Therefore:

$$dA^{-1} = - A^{-1} dA A^{-1} \qquad (4.8.4)$$

An obvious application of (4.8.4) arises, if there is some degree of uncertainty, concerning the precise magnitude of the elements of A, which might be the result of some kind of statistical estimation procedure.

Suppose for example:

$$A = \begin{bmatrix} 1 & 0.50 \\ 0.41 & 1 \end{bmatrix} \pm \begin{bmatrix} - & 0.10 \\ 0.10 & - \end{bmatrix}$$

If A has the estimated central value, we find:

$$A^{-1} = \begin{bmatrix} 1 & 0.50 \\ 0.41 & 1 \end{bmatrix}^{-1} = \begin{bmatrix} 1.25 & -0.63 \\ -0.52 & 1.26 \end{bmatrix}$$

The first-order approximation of the likely errors in the inverse
therefore is:

$$\Delta A^{-1} \;\cong\; -\begin{bmatrix} 1.25 & -0.63 \\ -0.52 & 1.26 \end{bmatrix}\; \Delta A \;\begin{bmatrix} 1.25 & -0.63 \\ -0.52 & 1.26 \end{bmatrix}$$

where ΔA may be any matrix, of which the diagonal elements are
zero, and the off-diagonal elements are up to 0.1 in absolute
value.

The differential expression $\partial A^{-1}/\partial A$ is not well manageable, as
there are n^4 partial derivatives of elements of A^{-1}, relative to
elements of A. However, we readily infer from (4.8.4), an
expression for a partial derivative of A^{-1}, relative to <u>one</u>
element of A.

If only one element of A is changing, then dA is:

$$dA = E_{ij}\, a_{ij} \tag{4.8.5}$$

where E_{ij} is an n by n matrix, of which the element $e_{ij} = 1$ is
the one non-zero element.

Under that assumption, dA^{-1} is:

$$dA^{-1} = -A^{-1}\, E_{ij}\, da_{ij}\, A^{-1}$$

and we then express $\partial A^{-1}/\partial a_{ij}$ as:

$$\partial A^{-1}/\partial a_{ij} = -A^{-1}\, E_{ij}\, A^{-1} \tag{4.8.6}$$

Answersheet on the reduced form

$$
\begin{array}{ccccc}
y_1 & y_2 & = & x_1 & x_2 & \Sigma
\end{array}
$$

$$
\begin{bmatrix} |\underline{2}| & -1 \\ -1 & 1 \end{bmatrix}
\qquad
\begin{bmatrix} 1 & 2 \\ 3 & 4 \end{bmatrix}
\qquad
\begin{array}{c} 4 \\ 7 \end{array}
$$

$$
\begin{array}{ccccc}
y_1 & y_2 & = & x_1 & x_2 & \Sigma
\end{array}
$$

$$
\begin{bmatrix} 1 & -0.5 \\ & |0.5| \end{bmatrix}
\qquad
\begin{bmatrix} 0.5 & 1 \\ 3.5 & 5 \end{bmatrix}
\qquad
\begin{array}{c} 2 \\ 9 \end{array}
$$

$$
\begin{array}{ccccc}
y_1 & y_2 & = & x_1 & x_2 & \Sigma
\end{array}
$$

$$
\begin{bmatrix} 1 & - \\ & 1 \end{bmatrix}
\qquad
\begin{bmatrix} 4 & 6 \\ 7 & 10 \end{bmatrix}
\qquad
\begin{array}{c} 11 \\ 18 \end{array}
$$

Therefore:

$$
y = \begin{bmatrix} 4 & 6 \\ 7 & 10 \end{bmatrix} x
\qquad\qquad
R = \begin{bmatrix} 4 & 6 \\ 7 & 10 \end{bmatrix}
$$

Answersheet on blockinversion

There are two ways of handling this particular problem.

<u>Method a</u>: Straightforward following of the example-illustration.

The T_1 tableau is:

$$
\begin{array}{cccccc}
x_1 & x_2 & = & y_1 & y_2
\end{array}
$$

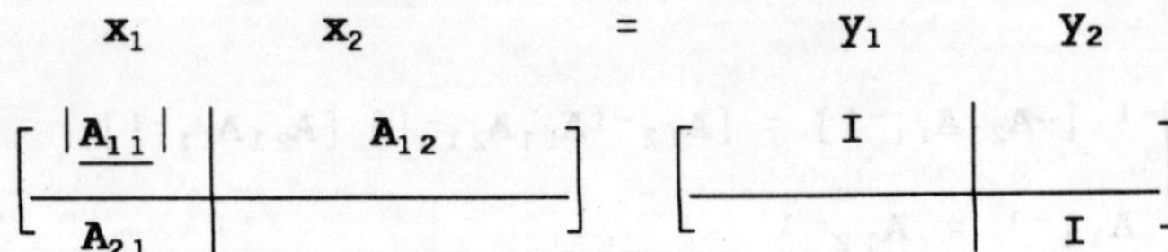

The corresponding S_1 tableau is:

$$
\begin{array}{cccc}
x_1 & x_2 & = & y_1 & y_2
\end{array}
$$

$$
\left[\begin{array}{c|c} I & A_{11}^{-1}A_{12} \\ \hline -A_{21} & -A_{21}A_{11}^{-1}A_{12} \end{array}\right]
\qquad
\left[\begin{array}{c|c} A_{11}^{-1} & \\ \hline -A_{21}A_{11}^{-1} & \end{array}\right]
$$

The equivalent explicit form of the formulae is:

$$
x_1 \;+\; A_{11}^{-1}A_{12}x_2 \;=\; A_{11}^{-1}y_1
$$

and, after pre-multiplication by $-A_{21}$:

$$
-A_{21}x_1 \;-\; A_{21}A_{11}^{-1}A_{12}x_2 \;=\; -A_{21}A_{11}^{-1}y_1
$$

We now develop the T_2 tableau:

$$
\begin{array}{cccc}
x_1 & x_2 & = & y_1 & y_2
\end{array}
$$

$$
\left[\begin{array}{c|c} I & A_{11}^{-1}A_{12} \\ \hline & \underline{-A_{21}A_{11}^{-1}A_{12}} \end{array}\right]
\qquad
\left[\begin{array}{c|c} A_{11}^{-1} & \\ \hline -A_{21}A_{11}^{-1} & I \end{array}\right]
$$

The inverse of the recursive product $A_{21}A_{11}^{-1}A_{12}$ is evaluated conform section 4.6 as:

$$
[A_{21}A_{11}^{-1}A_{12}]^{-1} = A_{12}^{-1}A_{11}A_{21}^{-1}
$$

We are therefore able to initiate the second block-inversion step, and develop the S_2 tableau:

$$
\begin{array}{cccc}
x_1 & x_2 & = & y_1 & y_2
\end{array}
$$

$$
\left[\begin{array}{c|c} & -A_{11}^{-1}A_{12} \\ \hline & I \end{array}\right]
\qquad
\left[\begin{array}{c|c} -A_{11}^{-1} & A_{21}^{-1} \\ \hline A_{12}^{-1} & -A_{12}^{-1}A_{11}A_{21}^{-1} \end{array}\right]
$$

Note the cancellation of several products of matrices and their inverses:

$$
[-A_{21}A_{11}^{-1}A_{12}]^{-1}\,[-A_{21}A_{11}^{-1}] = [A_{12}^{-1}A_{11}A_{21}^{-1}]\,[A_{21}A_{11}^{-1}]
$$

$$
= [A_{12}^{-1}A_{11}]\,A_{11}^{-1} = A_{12}^{-1}
$$

and:

$$
-A_{11}^{-1}A_{12}\,[-A_{12}^{-1}A_{11}A_{21}^{-1}] = A_{11}^{-1}\,[A_{11}A_{21}^{-1}] = A_{21}^{-1}
$$

as well as:

$$-\mathbf{A}_{11}^{-1}\mathbf{A}_{12}\ [-\mathbf{A}_{12}^{-1}] = \mathbf{A}_{11}^{-1}$$

The $\mathbf{T}_3$ tableau now is:

$$
\begin{array}{cc}
\mathbf{x}_1 \qquad\qquad \mathbf{x}_2 & = \qquad \mathbf{y}_1 \qquad\qquad \mathbf{y}_2
\end{array}
$$

$$
\left[\begin{array}{c|c}
\mathbf{I} & \\
\hline
 & \mathbf{I}
\end{array}\right]
\qquad
\left[\begin{array}{c|c}
 & \mathbf{A}_{21}^{-1} \\
\hline
\mathbf{A}_{12}^{-1} & -\mathbf{A}_{12}^{-1}\mathbf{A}_{11}\mathbf{A}_{21}^{-1}
\end{array}\right]
$$

<u>Method b</u>: Taking full advantage of the block-structure.

We first interchange the two block-rows/block-equations, to obtain the reordered $\mathbf{T}_1$ tableau:

$$
\begin{array}{cc}
\mathbf{x}_1 \qquad\qquad \mathbf{x}_2 & = \qquad \mathbf{y}_1 \qquad\qquad \mathbf{y}_2
\end{array}
$$

$$
\left[\begin{array}{c|c}
|\underline{\mathbf{A}_{21}}| & \\
\hline
\mathbf{A}_{11} & \mathbf{A}_{12}
\end{array}\right]
\qquad
\left[\begin{array}{c|c}
 & \mathbf{I} \\
\hline
\mathbf{I} &
\end{array}\right]
$$

The $\mathbf{T}_2$ successor tableau now is:

$$
\begin{array}{cc}
\mathbf{x}_1 \qquad\qquad \mathbf{x}_2 & = \qquad \mathbf{y}_1 \qquad\qquad \mathbf{y}_2
\end{array}
$$

$$
\left[\begin{array}{c|c}
\mathbf{I} & \\
\hline
 & |\underline{\mathbf{A}_{12}}|
\end{array}\right]
\qquad
\left[\begin{array}{c|c}
 & \mathbf{A}_{21}^{-1} \\
\hline
\mathbf{I} & -\mathbf{A}_{11}\mathbf{A}_{21}^{-1}
\end{array}\right]
$$

From which we obtain the $\mathbf{T}_3$ successor tableau:

$$
\begin{array}{cc}
\mathbf{x}_1 \qquad\qquad \mathbf{x}_2 & = \qquad \mathbf{y}_1 \qquad\qquad \mathbf{y}_2
\end{array}
$$

$$
\left[\begin{array}{c|c}
\mathbf{I} & \\
\hline
 & \mathbf{I}
\end{array}\right]
\qquad
\left[\begin{array}{c|c}
 & \mathbf{A}_{21}^{-1} \\
\hline
\mathbf{A}_{12}^{-1} & -\mathbf{A}_{12}^{-1}\mathbf{A}_{11}\mathbf{A}_{21}^{-1}
\end{array}\right]
$$

Either method leads to the result:

$$
\left[\begin{array}{c|c}
\mathbf{A}_{11} & \mathbf{A}_{12} \\
\hline
\mathbf{A}_{21} &
\end{array}\right]^{-1}
=
\left[\begin{array}{c|c}
 & \mathbf{A}_{21}^{-1} \\
\hline
\mathbf{A}_{12}^{-1} & -\mathbf{A}_{12}^{-1}\mathbf{A}_{11}\mathbf{A}_{21}^{-1}
\end{array}\right]
$$

CHAPTER V

DETERMINANTS AND RANK

5.1 Determinants and Minors

A determinant is a scalar function of a square matrix. It is conventional to indicate a determinant with the same symbol as the matrix, but with a bar '|' on both sides.

Thus, $|A|$ is the determinant of the matrix **A**, which must be assumed to be square. Unless otherwise stated, all matrices for which determinants are indicated in this chapter, will be assumed to be of order n by n.

The definition of the determinant is recursive:

$$|A| = \sum_{j=1}^{n} (-1)^{1+j} a_{1j} |A_{1j}| \qquad (n > 1) \qquad (5.1.1)$$

where $|A_{1j}|$ is the determinant of the (n-1) by (n-1) matrix, which is formed, by removing the leading row, and the jth column from A.

and

$$|A| = a_{11} \qquad (n = 1) \qquad (5.1.2)$$

The determinant of a square matrix, of which the ith row and the

j^{th} column have been removed, is called a <u>minor</u> (of the element a_{ij}).

We shall use the symbol $|A_{ij}|$ for the minor of a_{ij}. The corresponding matrix (block) is here referred to as the <u>minor-matrix</u>, and indicated as $\mathbf{A}_{ij}$. There is a potential possibility of confusion with block-rows and block-columns (i.e. belonging to the ith block-row and the jth block-column), but the context should make this clear.

If two rows and two columns have been removed, we speak of a <u>secondary minor</u>. or <u>second order minor</u>. The notation to be used here, presents the second order minor as a minor of a minor-matrix, i.e.

$||A_{ij}|_{rk}|$ is a minor of the minor-matrix $\mathbf{A}_{ij}$.

The indices refer, however, to the positions of these vectors in the parent-matrix A.

$||A_{ij}|_{rk}|$, $||A_{rj}|_{ik}|$ and $||A_{rk}|_{ij}|$ are all the same minor, i.e.

the determinant of a residual block of $\mathbf{A}$, obtained by the removal of the ith and the rth rows, and the jth and the kth columns.

One other notational complication which is useful to mention at this point, relates to the determinants of matrix expressions. When a capital letter occurs inside a '$|\ |$', it should be understood that we refer to a matrix. It was felt that the difference between a determinant and a matrix is enhanced, by dispensing with bold print of matrices. To clearly distinguish vectors from single elements of matrices, it is however still necessary to print vectors in bold, in '$[\ ]$', as well as in '$|\ |$' expressions.

One of the implications of the recursive definition of a determinant, is that <u>proofs</u> of theorems relating to determinants are often developed recursively as well:

For a 1 by 1 matrix, i.e. a single number, very few theorems have any meaningful content, and are therefore true in the same way that the words 'I spoke nothing but the truth', are true whenever one has not spoken at all. At the same time, it is often easier to prove a theorem concerning determinants, when it may be assumed to be true for the minor-matrices figuring in (5.1.1), or other recursive relationships concerning determinants, which we shall be developing.

The theorem itself, assumed true for a k by k matrix, can then be applied to the purpose of providing its own proof for a k+1 by k+1 matrix. Once it has been shown that the truth of a theorem for a matrix of order k by k, implies the truth of the same theorem for a matrix of order k+1 by k+1, the general proof follows: It is true for a 1 by 1 matrix (virtually every plausible theorem is true for a 1 by 1 matrix), therefore, it is true for a 2 by 2 matrix, therefore it is true for a 3 by 3 matrix .... etc. Such a type of proof is known as a <u>proof by recursive induction</u>.

That type of proof is occasionally used in other topics of mathematics (e.g. sums of series), but with determinants it is more or less the normal method of proof.

Example of the calculation of a determinant:

$$A = \begin{bmatrix} 1 & 2 & - \\ 3 & -4 & 5 \\ - & 6 & - \end{bmatrix}$$

$$|A| = 1 \times \begin{vmatrix} -4 & 5 \\ 6 & - \end{vmatrix} - 2 \times \begin{vmatrix} 3 & 5 \\ - & - \end{vmatrix} + 0 \times \begin{vmatrix} 3 & -4 \\ - & 6 \end{vmatrix}$$

To actually evaluate $|A|$, we now need to calculate the minors, at least insofar as they are to be multiplied by a non-zero number.

$$|A_{11}| = -4 \times |-| - 5 \times |6| = 0 - 30 = -30.$$

$$|A_{12}| = 3 \times |-| - 5 \times |-| = 0 - 0 = 0.$$

There is no need to evaluate $|A_{13}|$, as this minor will in any case be multiplied by $a_{13} = 0$.

Therefore:

$$|A| = a_{11} \times |A_{11}| - a_{12} \times |A_{12}|$$

$$= 1 \ (-30) - 2 \ 0 = -30.$$

5.2 **A determinant can be developed by any row**

Theorem:

For an n by n matrix **A**,

$$\sum_{j=1}^{n} a_{ij}\,(-1)^{i+j}\,|A_{ij}| \;=\; |A| \qquad\qquad (5.2.1)$$

holds for i = 1, 2, ... n (i.e. A can be developed by any row, not just the first).

Proof:

For i=1, the theorem coincides with the definition of $|A|$, and no proof is needed.

q.e.d. for n=1.

Now suppose the theorem to be true for matrices of orders 2 by 2, 3 by 3, ... (n-1) by (n-1), in addition to 1 by 1 as already proved.

Application of the theorem, as assumed true for the minors of **A**, allows us to express each $|A_{1j}|$ as:

$$|A_{1j}| \;=\; \sum_{k=1}^{j-1} a_{ik}\,(-1)^{i+k-1}\,||A_{1j}|_{ik}|$$

$$+\; \sum_{k=j+1}^{n} a_{ik}\,(-1)^{i+k-2}\,||A_{1j}|_{ik}| \qquad\qquad (5.2.2)$$

(The terms -1 and -2 in the exponents i+j-1 and i+j-2 arise, because the minor-matrix $[A_{1j}]$ misses the row which corresponds to the leading row of **A**, hence the row-indices in $[A_{1j}]$ are all one less than in **A**, and for k > j a similar consideration applies to the column-indices, on account of the removal of the jth column.)

In the interest of streamlining notation, we introduce n^2 symmetry operators, which have the value:

$s_{jk} = -1$, if j > k; $s_{jk} = 0$, of j = k; and $s_{jk} = 1$, if j < k

$(j = 1, 2, \ldots\ldots n, \ k = 1, 2, \ldots\ldots n)$.

We also introduce an artificial factor $||A_{1j}|_{ij}|$, which can have any value, as it will be multiplied by $s_{jj} = 0$.

These notational devices permit us to express $|A_{1j}|$ in a more compact form.

We multiply the first group of terms in (5.2.2) by $(-1)\ s_{jk} = 1$, thereby causing the exponent of (-1) to increase from $i+k-1$ to $i+k$, and the second group of terms by $(-1)^2\ s_{jk} = 1$, also causing the exponent to become $i+k$, and obtain:

$$|A_{1j}| \ = \ \sum_{k=1}^{n} a_{ik}\ (-1)^{i+k}\ s_{jk}\ ||A_{1j}|_{ik}| \qquad\qquad (5.2.3)$$

Application of the same notation to the development of $|A_{ij}|$ by its leading row results in:

$$|A_{ij}| \ = \ \sum_{k=1}^{n} a_{1k}\ (-1)^{1+k}\ s_{jk}\ ||A_{1j}|_{ik}| \qquad\qquad (5.2.4)$$

Substitution of the righthand side of (5.2.3) for $|A_{1j}|$ into (5.1.1), the definition of $|A|$, now gives rise to::

$$|A| = \sum_{j=1}^{n} a_{1j}\ (-1)^{1+j} \sum_{k=1}^{n} a_{ik}\ (-1)^{i+k}\ s_{jk}\ ||A_{1j}|_{ik}|$$

$$= \sum_{k=1}^{n} a_{ik}\ (-1)^{i+k} \sum_{j=1}^{n} a_{1j}\ (-1)^{1+j}\ s_{jk}\ ||A_{1j}|_{ik}| \qquad (5.2.5)$$

Comparison of (5.2.5) and (5.2.4) reveals that the last members of (5.2.5) can be rewritten as:

$$|A| \qquad = \sum_{k=1}^{n} a_{ik}\ (-1)^{i+k} \sum_{j=1}^{n} a_{1j}\ (-1)^{1+j}\ s_{jk}\ ||A_{1j}|_{ik}|$$

$$|A| \qquad = \sum_{k=1}^{n} a_{ik}\ (-1)^{i+k}\ |A_{ik}| \qquad\qquad (5.2.6)$$

and recursive induction is shown.

q.e.d.

Example:

$$A = \begin{bmatrix} 1 & 2 & 3 \\ 4 & 5 & 6 \\ 7 & 8 & 9 \end{bmatrix}$$

The development of $|A|$, by its definition, i.e. according to the minors of the leading row, is as follows:

$$|A| = 1 \begin{vmatrix} 5 & 6 \\ 8 & 9 \end{vmatrix} - 2 \begin{vmatrix} 4 & 6 \\ 7 & 9 \end{vmatrix} + 3 \begin{vmatrix} 4 & 5 \\ 7 & 8 \end{vmatrix}$$

$$= 1 \quad (-3) \quad - 2 \quad (-6) \quad + 3 \quad (-3)$$
$$= -3 \quad\quad\quad + 12 \quad\quad\quad - 9 \quad\quad = 0.$$

We now also develop $|A|$ by the second row:

$$|A| = -4 \times \begin{vmatrix} 2 & 3 \\ 8 & 9 \end{vmatrix} + 5 \times \begin{vmatrix} 1 & 3 \\ 7 & 9 \end{vmatrix} - 6 \times \begin{vmatrix} 1 & 2 \\ 7 & 8 \end{vmatrix}$$

$$= -4 \quad (-6) \quad + 5 \quad (-12) \quad - 6 \quad (-6)$$
$$= 24 \quad\quad\quad - 60 \quad\quad\quad + 36 \quad\quad = 0.$$

5.3 A determinant can be developed by a column

Theorem

$$\sum_{i=1}^{n} a_{i1} (-1)^{i+1} |A_{i1}| = |A| \qquad\qquad (5.3.1)$$

Proof:

For n=1, the theorem has no content, and no proof is required in that case.

Now assume that the theorem is true for matrices of order 2 by 2, 3 by 3, ... n-1 by n-1, and therefore for the minor-matrices of A.

Therefore, for the minors of the elements of the first row of **A**, we have by assumption:

$$|A_{1j}| = \sum_{i=2}^{n} a_{i1} (-1)^i ||A_{1j}|_{i1}| \qquad (5.3.2)$$

(j = 2, 3, ... n)

(The exponent of (-1) is only i, not i+1, because the leading row of **A** is not part of the minor-matrices, and $|A_{11}|$ cannot at all be developed by elements from the leading column of **A**, because no such elements are contained in its minor-matrix.)

We then develop $|A|$ by the minors of its leading row as:

$$|A| = \sum_{j=1}^{n} a_{1j} (-1)^{1+j} |A_{1j}|$$

$$= a_{11}|A_{11}| + \sum_{j=2}^{n} a_{1j} (-1)^{1+j} \sum_{i=2}^{n} a_{i1} (-1)^i ||A_{1j}|_{i1}| \qquad (5.3.3)$$

On replacement of the $||A_{1j}|_{i1}|$ by the identical minors $||A_{i1}|_{1j}|$, and interchanging the summations over i and j, we obtain from (5.3.3):

$$|A| = \sum_{j=1}^{n} a_{1j} (-1)^{1+j} |A_{1j}|$$

$$= a_{11}|A_{11}| + \sum_{i=2}^{n} a_{i1} (-1)^i \sum_{j=2}^{n} a_{1j} (-1)^{1+j} ||A_{i1}|_{1j}|$$

$$= a_{11} |A_{11}| + \sum_{i=2}^{n} a_{i1} (-1)^{1+i} \sum_{j=2}^{n} a_{1j} (-1)^{j} ||A_{i1}|_{1j}|$$

$$= a_{11} |A_{11}| + \sum_{i=2}^{n} a_{i1} (-1)^{1+i} |A_{i1}|$$

$$= \sum_{i=1}^{n} a_{i1} (-1)^{1+i} |A_{i1}| \qquad (5.3.4)$$

q.e.d.

Theorem

$$\sum_{i=1}^{n} a_{ik} (-1)^{i+k} |A_{ik}| = |A| \qquad (5.3.5)$$

(**A** can be developed by any k^{th} column, not just the leading one)

The proof of this theorem is analogous to the similar theorem (5.2.1), concerning the development of $|A|$ by any i^{th} row.

q.e.d.

Theorem

$$|A| = |A'|$$

(A square matrix and its transpose, have identical determinants)

Proof:

For n=1 we have $|A| = |A'| = a_{11}$

q.e.d. for n=1.

For n > 1, assume the theorem to be true, for matrices of order 2 by 2, 3 by 3, ... (n-1) by (n-1), and therefore for the minor-matrices of **A**, and of **A'**.

The development of $|A|$ by its jth column, and of **A'** by its jth

row are then identical.

The general proof now follows by recursive induction.

q.e.d

Exercise: Develop $\begin{vmatrix} 1 & 2 & 3 \\ 4 & 5 & 6 \\ 7 & 8 & 9 \end{vmatrix}$

by its leading row, and also by its third column.

Also develop the transpose $\begin{vmatrix} 1 & 4 & 7 \\ 2 & 5 & 8 \\ 3 & 6 & 9 \end{vmatrix}$ by its third row.

The answers should obviously be all the same (zero).

5.4 The non-recursive development of a determinant

There are circumstances, when it is useful to develop a
determinant, not recursively according to its definition or
according to one of the other recursive ways of developing a
determinant outlined in the two previous sections, but as the
summation of n! = 1 × 2 × 3 × ..... n separate products of n
elements of the matrix, each with the appropriate sign-
correction.

The appropriate sign correction is then found <u>with</u> the relevant
recursive product of n elements of the matrix, as the determinant
of a matrix, of which the position of the relevant elements is
as those of a permutation-operator.

We shall refer to a square matrix **A**, for which $a_{rk} \neq 0$ implies
$a_{ik} = 0$ for $i \neq r$ and a_{rj} for $j \neq k$, as a <u>permutoid</u>. The term
'permutoid' is a reminder of its relationship with the
permutation operator. Note, however, that no element of a
permutoid is required to be non-zero, or to attain any particular
numerical value.

$$\mathbf{P} = \begin{bmatrix} - & 1 & - \\ 1 & - & - \\ - & - & 1 \end{bmatrix}$$

is a permutation operator, and by implication a permutoid, and

$$\mathbf{A} = \begin{bmatrix} - & 1 & - \\ 2 & - & - \\ - & - & 3 \end{bmatrix}$$

is the sort of matrix we would in first instance think of, when speaking of a permutoid.

In fact, $\mathbf{C} = \begin{bmatrix} - & 1 & - \\ 2 & - & - \\ - & - & 0 \end{bmatrix}$

also is a permutoid, as is, incidentally, a null matrix.

We may now obtain the determinant of an n by n matrix $\mathbf{A}$, as the sum of the determinants of all the n! permutoids, that may be obtained as the elementswise products of $\mathbf{A}$ and all the permutations of the identity matrix.

Example:

$$\begin{vmatrix} 1 & 2 & - \\ 3 & -4 & 5 \\ - & 6 & 7 \end{vmatrix} =$$

$$\begin{vmatrix} 1 & - & - \\ - & -4 & - \\ - & - & 7 \end{vmatrix} + \begin{vmatrix} - & 2 & - \\ 3 & - & - \\ - & - & 7 \end{vmatrix} + \begin{vmatrix} - & - & - \\ - & -4 & - \\ - & - & - \end{vmatrix} +$$

$$\begin{vmatrix} 1 & - & - \\ - & - & 5 \\ - & 6 & - \end{vmatrix} + \begin{vmatrix} - & 2 & - \\ - & - & 5 \\ - & - & - \end{vmatrix} + \begin{vmatrix} - & - & - \\ 3 & - & - \\ - & 6 & - \end{vmatrix}$$

$$= -28 \quad -42 \quad + 0 \quad - 30 \quad + 0 \quad = \quad -100.$$

Exactly the same figures are also obtained, if $|A|$ is developed according to its definition, as may be seen by reordering the terms in the illustration-example, as follows:

$$
\begin{vmatrix} 1 & - & - \\ - & -4 & - \\ - & - & 7 \end{vmatrix}
+
\begin{vmatrix} 1 & - & - \\ - & - & 5 \\ - & 6 & - \end{vmatrix}
= 1
\begin{vmatrix} -4 & 5 \\ 6 & 7 \end{vmatrix}
= a_{11}\,|A_{11}|
$$

$$
\begin{vmatrix} - & 2 & - \\ 3 & - & - \\ - & - & 7 \end{vmatrix}
+
\begin{vmatrix} - & 2 & - \\ - & - & 5 \\ - & - & - \end{vmatrix}
= -2
\begin{vmatrix} 3 & 5 \\ - & 7 \end{vmatrix}
= -a_{12}\,|A_{12}|
$$

where the minus sign arises out of the development of the two permutoids by their leading rows.

$$
\begin{vmatrix} - & - & - \\ 3 & - & - \\ - & 6 & - \end{vmatrix}
+
\begin{vmatrix} - & - & - \\ - & -4 & - \\ - & - & - \end{vmatrix}
= 0
\begin{vmatrix} 3 & -4 \\ 6 & - \end{vmatrix}
= a_{13}\,|A_{13}|
$$

This way of grouping the determinants of the permutoids amounts to a proof of the statement with which we opened this section:

The group of permutoids, which is obtained as the elementswise product of **A** and the permutation-operators for which $p_{1j}=1$ applies, all have a_{1j} as the only non-zero element in the leading row. And if the statement is true for n-1 by n-1 matrices, then that group of determinants of permutoids adds up to $(-1)^{1+j}|A_{1j}|$.

It is basically a development of $|A|$ according to its definition.

5.5 **Proportionality of vectors**

The following theorem is here stated as a separate theorem, although it is hardly more than a corollary of the theorem from section 5.2.

Theorem

If the n by n matrices **B** and **A** are related by:

$$b_{rj} = \beta\, a_{rj} \qquad (j = 1, 2, \ldots n)$$

and

$$b_{ij} = a_{ij} \qquad (i = 1, 2, \ldots r-1, r+1, r+2, \ldots, n,$$
$$j = 1, 2, \ldots n)$$

then $|B| = \beta\, |A|$.

(Multiplication of a row by a multiplier β, implies multiplication of the determinant by β.)

Proof:

Develop both $|B|$ by its rth row, to obtain:

$$|B| = \sum_{j=1}^{n} b_{rj}\, (-1)^{r+j}\, |B_{rj}|$$

$$= \sum_{j=1}^{n} \beta\, a_{rj}\, (-1)^{r+j}\, |A_{rj}| = \beta\, |A|$$

$$= \beta \sum_{j=1}^{n} a_{rj}\, (-1)^{r+j}\, |a_{rj}| = \beta\, |A|$$

q.e.d.

The similar theorem concerning the proportionality of two columns (and its proof by developing the determinants by the relevant columns), will be obvious.

Theorem

If the n by n matrices **B** and **A** are related by:

$$b_{rj} = a_{rj} + \beta\, a_{hj} \qquad (j = 1, 2, \ldots\ldots n)$$

and

$$b_{ij} = a_{ij} \qquad (i = 1, 2, \ldots\ldots r-1, r+1, r+2, \ldots\ldots, n,$$
$$j = 1, 2, \ldots\ldots n)$$

then $|B| = |A|$.

(Adding a β-fold multiple of the hth row to the rth row leaves the determinant unaffected.)

Proof:

We first deal with some special cases:

Case 1) n=1:

The assumed operation is not possible. Therefore, the theorem has no content for n=1, and no proof is needed.

q.e.d. for case 1.

Case 2) n=2:

$$|B| = a_{11} (a_{22} + \beta a_{12}) - a_{12} (a_{21} + \beta a_{11})$$

$$= a_{11} \times a_{22} + a_{11} \times \beta a_{12} - a_{12} \times a_{21} - a_{12} \times \beta a_{11}$$

$$= a_{11} \times a_{22} \qquad\qquad - a_{12} \times a_{21} = |A|$$

The proof for h=2, r=1 is analogous.

q.e.d. for case 2.

Case 3) n > 2:

We introduce the provisional assumption, that the theorem is true for matrices of order 3 by 3, 4 by 4, ... (n-1) by (n-1), in addition to cases 1 and 2 as already proved.

A proof by recursive induction then arises, by developing $|A|$ and $|B|$ by any q^{th} row ($q \neq h$, $q \neq r$). Application of the theorem, as assumed true for the minors $|A_{qj}|$ and $|B_{qj}|$, then provides the proof by recursive induction.

q.e.d.

The similar theorem, concerning the addition or subtraction of a multiple of a column, and its proof, will be obvious.

These theorems are useful, to evaluate a determinant of a matrix of some size, in which some vectors have a typical structure.

Example:

$$A = \begin{bmatrix} 1 & 1 & 1 & 1 \\ 1 & 2 & 1 & 2 \\ 1 & 1 & 2 & 3 \\ 1 & 1 & 1 & 4 \end{bmatrix}$$

On subtracting the leading column from the other columns, we evaluate $|A|$ as:

$$|A| = \begin{vmatrix} 1 & - & - & - \\ 1 & 1 & - & 1 \\ 1 & - & 1 & 2 \\ 1 & - & - & 3 \end{vmatrix} = 1 \times \begin{vmatrix} 1 & - & 1 \\ - & 1 & 2 \\ - & - & 3 \end{vmatrix}$$

and we have no difficulty in establishing $|A| = 3$.

Theorem

If the n by n matrices **A** and **B** are related by:

$b_{rj} = a_{hj}$ (j = 1, 2, ... n)

and

$b_{hj} = a_{rj}$ (j = 1, 2, ... n, r ≠ h)

and

$b_{ij} = a_{ij}$ (j = 1, 2, ... n, all i satisfying i ≠ h, i ≠ r),

then: $|B| = - |A|$ is true.

(Interchanging two rows of a matrix causes its determinant to change in sign, while leaving its absolute value unaffected.)

Proof:

For n = 1, the stated conditions cannot be met, and no proof is needed, as the theorem has no content for n=1.

q.e.d. for n=1.

For n = 2, $|B| = a_{21} \times a_{12} - a_{22} \times a_{11} = - |A|$

q.e.d. for n = 2.

For n > 2, an assumption, that the theorem is true for the minor-matrices of **A** and **B**, leads to the evaluation of $|B|$ by any i^{th} row (i ≠ h, i ≠ r), as:

$$|B| = \sum_{j=1}^{n} (-1)^{i+j} |B_{ij}| = \sum (-1)^{i+j} (-|A_{ij}|) = -|A|$$

The proof now follows by recursive induction.

q.e.d.

Corollary:

If **A** has two equal rows, $|A| = 0$ applies.

Furthermore, since multiplication of a row by a factor β, also multiplies the determinant by β, we have the following further

Corollary:

If two rows of the n by n matrix **A** are proportional to each other, then $|A| = 0$ applies.

The generalization of this theorem and its corollaries, to the interchanging, the equality, or the proportionality of two columns, will be obvious.

5.6 Determinant and inversion by row operations

Recall, from chapter IV, the process of inversion by row operations. That operation implies a calculation of the determinant.

To invert a square matrix **A**, i.e. to resolve the **Ax** = **y** system, we start with a calculation tableau, of which the lefthand square block, to which we shall refer as **L**, initially is **A**, and the righthand block, to which we shall refer as **R**, initially is I_n.

Per inversion step, we then perform operations (a) to (c) as listed below:

(a) If necessary, interchange two rows. This operation, if performed, causes the determinants of both blocks to change in sign, while maintaining the same absolute value.

(b) Multiply the pivotal row by the reciprocal of the pivot. This causes the determinants of both blocks to be divided by the pivot.

(c) Add or subtract suitable multiples of the pivotal row to (or from) the other rows. This is the actual elimination operation, and it leaves the determinants of both blocks unchanged.

At the end of the inversion operation, **L** is an identity matrix,
of which the determinant is 1=one. Therefore, if no row-swaps and
corresponding changes in the sign of the determinants took place,
the determinant of the initial lefthand block, must have been the
recursive product of the numbers by which it was divided in
operation (b).

Example:

$$\mathbf{A} = \begin{bmatrix} 1 & 2 & - \\ 3 & -4 & 5 \\ - & 6 & - \end{bmatrix}$$

The corresponding calculation tableaus are:

x_1	x_2	x_3	$=$	y_1	y_2	y_3	Σ
$\boxed{\underline{1}}$	2	-		1	-	-	4
3	-4	5		-	1	-	5
	6			-	-	1	7

$$|L| = |A| \qquad\qquad |R| = |I| = 1$$

x_1	x_2	x_3	$=$	y_1	y_2	y_3	Σ
1	2	-		1	-	-	4
-	$\boxed{\underline{-10}}$	5		-3	1	-	-7
-	6	-		-	-	1	7

$$|L| = |A| \qquad\qquad |R| = |I| = 1$$

(As yet, division of both determinants, by the pivot, has made no
change, because the pivot was 1)

x_1	x_2	x_3	$=$	y_1	y_2	y_3	Σ
1	-	1		0.4	0.2	-	2.6
-	1	-0.5		0.3	-0.1	-	0.7
-	-	$\boxed{\underline{3}}$		-1.8	0.6	1	2.8

$$|L| = |A|/(-10) \qquad\qquad |R| = 1 / (-10) = -0.1$$

(The pivotal row was divided by -10)

$$
\begin{array}{ccccccccc}
x_1 & & x_2 & & x_3 & = & y_1 & y_2 & y_3 & \Sigma \\
\begin{bmatrix} 1 & - & - \\ - & 1 & - \\ - & - & 1 \end{bmatrix} & & & & & & \begin{bmatrix} 1 & - & -0.33 \\ - & - & 0.17 \\ -0.6 & 0.2 & 0.33 \end{bmatrix} & & & \begin{matrix} 1.67 \\ 1.17 \\ 0.93 \end{matrix}
\end{array}
$$

$$|L| = (|A|/(-10))/3 \qquad\qquad |R| = -0.1 \; / \; 3 = -1/30$$

(The pivotal row was divided by 3)

Since $|L|$ now clearly is $|L| = |I| = 1$, we conclude:

$$|L| = |A| \;/(-30) = 1, \text{ therefore: } \rightarrow |A| = -30.$$

(The determinant of the same matrix was calculated by its definition in section 5.1)

Obviously, this result is generally valid: The recursive product of the pivots used to invert a matrix, with the sign changed every time two rows need to be interchanged, has the determinant of the inverted matrix as end-product.

5.7 Failure to invert = singularity

We began our analysis of the relation between the determinant and the calculation of the inverse, by assuming that the calculation of the inverse, by the method of row operations, as outlined, is indeed possible.

We now consider the situation which arises, if the inversion procedure breaks down, because the kth pivotal column does not contain a non-zero element, that could be used as pivot. We have dealt with the case of a zero on the main diagonal, while there still are non-zero elements lower down in the kth column, in section 4.5. But let us now assume that there is no non-zero element further down either.

If this happens immediately with the leading column, for k=1, it is clear that this is a matrix of which the determinant is zero.

But, as will be illustrated below, the same problem may indeed

arise at a later stage, possibly even when there were no zero
vectors in the original matrix.

It will be seen below, that failure to invert, whether
immediately at the first step, (the leading column of **A** being a
null vector), or after n_1 inversion steps ($0 < n_1 < n$) implies
both a zero determinant, and singularity as defined in section
4.1, i.e. row dependence.

At this stage, we first of all illustrate failure to invert
following partial inversion.

Example:

$$\mathbf{A} = \begin{bmatrix} 1 & 1 & 1 & 1 \\ 2 & 2 & 2 & - \\ - & 2 & 3 & 3 \\ 4 & 4 & 4 & 2 \end{bmatrix}$$

We develop the following calculation tableaus:

x_1	x_2	x_3	x_4	=	y_1	y_2	y_3	y_4	Σ
|1|	1	1	1		1	-	-	-	5
2	2	2	-		-	1	-	-	7
-	2	3	3		-	-	1	-	9
4	4	4	2		-	-	-	1	15

x_1	x_2	x_3	x_4	=	y_1	y_2	y_3	y_4	Σ
1	1	1	1		1	-	-	-	5
-	-	-	-2		-2	1	-	-	-3
-	2	3	3		-	-	1	-	9
-	-	-	-2		-4	-	-	1	-5

Before we can make the second inversion step, we need to
interchange the second and third rows, to obtain the following
tableaus:

$$
\begin{array}{cccc}
x_1 & x_2 & x_3 & x_4
\end{array} \quad = \quad
\begin{array}{cccc}
Y_1 & Y_2 & Y_3 & Y_4
\end{array} \quad \Sigma
$$

$$
\left[\begin{array}{cccc}
1 & 1 & 1 & 1 \\
- & |\underline{2}| & 3 & 3 \\
- & - & - & -2 \\
- & - & - & -2
\end{array}\right]
\qquad
\left[\begin{array}{cccc}
1 & - & - & - \\
- & - & 1 & - \\
-2 & 1 & - & - \\
-4 & - & - & 1
\end{array}\right]
\begin{array}{c}
5 \\ 9 \\ -3 \\ -5
\end{array}
$$

$$
\begin{array}{cccc}
x_1 & x_2 & x_3 & x_4
\end{array} \quad = \quad
\begin{array}{cccc}
Y_1 & Y_2 & Y_3 & Y_4
\end{array} \quad \Sigma
$$

$$
\left[\begin{array}{cccc}
1 & - & -0.5 & -0.5 \\
- & 1 & 1.5 & 1.5 \\
- & - & - & -2 \\
- & - & - & -2
\end{array}\right]
\qquad
\left[\begin{array}{cccc}
1 & - & -0.5 & - \\
- & - & 0.5 & - \\
-2 & 1 & - & - \\
-4 & - & - & 1
\end{array}\right]
\begin{array}{c}
-0.5 \\ 4.5 \\ -3 \\ -5
\end{array}
$$

At this point, the inversion algorithm ends, with failure to invert, following the partial inversion of a 2 by 2 block A_{11} (after re-ordering that block to the top lefthand corner, by interchanging the second and third rows).

Theorem:

For every n by n matrix **A** one of the following conditions applies:

Either: a)

    a1) **A** is singular,

    a2) $|A| = 0$, and

    a3) **A** cannot be inverted by row-operations,

Or: b)

    b1) **A** can be inverted by row operations, and:

    b2) For every $0 < n_1 < n$, there is a (trivial or non-trivial) re-ordering of the rows of **A**, (requiring r times the interchanging of two rows of **A**), for which **A** partitions as:

$$
\mathbf{A} = \left[\begin{array}{c|c}
\mathbf{A}_{11} & \mathbf{A}_{12} \\
\hline
\mathbf{A}_{21} & \mathbf{A}_{22}
\end{array}\right]
\tag{5.7.1}
$$

where $\mathbf{A}_{11}$ is square and invertible, and for which:

b2)

$$|A| = |A_{11}| \times |A_{22} - A_{21} A_{11}^{-1} A_{12}| \quad \neq 0$$

$$(5.7.2)$$

applies.

Proof:

The following cases arise:

Case 1): $a'_1 = [0]'$ (the leading row is a null vector)

Then $e'_1 A = [0]'$ states a1), the singularity of A.

a1) follows immediately, on development of $|A|$, by the leading row.

Since a singular matrix does not have an inverse, it cannot be calculated (by row operations or otherwise), and a3) is implied.

q.e.d. for case 1.

For n=1, the theorem is true by triviality.
-Either a) for $A = [0]$, or for $A = [a_1]$, $a_{11} =/ 0$, b) is true by way of not stating anything at all, the set of numbers n_1, for which b) has content, being empty.-

We now consider the non-trivial

case 2) $n > 1$, $a'_1 \neq [0]'$

Then b2) is stated for A' at least for $n_1 = 1$, and we will use the notation used in (5.7.1), even if it may be the case that the partitioning requires column reordering (= row-reordering of A')

We now invoke the provisional assumption that the theorem is true for all matrices of orders 1 by 1, to (n-1) by (n-1), and therefore for both $[A_{22} - A_{21} A_{11}^{-1} A_{12}]$ as well as for its transpose (of which the determinants are identical).

We now first consider:

Subcase 2a):

$$|\mathbf{A}_{22}-\mathbf{A}_{21}\mathbf{A}_{11}{}^{-1}\mathbf{A}_{12}| = |\mathbf{A}_{22}{}'-\mathbf{A}_{12}{}'\mathbf{A}_{11}{}'^{-1}\mathbf{A}_{21}| = 0 \qquad (5.7.3)$$

Since successful completion of the inversion of $\mathbf{A}'$ by row operations would imply the inversion of $[\mathbf{A}_{22}{}'-\mathbf{A}_{12}{}'\mathbf{A}_{11}{}'^{-1}\mathbf{A}_{21}]$, we must assume that an attempt to complete the inversion of $\mathbf{A}'$ will come to the abnormal ending discussed earlier in this section.

<u>Denoting</u> $\mathbf{B} = \mathbf{A}'$, the attempted inversion $\mathbf{B} = \mathbf{A}'$ ends with the following calculation tableau:

$$
\begin{array}{ccc}
\mathbf{w}_1 \quad \mathbf{w}_2 \qquad\qquad \mathbf{w}_3 & = & \mathbf{u}_1 \quad \mathbf{u}_2
\end{array}
$$

$$
\left[\begin{array}{c|c}
\mathbf{I} \quad \mathbf{B}_{11}{}^{-1}\mathbf{b}_{12} & \mathbf{B}_{11}{}^{-1}\mathbf{B}_{13} \\
\hline
- & \mathbf{B}_{23} - \mathbf{B}_{21}\mathbf{B}_{11}{}^{-1}\mathbf{B}_{13}
\end{array}\right]
\qquad
\left[\begin{array}{c|c}
\mathbf{B}_{11}{}^{-1} & \\
\hline
-\mathbf{B}_{21}\mathbf{B}_{11}{}^{-1} & \mathbf{I}
\end{array}\right]
$$

$$
|L| = 0 \qquad\qquad\qquad |R|
$$

where $|L| = 0$ arises on developing $|L|$ by its leading column and invoking the provisional assumption in relation to the (n-1) by (n-1) minor-matrix associated with $l_{11} = 1$.

We saw in the previous section, that $|L|$ is in absolute value equal to $|\mathbf{A}'|$ (or $|\mathbf{A}|$), divided by the recursive product of the pivots used to invert $\mathbf{A}_{11}{}'$.

Therefore: $|\mathbf{A}'| = |\mathbf{A}| = 0$, a2).

This calculation tableau also provides a solution:

$$\mathbf{w}_1 = \mathbf{B}_{11}{}^{-1}\mathbf{b}_{12}, \ \mathbf{w}_2 = -1$$

(and $\mathbf{w}_3 = [0]$, $\mathbf{u}_1 = [0]$, $\mathbf{u}_2 = [0]$)

for the $\mathbf{B}\,\mathbf{w} = [0]$ system, and therefore a vector $\mathbf{w}' \neq [0]'$, which satisfies $\mathbf{w}'\mathbf{A} = [0]'$.

Therefore a1), the singularity (rowdependence) of $\mathbf{A}$ is stated.

Since a singular matrix does not have an inverse, a3) is then also implied.

Hence the assumed (5.7.3) implies all of a).

q.e.d. for case 2a).

Subcase 2b)

$$|\mathbf{A}_{22}-\mathbf{A}_{21}\mathbf{A}_{11}^{-1}\mathbf{A}_{12}| = |\mathbf{A}_{22}'-\mathbf{A}_{12}'\mathbf{A}_{11}'^{-1}\mathbf{A}_{21}| \neq 0 \qquad (5.7.4)$$

The theorem, assumed true for $[\mathbf{A}_{22}'-\mathbf{A}_{12}'\mathbf{A}_{11}'^{-1}\mathbf{A}_{21}]$ then implies that the inversion of $\mathbf{B} = \mathbf{A}'$ by row-operations, can be completed, and (see the previous section) $|\mathbf{B}| = |\mathbf{A}'| = |\mathbf{A}| \neq 0$ is implied.

Now assume, contrary to what the theorem states, that $\mathbf{A}$ cannot be inverted by row operations. That would establish, (for $\mathbf{A}$ with its columns in their natural ordering):

$$|\mathbf{L}| = 0,$$

and hence:

$$|\mathbf{A}| = 0, \text{ contradicting } |\mathbf{A}'| = |\mathbf{A}| \neq 0 \text{ as already proved.}$$

Therefore successful completion of the inversion of $\mathbf{A}$, b1) must be assumed instead.

If b1) is true, b2) is implied, with n_1 being the number of inversion steps already made:

Since row-swaps change the signs of both $|\mathbf{L}|$, and of the ratio between $|\mathbf{A}_{11}|$ and the recursive product of the pivots, we conclude from a development of $|\mathbf{L}|$ by its leading column:

$$|\mathbf{L}| = |\mathbf{A}| / |\mathbf{A}_{11}| = |\mathbf{A}_{22}-\mathbf{A}_{21}\mathbf{A}_{11}^{-1}\mathbf{A}_{12}| \qquad (5.7.5)$$

and (5.7.2), i.e. b3) follows.

q.e.d. for subcase 2b)

q.e.d. for case 2

q.e.d.

It is also clear, both from this theorem and its proof, and from the previous section, that the equality:

$$|A| = |A_{11}| \times |A_{22} - A_{21} A_{11}{}^{-1} A_{12}|$$

applies for every invertible block A_{11}, irrespective of whether or not **A** is invertible: If **A** is invertible we have (5.7.2), otherwise we have (5.7.3).

The following statements are thus seen to be equivalent:

A is singular
(defined originally as $\mathbf{w'A} = [0]'$, for some $\mathbf{w'} \neq [0]'$).

$|A| = 0$

$\mathbf{Ax} = [0]$, $\mathbf{x} \neq [0]$, (follows from $|A'| = 0$)

A cannot be inverted by row operations.

In addition, we saw in section 4.4, that a singular matrix does not have an inverse, ($\mathbf{w'A} = [0]'$, implies $\mathbf{w'AB} = [0]'$ therefore either $\mathbf{w'} = [0]'$, or no matrix **B** satisfying $\mathbf{AB} = I$ can exist). Conversely, if a matrix is not singular, $|A| \neq 0$ must apply, and $\mathbf{A}^{-1}$ can be calculated by row operations.

Therefore, the statement:

A does not have an inverse, is also equivalent to the four listed above.

Note, however, that the number of inversion-steps, after which an attempt to invert by row operations will result in failure to invert, may be different, if we try to invert **A**, or **A'**.

For the example used in this section,

$$\mathbf{A} = \begin{bmatrix} 1 & 1 & 1 & 1 \\ 2 & 2 & 2 & - \\ - & 2 & 3 & 3 \\ 4 & 4 & 4 & 2 \end{bmatrix} \quad \text{therefore} \quad \mathbf{A'} = \begin{bmatrix} 1 & 2 & - & 4 \\ 1 & 2 & 2 & 4 \\ 1 & 2 & 3 & 4 \\ 1 & - & 3 & 2 \end{bmatrix}$$

row-permutation, as prescribed by the algorithm of inversion by row-operations, would identify a 3 by 3 invertible block

$$A^*_{11}{}' = \begin{bmatrix} 1 & 2 & - \\ 1 & 2 & 2 \\ 1 & - & 3 \end{bmatrix}$$

and it would take a step more, before the attempt to invert A' would be revealed as a failure, while the inversion of A failed already after two steps.

Now that we have shown that an inverse, if it exists, can always be calculated by row operations, we are able to draw some further inferences from our analysis of the previous section.

From the division of both sides of the calculation-tableau, by the recursive product of the pivots (with or without sign-swap, which affects both sides in the same way), we infer

$$|A^{-1}| = 1 / |A|$$

as illustrated in the previous section, to be generally valid: For $|A| = 0$, there is no inversion by row-operations, and no inverse either.

Application of the same logic to a reduced form system:

$$A\,y = B\,x$$

which, for $|A| \neq 0$ resolves as

$$y = A^{-1}\,B\,x$$

leads in the case of a square matrix B to the result:

$$|[A^{-1}\,B]| = |B| / |A| \tag{5.7.6}$$

Application of this last result to the $A\,y = [A\ B]\,x$ calculation tableau leads, for $|A| \neq 0$ to:

$$|[A^{-1}\,A\,B]| = |B| = |[A\ B]| / |A|$$

and therefore

$$|[A\ B]| = |A| \times |B| \tag{5.7.7}$$

For a singular A, $w'A = [0]'$, $w' \neq [0]'$ implies:

$\mathbf{w}'[\mathbf{A}\ \mathbf{B}] = [0]'$, $\mathbf{w}' \neq [0]'$, and hence the singularity of $[\mathbf{A}\ \mathbf{B}]$, and therefore both $|\mathbf{A}| = 0$ and $|[\mathbf{A}\ \mathbf{B}]| = 0$.

The result:

$$|[\mathbf{A}\ \mathbf{B}]| = |\mathbf{A}| \times |\mathbf{B}| = |\mathbf{B}| \times |\mathbf{A}| = |[\mathbf{B}\ \mathbf{A}]|$$

therefore applies for all square matrices, whether invertible or singular.

The determinant of the product of two or more square matrices is the product of their determinants.

Obviously, $|[\mathbf{A}\ \mathbf{B}]| = 0$ then implies $|\mathbf{A}| = 0$, or $|\mathbf{B}| = 0$, or both.

Exercise:

$$\text{For} \quad \mathbf{A} = \begin{bmatrix} - & 1 & 2 \\ 3 & -4 & 5 \\ -6 & 7 & -8 \end{bmatrix}, \quad \mathbf{B} = \begin{bmatrix} - & 1 & 2 \\ 3 & 4 & 5 \\ 6 & 7 & 8 \end{bmatrix}, \quad \mathbf{C} = \begin{bmatrix} 1 & -2 & 3 \\ -4 & 5 & -6 \\ 7 & -8 & 9 \end{bmatrix}$$

calculate the determinants, and if existing, the inverses, by inversion or attempt to invert by row-operation, and also independently, by developing the determinants according to the definition of a determinant. (This exercise is self-checking, but an answersheet -for $\mathbf{A}$ only- is provided at the end of the chapter anyhow).

5.8 Co-factors and the adjoint

The term $(-1)^{i+j}\ |A_{ij}|$, as occurring in the development of a determinant, is called the co-factor of the corresponding element. The transposed matrix of co-factors is called the adjoint.

Theorem:

If $B = A^{-1}$ exists,   then:

$$B = A^{-1} = [(-1)^{i+j} |A_{ij}|]' / |A| \qquad (5.8.1)$$

(The inverse is the adjoint, divided by the determinant.)

Proof:

We first consider the special case $i=j=n$, (i.e. b_{nn}, the n,n element of the inverse):

Recall section 5.7.   We partition A as follows:

$$A = \left[\begin{array}{c|c} A_{11} & a_{1n} \\ \hline a'_{n1} & a_{nn} \end{array} \right] \qquad (5.8.2)$$

If A_{11} is invertible, we apply (5.7.2), as follows:

$$|A| = |A_{11}| \cdot |a_{nn} - a'_{n1}A_{11}^{-1}a_{1n}| =$$

$$|A_{11}| (a_{nn} - a'_{n1}A_{11}^{-1}a_{1n}) \qquad (5.8.3)$$

The factor $(a_{nn} - a'_{n1}A_{11}^{-1}a_{1n})$ on the righthand side of (5.8.3) is also the last pivot, to be used in the inversion of A by row-operations. For $|A| \neq 0$, this factor must be non-zero.

Therefore:

$$(a_{nn} - a'_{n1}A_{11}^{-1}a_{1n})^{-1} = b_{nn} \qquad (5.8.4)$$

Row-permutation during inversion of A_{11} changes the orderings, and by implication the signs of both $|A|$ and $|A_{11}|$.

Therefore by (5.8.3) and (5.8.4):

$$b_{nn} = |A_{11}| / |A| \qquad (5.8.5)$$

q.e.d. for $i=j=n$, $|A_{11}| \neq 0$.

Now assume, by contra-assumption: $|A_{11}| = 0$, $b_{nn} \neq 0$.

We partition the inverse system $\mathbf{x} = \mathbf{B}\mathbf{y}$, as follows:

$$\mathbf{x}_1 \quad = \quad \mathbf{B}_{11}\mathbf{y}_1 + \mathbf{b}_{1n}\mathbf{y}_n \tag{5.8.6}$$

$$\mathbf{x}_n \quad = \mathbf{b'}_{n1}\mathbf{y}_1 + \mathbf{b}_{nn}\mathbf{y}_n \tag{5.8.7}$$

Therefore, substituting $b_{nn}^{-1}x_n - b_{nn}^{-1}\mathbf{b'}_{n1}\mathbf{y}_1$ for y_n, conform (5.8.7) into (5.8.6):

$$\mathbf{x}_1 \quad = [\mathbf{B}_{11} - \mathbf{b}_{1n}\mathbf{b}_{nn}^{-1}\mathbf{b'}_{n1}]\mathbf{y}_1 + \mathbf{b}_{1n}\mathbf{b}_{nn}^{-1}\mathbf{x}_n \tag{5.8.8}$$

The solvability of $\mathbf{B}\mathbf{x} = \mathbf{y}$ now implies the invertability of

$$[\mathbf{B}_{11} - \mathbf{b}_{1n}\mathbf{b}_{nn}^{-1}\mathbf{b'}_{n1}] = \mathbf{A}_{11}^{-1} \tag{5.8.9}$$

and hence $|A_{11}| \neq 0$.

Therefore $b_{nn} \neq 0$ cannot be true for $|A_{11}| = 0$.

q.e.d. for $i=j=n$, $|A_{11}| = 0$.

q.e.d. for $i=j=n$.

The general proof now follows on reordering the $\mathbf{A}\mathbf{x} = \mathbf{y}$ system, causing the determinant of the re-ordered matrix $\mathbf{A}^*$ to change sign conform (5.8.1)

q.e.d.

Note the relation between the reordering of the $\mathbf{A}\mathbf{x} = \mathbf{y}$ system, and the transposition of the matrix of co-factors. To get the i,j element of $\mathbf{A}$ in the n,n position, we need to reorder as follows: To get $a_{i,j}$ in the $a^*_{n,j}$ position, we need to reorder the rows of $\mathbf{A}$, and the elements of $\mathbf{y}$, placing y_i in position n. The corresponding reordering of $\mathbf{B} = \mathbf{A}^{-1}$ is a column-reordering, i.e. while the ith row of $\mathbf{A}$ comes at the end, the jth column of $\mathbf{B}$ is moved to the end. Similarly, column-reordering in $\mathbf{A}$ corresponds to a reordering of the vector $\mathbf{x}$, and of the rows of $\mathbf{B}$.

Example:

$$\mathbf{A} = \begin{bmatrix} 2 & 5 & 1 \\ 11 & -7 & 1 \\ -5 & 3 & 1 \end{bmatrix}$$

We now wish to obtain the $b_{2,1}$ element of the inverse.

To be able to apply the i=j=n case, we need to have x_2 at the end of the vector $\mathbf{x}$, as x^*_n, and y_1 at the end of the vector $\mathbf{y}$, as y_n.

The corresponding re-ordering of the $\mathbf{Ax} = \mathbf{y}$ system is as follows:

Initial system:

$$2x_1 + 5x_2 + x_3 = y_1$$
$$11x_1 - 7x_2 + 8x_3 = y_2$$
$$-5x_1 + 3x_2 + x_3 = y_3$$

Re-ordered system:

$$11x_1 + 8x_3 - 7x_2 = y_2$$
$$-5x_1 + x_3 + 3x_2 = y_3$$
$$2x_1 + x_3 + 5x_2 = y_1$$

Accordingly, the minor-matrix, which is associated with the co-factor to be used in calculating b_{21} is:

$$|A_{12}| = \begin{vmatrix} 11 & 8 \\ -5 & 1 \end{vmatrix} = 11 - (-40) = 51$$

Correspondingly, no change in the minor-matrix arises: We simply put the ith row and the jth column at the end, and the indices of the other rows above position i, and of the columns above position j drop one position. Correspondingly, the non-leading elements of the re-ordered vectors $\mathbf{x}^*$ (x_1 and x_2), and of $\mathbf{y}^*$ (y_2 and y_3), stay in their natural ordering. Hence no sign-swap of the minor, only between $|A|$ and $|A^*|$.

The older traditional approach in linear algebra, for example Aitken [2], Chapter III, uses (5.8.1) as the definition of the inverse, i.e. the inverse is defined as the matrix which conforms to (5.8.1), and the other properties which we discussed so far, then follow as results.

The theory of determinants is older than the use of matrix notation, and there are authors, e.g. Maurer [30], who appear to think that the equivalence with inversion by row-operations is a recent discovery. This is not so. Vandermonde [47] (as referred to by Muir [35], Vol I, p. 19), was one of the first people to

develop the use of determinants and determinantal notation. He
did so, in a paper entitled Note on elimination (i.e. Gaussian
elimination by row-operations). And there were other people who
clearly were well aware of the close relationship between
determinants, elimination and inversion by row-operations, see
for example Aitken [1], [2], pp. 45-48, as well as Mehmke [31],
[32].

It was nevertheless at one stage traditional, to start with
determinants, and then proceed to using them in solving linear
equations. More recently, a number of authors e.g. Hohn [25], p.
27, 271, and indeed some early pioneers of matrix algebra (e.g.
Bodewig [6], p. 16), have, like myself, found the move in the
other direction more naturally explainable.

Exercise:

For $\mathbf{A} = \begin{bmatrix} 2 & 4 & 6 \\ 1 & 1 & 2 \\ 3 & 3 & 3 \end{bmatrix}$

calculate the determinant, and the inverse, by the following
independent methods:

(1) Calculate $|\mathbf{A}|$ according to its definition, (development by
the leading row)

(2) Developing $|\mathbf{A}|$ by the second column.

(3) Then calculate $\mathbf{A}^{-1}$ according to (5.8.1).

(4) Invert $\mathbf{A}$ by row-operations, and cross-check that both the
inverse and the determinant match the results obtained under (1)
to (3).

5.9 **The adjoint all integer elimination method**

For a square matrix of not too large order, it is a practical proposition to invert while avoiding rounding altogether. The advantage of such a procedure, if possible, are most obvious if we are concerned with the inversion of a matrix whose elements are all integer numbers. This section is devoted to outlining a method of inversion, in which we store the adjoint, rather than the inverse of the already 'inverted' top lefthand part of the matrix which is to be inverted.

For $\mathbf{A} = \begin{bmatrix} 2 & 5 & 1 \\ 11 & -7 & 8 \\ -5 & 3 & 1 \end{bmatrix}$

and therefore the $\mathbf{Ax} = \mathbf{y}$ system:

$$2\,x_1 + 5\,x_2 + x_3 = y_1$$
$$11\,x_1 - 7\,x_2 + 8\,x_3 = y_2$$
$$-5\,x_1 + 3\,x_2 + x_3 = y_3$$

we develop a series of calculation tableaus, as follows:

$\mathbf{T}_1$:

x_1	x_2	x_3	=	y_1	y_2	y_3	Σ	Ratio
|2|	5	1		1	–	–	9	1
11	-7	8		–	1	–	13	1
-5	3	1		–	–	1	–	1

As yet, the tableau is identical to the set-up tableau for the standard method of elimination by row-operations, i.e. the Gauss-Jordan method. The one difference is the additional column, which indicates the ratio between the entries in the various rows of the calculation tableau, and the similar Gauss-Jordan tableau.

Instead of dividing the pivotal row by the pivot, we multiply the other rows by that number, to obtain the following intermediate tableau:

U_1 :

x_1	x_2	x_3	=	y_1	y_2	y_3	Σ	Ratio
|2|	5	1		1	–	–	9	2
22	-14	16		–	2	–	26	2
-10	6	2		–	–	2	–	2

This intermediate tableau represents a slightly different stage
of calculation, compared to the intermediate tableau which we
indicated by the letter **S** in Chapters I and IV, when illustrating
the standard method of elimination by row-operations.

The equivalent tableau with unity coefficients on the lefthand
side, the one which is obtained immediately after division of
the pivotal row by the pivot, i.e.:

G_1 :

x_1	x_2	x_3	=	y_1	y_2	y_3	Σ	Ratio
1	2.5	0.5		0.5	–	–	0.5	1
11	-7	8		–	1	–	13	1
-5	3	1		–	–	1	–	1

Here, the ratio is one by definition, as the ratio column indicates the ratio between the all-integer tableau and the corresponding unit vectors tableau.

Now returning to tableau U_1, we subtract the leading pivotal row
from the other rows, with multiples of 11 and -5, to obtain the
successor tableau:

T_2 :

x_1	x_2	x_3	=	y_1	y_2	y_3	Σ	Ratio
2	5	1		1	–	–	9	2
–	|-6|	5		-11	2	–	-73	2
–	31	7		5	–	2	45	2

The multipliers, used in adding multiples of the pivotal row to
the other rows of the tableau (-11 and 5), are precisely the same
as in the conventional method of elimination by row operations.

Note the different ways in which the figure of 2 in the ratio column was arrived at: Rows 2 and 3 were multiplied by 2, but for the leading row, it was a question of not dividing by 2.

In the usual procedure, we obtain the determinant of the next $\mathbf{A}_{11}$ block, by multiplying the determinant of the block as already inverted, by the next pivotal element. (This assumes that the matrix is ordered in a way which permits inversion along the diagonal.) As long as the ratio is $|\mathbf{A}_{11}|$, it follows that we always have the determinant of the next $\mathbf{A}_{11}$ block as the tableau element in the pivotal position.

It is is readily verified, that this is indeed the case:

$$\begin{vmatrix} 2 & 5 \\ 11 & -7 \end{vmatrix} = 2 \times (-7) \quad - \quad 5 \times 11 \quad = \quad -14 - 55 = -69.$$

We now again multiply the non-pivotal rows by the pivotal element, instead of dividing the pivotal row by its pivotal diagonal entry, and we obtain the intermediate tableau:

$\mathbf{U}_2$:

x_1	x_2	x_3	=	y_1	y_2	y_3	Σ	Ratio
-138	-345	-69		-69	-	-	-621	-138
-	$\underline{|-69|}$	5		-11	2	-	-73	-69
-	-2139	-483		-345	-	-138	-3105	-138

As before, the change in the ratio with the corresponding unit vectors tableau arises, in the pivotal row by not dividing, and in the other rows by multiplication. From the second step onwards, the ratio in the pivotal row is not in all stages of calculation the same as in the other rows. In the pivotal row, the inhibited division would not have been by $|\mathbf{A}_{11}|$, but by the pivot. The second row would not have been multiplied by 2, and that pivotal entry would have been -34.5 , instead of the -69 which is there in the all-integer tableau.

We now subtract the second pivotal row 5 times from the leading row, and 31 times from the third row, thereby 'sweeping out' the second column, i.e causing all its entries, except the diagonal entry of -69 to vanish:

V_2 :

	x_1	x_2	x_3	=	y_1	y_2	y_3	Σ	Ratio
	-138	$-$	-94		-14	-10	$-$	-256	-138
	$-$	-69	5		-11	2	$-$	-73	-69
	$-$	$-$	-638		-4	-62	-138	-842	-138

But clearly, the -69, which is $|A_{11}|$, is a sufficient number, to ensure the integer nature of the entire calculation tableau. Reference to section 4.7 indicates that all blocks of the Gauss-Jordan calculation tableau are obtainable as matrix-expressions involving blocks of $\mathbf{A}$, and $\mathbf{A}_{11}^{-1}$.

Multiplication by $|A_{11}|$ converts these expressions into corresponding expressions, with $\mathbf{A}_{11}^{-1}$ replaced by the adjoint.

This consideration permits us to normalize the calculation tableau, dividing the non-pivotal rows by the <u>previous</u> multiplier (i.e. 2), to obtain the successor tableau:

T_3 :

	x_1	x_2	x_3	=	y_1	y_2	y_3	Σ	Ratio
	-69	$-$	-47		-7	-5	$-$	-128	-69
	$-$	-69	5		-11	2	$-$	-73	-69
	$-$	$-$	$|-319|$		-2	-31	-69	-421	-69

We are now ready to prepare the last inversion step. We multiply rows 1 and 3 by -319 ($= |A_{11}| = |A|$), and develop the following intermediate tableau:

U_3 :

	x_1	x_2	x_3	=	y_1	y_2	y_3	Σ	Ratio
	22011	$-$	14993		2233	1595	$-$	40832	22011
	$-$	22011	-1595		3509	-638	$-$	23287	22011
	$-$	$-$	$|-319|$		-2	-31	-69	-421	-319

As before, the change in ratio in the pivotal row arises on account of <u>not</u> dividing (by the conventional pivot, which would have been $-319/(-69) = 4.628$), while the change in ratio in the other rows is due to the multiplication by -319.

Addition of 47 times the third row to the first row, and subtraction of 5 times the third row from the second row of the tableau now leeds to the following further intermediate tableau:

V_3 :

x_1	x_2	x_3	=	y_1	y_2	y_3	Σ	Ratio
22011	–	–		2139	138	–3243	21045	22011
–	22011	–		3519	–438	345	25392	22011
–	–	|–319|		–2	–31	–69	–421	–319

At which point we may divide the two leading rows by the excess ratio-factor, i.e. by –69, to obtain:

T_4 :

x_1	x_2	x_3	=	y_1	y_2	y_3	Σ	Ratio
–319	–	–		–31	–2	47	–305	–319
–	–319	–		–51	7	–5	–368	–319
–	–	–319		–2	–31	–69	–421	–319

If we want the inverse, we now need to divide by $|A| = -319$:

$$A^{-1} = \begin{bmatrix} 2 & 5 & 1 \\ 11 & -7 & 8 \\ -5 & 3 & 1 \end{bmatrix}^{-1} =$$

$$\begin{bmatrix} -31 & -2 & 47 \\ -51 & 7 & -5 \\ -2 & -31 & -69 \end{bmatrix} \times (-319^{-1}) = \begin{bmatrix} 0.097 & 0.006 & -0.147 \\ 0.160 & -0.022 & 0.016 \\ 0.006 & 0.097 & 0.216 \end{bmatrix}$$

The method, as outlined in this section, is not strictly limited to the inversion of matrices whose coefficients are integer numbers. Rather, the one computational advantage of this particular method, is the avoidance of rounding errors, and that is most desirable and attainable (if it is), in the context of a system with integer coefficients.

That computational advantage is in any case restricted by the practical limits on storing integer numbers of high absolute values. For example, for the typical 8-bit micro (maximum integer number: maxint = $2^{15} - 1$ = 32767), this 3 by 3 example already comes near to that limit.

This section was written, not so much to offer help with computation, but to drive home the relation between the adjoint and the inverse: the adjoint all integer elimination method calculates the adjoint.

5.10 Rank

The rank r of an m by n matrix **A** is the largest order-parameter, of any r by r invertible block, that can be extracted from the matrix, if necessary by reordering the rows and/or the columns of **A**.

Thus, $\mathbf{A} = \begin{bmatrix} 1 & 1 & 1 & 1 \\ 2 & 2 & 2 & 2 \\ - & 3 & 3 & 3 \\ - & 4 & 4 & 4 \\ 5 & 5 & 5 & 5 \end{bmatrix}$ is of order 5 by 4,

but its rank is only 2.

The blocks $\begin{bmatrix} 1 & 1 \\ - & 3 \end{bmatrix}$, $\begin{bmatrix} 2 & 2 \\ - & 3 \end{bmatrix}$, $\begin{bmatrix} 1 & 1 \\ - & 4 \end{bmatrix}$, $\begin{bmatrix} 2 & 2 \\ - & 4 \end{bmatrix}$, $\begin{bmatrix} - & 3 \\ 5 & 5 \end{bmatrix}$, and $\begin{bmatrix} - & 4 \\ 5 & 5 \end{bmatrix}$

are square and non-singular. But all blocks of order 3 by 3, or 4 by 4, e.g.

$$\begin{bmatrix} 1 & 1 & 1 \\ 2 & 2 & 2 \\ - & 3 & 3 \end{bmatrix},$$

contain proportional rows and identical columns.

Obviously the rank of a matrix is at most equal to its smallest order parameter. If so, we say that such a matrix is of <u>full</u> rank.

Thus, $\mathbf{A} = \begin{bmatrix} 1 & -2 & - & - \\ -2 & 1 & - & - \end{bmatrix}$

is of full rank 2, despite the zero block further to the right.

Since we cannot have a negative rank, the lowest possible rank is zero, in which case the matrix contains nothing but zeros.

We admit zero as a possible value for the rank of a matrix, and we also accept zero as a possible order parameter.

The rank can therefore be zero, either because all the elements are zero-valued, or because one of the order parameters is zero, in which case there are no elements at all, and by implication no non-zero elements.

Before we develop the theory concerning rank any further, we discuss an algorithm establishing the rank of a matrix. The method is closely similar to the algorithm for inversion. Like the method for inversion by row operations, it operates on a partitioned system, and a square block A_{11} is inverted.

The differences with the usual method of elimination by row-operations are:

(1) We swap columns as well as rows, and
(2) It applies to non-square matrices, as well as to square ones.

Example:

$$A = \begin{bmatrix} 1 & 1 & 1 & 1 & - \\ - & 2 & 4 & 3 & -1 \\ 2 & 2 & 2 & - & 2 \\ 4 & 4 & 4 & 2 & 2 \end{bmatrix}$$

The following calculations are performed on the $Ax = [0]$ system:

We have no difficulty in identifying an initial 1 by 1 invertible block A_{11} in its natural ordering, and we make an inversion step.

	x_1	x_2	x_3	x_4	x_5	Σ
	$\underline{|1|}$	1	1	1	-	4
	-	2	4	3	-1	8
	2	2	2	-	2	8
	4	4	4	2	2	16

The successor tableau is:

	x_1	x_2	x_3	x_4	x_5	Σ
	1	1	1	1	–	4
	–	2	4	3	–1	8
	–	–	–	–2	2	–
	–	–	–	–2	2	–

We now re-define the partitioning, by adding the second column and the second row to A_{11}, and make the next inversion step:

	x_1	x_2	x_3	x_4	x_5	Σ
	1	1	1	1	–	4
	–	$\underline{|2|}$	4	3	–1	8
	–	–	–	–2	2	–
	–	–	–	–2	2	–

The successor tableau then becomes:

	x_1	x_2	x_3	x_4	x_5	Σ
	1	–	–1	–0.5	0.5	–
	–	1	2	0.5	–0.5	4
	–	–	–	–2	2	–
	–	–	–	–2	2	–

It is at this stage actually clear, that the remaining bottom righthand block:

$$[A_{22} - A_{21}A_{11}^{-1}A_{12}] = \begin{bmatrix} - & -2 & 2 \\ - & -2 & 2 \end{bmatrix}$$

is of rank 1, and since a 2 by 2 block

$$A_{11} = \begin{bmatrix} 1 & 1 \\ - & 2 \end{bmatrix}$$

has been inverted so far, and only one further inversion step can be made, the rank of A must be $r = 2 + 1 = 3$.

The logical end of a rank-calculation algorithm is, however, the reduction of the remainder block $[A_{22} - A_{21}A_{11}^{-1}A_{12}]$ to a nil block, and this requires a re-ordering, in order to form a 3 by 3 matrix A_{11}.

To this purpose, we interchange the x_3 and x_4 columns. The tableau now becomes:

$$
\begin{array}{ccccc}
x_1 & x_2 & x_4 & x_3 & x_5 & \Sigma
\end{array}
$$

$$
\left[
\begin{array}{ccccc}
1 & - & -0.5 & -1 & 0.5 \\
- & 1 & 1.5 & 2 & -0.5 \\
- & - & |\underline{-2}| & - & 2 \\
- & - & -2 & - & 2
\end{array}
\right]
\begin{array}{c}
- \\
4 \\
- \\
-
\end{array}
$$

We now make one more inversion-step, and obtain the final tableau:

$$
\begin{array}{ccccc}
x_1 & x_2 & x_4 & x_3 & x_5 & \Sigma
\end{array}
$$

$$
\left[
\begin{array}{ccccc}
1 & - & - & -1 & - \\
- & 1 & - & 2 & 1 \\
- & - & 1 & - & -1 \\
- & - & - & - & -
\end{array}
\right]
\begin{array}{c}
- \\
4 \\
- \\
-
\end{array}
$$

At this point, the rank of **A** is seen to be the rank of the leading block-row of this final tableau, r = 3.

These calculations make an implicit use of the following

Theorem:

If the m by n matrix **A** is of rank r, and permits partitioning as

$$
A = \left[
\begin{array}{c|c}
A_{11} & A_{12} \\
\hline
A_{21} & A_{22}
\end{array}
\right]
\tag{5.10.1}
$$

where A_{11} is of order k by k and invertible,
then:

$[A_{22} - A_{21}A_{11}^{-1}A_{12}]$ is of rank $q = r-k$.

Proof:

We consider the further partitioning of A, conform:

$$A = \begin{bmatrix} A_{11} & A_{12} & A_{13} \\ \hline A_{21} & A_{22} & A_{23} \\ \hline A_{31} & A_{32} & A_{33} \end{bmatrix} \qquad (5.10.2)$$

where A_{11} is A_{11} as in (5.10.1), and the orders of the diagonal blocks are $k=n_1$, n_2 and $n_3=n-n_1-n_2$. and A_{22}.

We can always re-order the rows of A between the second and the third block-row, and the columns between the second and third block-column, and we may therefore assume, without loss of generality, $n_2 = q$.

The partitioning conform (5.10.2) relates to the calculation-tableau which arises in relation to the partial inversion by row operations of the $Ax=[0]$ system.

The current form of this calculation tableau is indicated as C, and the block-inversion-step is summarized below, as follows:

$$
\begin{array}{ccc}
 & A & \\
x_1 \quad x_2 \quad x_3 & \qquad\qquad & x_1 \qquad x_2 \qquad x_3
\end{array}
$$

$$
\begin{bmatrix} \underline{|A_{11}|} & A_{12} & A_{13} \\ \hline A_{21} & A_{22} & A_{23} \\ \hline A_{31} & A_{32} & A_{33} \end{bmatrix}
\qquad
\begin{bmatrix} I & A_{11}^{-1}A_{12} & A_{11}^{-1}A_{13} \\ \hline A_{22}-A_{21}A_{11}^{-1}A_{12} & A_{23}-A_{21}A_{11}^{-1}A_{13} \\ \hline A_{32}-A_{31}A_{11}^{-1}A_{12} & A_{33}-A_{31}A_{11}^{-1}A_{13} \end{bmatrix}
$$

Successful inversion of any block $[A_{22}-A_{21}A^{-1}A_{12}]$ implies the inversion of the composite block:

$$A^{*}{}_{11} = \begin{bmatrix} A_{11} & A_{12} \\ \hline A_{21} & A_{22} \end{bmatrix} \qquad (5.10.3)$$

and

$$r \geq q + k \qquad (5.10.4)$$

is shown.

Since (see section 5.6), a pivoting operation divides the determinant of any block to which the pivot belongs by the value of the pivot, the rank of **T** is also r, the rank of **A**.

Thus, for k = 1,

$$T_{22} = \left[\begin{array}{c|c} A_{22}-A_{21}A_{11}{}^{-1}A_{12} & A_{23}-A_{21}A_{11}{}^{-1}A_{13} \\ \hline A_{32}-A_{31}A_{11}{}^{-1}A_{12} & A_{33}-A_{31}A_{11}{}^{-1}A_{13} \end{array} \right] \qquad (5.10.5)$$

must be assumed to be or rank

$$q \leq r-1$$

(there is only one row and one column of **A** which is not included in the corresponding block of **A**)

and therefore

$$q+1 \leq r \qquad (5.10.6)$$

Since $r \geq q + k$ (= q+1 in the k=1 case), was already proved irrespective of recursive induction, this completes the proof for the k=1 case. An assumption that the theorem is true for T_{22}, now immediately leads to a proof by recursive induction.

q.e.d.

The rank calculation algorithm, as outlined above, ends either with a block $[A_{22} - A_{21}A_{11}{}^{-1}A_{12}] = [0]$, or with finding the matrix to be of full rank, and in either case the order of the successfully inverted block is the rank of the matrix.

5.11 **Rank and linear (in)dependence**

The column vector **x** is said to be a <u>combination</u> of n other column
vectors $\mathbf{v}_j$, if there exist numbers w_j (j = 1, 2, ... n), not all
of them zero, for which:

$$\mathbf{x} = \sum_{j=1}^{n} w_j \, \mathbf{v}_j \qquad\qquad (5.11.1)$$

holds.

An equivalent definition, grouping the columns $\mathbf{v}_j$ in a matrix is:

The column vector **x** is a <u>combination</u> of the columns of the m by n
matrix **A**, if there exists a column vector $\mathbf{w} \neq [0]$, for which:

$$\mathbf{x} = \mathbf{Aw} \qquad\qquad (5.11.2)$$

holds.

Note that this definition includes the trivial case, where one w_j
is 1, and all the others are zero, i.e. the column is a
combination of itself with nothing else.

Similarly, in the matrix notation version of the definition, the
term combination includes the columns of **A** itself.

A set of columns, of which none can be expressed as a combination
of the others, is said to be <u>independent</u> from each other.
Equivalently, we may say, that some columns are independent from
each other, if they do not permit to express the null vector as a
combination of them:

$$\mathbf{v}_1 = 2\mathbf{v}_2 + 3\mathbf{v}_3 \text{ implies: } -\mathbf{v}_1 + 2\mathbf{v}_2 + 3\mathbf{v}_3 = [0].$$

The generalization of these definitions to combinations of rows,
and independent rows, will be obvious.

The following theorem is at this stage given in a provisional
form, and we shall be discussing extensions of it. The reason is,
that the proof is simplified, if the case of a non-trivial
partitioning is dealt with separately from the various cases
where no meaningful partitioning exists.

Theorem (Rank completion theorem):

If:

$$A = \begin{bmatrix} A_{11} & A_{12} \\ A_{21} & A_{22} \end{bmatrix}$$

(5.11.3)

is of order m by n, and of rank r, $(0 < r < m, 0 < r < n)$, and A_{11} is of order r by r and invertible,

then:

(1) There is at least one n by n-r matrix C, of full rank n-r, which satisfies:

$$A\,C = [0]$$

(5.11.4)

and

(2) Any column $c^* \neq [0]$, which satisfies:

$$A\,c^* = [0]$$

(5.11.5)

is a combination of the columns of C.

Proof:

If $[A_{22} - A_{21}A_{11}^{-1}A_{12}] \neq [0]$ were to hold, this would prove that the rank of A was in excess of r, the order of A_{11}. (We could use a non-zero element in $[A_{22} - A_{21}A_{11}^{-1}A_{12}]$ as 'pivot', to invert a larger block A_{11}, adding a column and a row to it. -See sections 5.6 and 5.7-)

Therefore:

$$[A_{22} - A_{21}A_{11}^{-1}A_{12}] = [0]$$

(5.11.6)

Therefore, for

$$C = \begin{bmatrix} -A_{11}^{-1}A_{12} \\ \hline I \end{bmatrix}$$

(5.11.7)

we find:

$$\mathbf{A}\,\mathbf{C} = \begin{bmatrix} \mathbf{A}_{11} & \mathbf{A}_{12} \\ \mathbf{A}_{21} & \mathbf{A}_{22} \end{bmatrix} \begin{bmatrix} -\mathbf{A}_{11}^{-1}\mathbf{A}_{12} \\ \mathbf{I} \end{bmatrix} =$$

$$\begin{bmatrix} -\mathbf{A}_{11}\mathbf{A}_{11}^{-1}\mathbf{A}_{12} + \mathbf{A}_{12} \\ -\mathbf{A}_{21}\mathbf{A}_{11}^{-1}\mathbf{A}_{12} + \mathbf{A}_{22} \end{bmatrix} = \begin{bmatrix} 0 \\ \hline 0 \end{bmatrix} \qquad (5.11.8)$$

q.e.d. ad (1).

From (5.11.5) we extract the leading block-equation:

$$\mathbf{A}_{11}\mathbf{c}^*{}_1 + \mathbf{A}_{12}\,\mathbf{c}^*{}_2 = [0] \qquad\qquad (5.11.9)$$

For $\mathbf{c}^*{}_2 = [0]$, (5.11.9) reduces to $\mathbf{A}_{11}\mathbf{c}^*{}_1 = [0]$.

Since $\mathbf{A}_{11}$ was assumed to be invertible, this implies $\mathbf{c}^*{}_1 = [0]$, and therefore $\mathbf{c}^* = [0]$.

However, $\mathbf{c}^* \neq [0]$ was assumed, and $\mathbf{c}^*{}_2 \neq [0]$ needs to be assumed instead.

The corresponding vector $\mathbf{c}^*{}_1$ is then expressed by (5.11.9) as:

$$\mathbf{c}^*{}_1 = -\mathbf{A}_{11}^{-1}\mathbf{A}_{12}\mathbf{c}^*{}_2 \qquad\qquad (5.11.10)$$

Therefore:

$$\mathbf{c}^* = \begin{bmatrix} -\mathbf{A}_{11}^{-1}\mathbf{A}_{12} \\ \mathbf{I} \end{bmatrix} \mathbf{c}^*{}_1 = \mathbf{C}\,\mathbf{c}^*{}_1 \qquad (5.11.11)$$

q.e.d. ad (2).

q.e.d.

A more compact, and, as we shall see below, more general formulation of the same theorem is:

If the m by n matrix $\mathbf{A}$ is of rank r ($r \leq m$, $r \leq n$), then there exist n-r independent columns $\mathbf{c}_j \neq [0]$, satisfying $\mathbf{A}\,\mathbf{c}_j = [0]$.

We now review the equivalence, or otherwise, of this one-sentence formulation of the rank completion theorem, with the version of the theorem, as stated above.

If the partitioning conform (5.11.3) exists, then $r > 0$, $r < m$, and $r < n$ are implied, at least insofar as we restrict ourselves to non-trivial partitionings.

In that case, the n-r vectors are the columns of **C**.

A transformation of **C** (which may be any post-multiplication of **C** by a non-singular matrix), will also do, and confirms both formulations of the theorem.

The one-sentence formulation also applies to a number of special cases, where no partitioning conform (5.10.3) exists.

For r=0, **A** is a null matrix.

Then $\mathbf{C} = \mathbf{I}_n$, n independent vectors $\mathbf{c}_j$ being (for example), the n unit vectors $\mathbf{e}_1$ to $\mathbf{e}_n$ will do. But the original form of the theorem is not stated as being applicable, and its proof does not cover this case. There is no inverse $\mathbf{A}_{11}^{-1}$.

For r=n (**A** is of full rank n), there are indeed n-r = 0 vectors $\mathbf{c}_j \neq [0]$, for which $\mathbf{Ac}_j = [0]$ holds. The theorem, as originally formulated, does not strictly cover this case, but could be extended to it, if a block $[\mathbf{A}_{22} - \mathbf{A}_{21}\mathbf{A}_{11}^{-1}\mathbf{A}_{12}]$ is understood to be a null matrix, if it is of zero order.

For $r = m < n$, the matrix **C** also exists, as do the n-m independent vectors. Again, the way the theorem and its proof were originally formulated, do not strictly cover this case, but could be extended to it, by including blocks of zero order.

The following corollary will be obvious:

If the m by n matrix **A** is of rank r, there are m-r independent row-vectors $\mathbf{w}'_i \neq [0]$, for which $\mathbf{w}'_i\mathbf{A} = [0]$ applies.

Hence the more general formulation of the rank completion theorem is:

If the m by n matrix **A** is of rank r, then there are n-r independent columns $\mathbf{c}_j \neq [0]$, satisfying $\mathbf{Ac}_j = [0]$, and m-r independent rows $\mathbf{w}'_i \neq [0]$, satisfying $\mathbf{w}'_i\mathbf{A} = [0]$.

The rank of a matrix can also be revealed by the number of independent columns which can <u>not</u> be expressed as a combination of its columns.

Theorem:

If:

no combination of the columns of the m by k matrix **A**, which is of full rank k (= there are k independent columns a_j), can be expressed as a combination of the columns of the m by n matrix **B**,

then:

the rank r of **B** is not more than m-k.

Proof:

We first deal with some special, somewhat trivial cases:

Case 1): **A** is of full rank m

In that case all non-zero vectors b_j can be expressed as:

$$A\ x = b$$

This case only arises for **B** = [0].

q.e.d. for case 1

Case 2): k=1

The theorem then says, that the inability to resolve, for all square blocks B_1 (and all column-reorderings of **B**), the $B_1 x = a$ system, proves the singularity of all such blocks B_1. Which is true.

q.e.d. for case 2.

We now consider the non-trivial case m > r, k > 1 (Case 3):

Consider any attempt, successful or otherwise, to (part)-resolve the $\mathbf{A}\,\mathbf{x} = \mathbf{B}\,\mathbf{y}$ system, by completing, or attempting to complete, $\mathbf{A}$ into an invertible m by m block $[\mathbf{A} \mid \mathbf{B}_1]$.

The following sub-cases arise:

Subcase 3a):

All non-zero columns of $\mathbf{B}$ are combinations of the columns of $\mathbf{A}$.

Every system $\mathbf{A}\,\mathbf{x} = \mathbf{b}_j$ contradicts what the theorem assumes.

q.e.d. for subcase 3a.

Subcase 3b):

 $[\mathbf{A}_1 \mid \mathbf{B}_1]$ is of order m by m and invertible.

Then either:

 3ba: Another non-zero column of $\mathbf{B}$ permits the expression:

$\mathbf{A}_1\,\mathbf{x}_1 = \mathbf{b}_2 - \mathbf{B}_1\,\mathbf{y}_2$, contrary to what the theorem assumes,

or 3bb: $\mathbf{B}_2 = [0]$

q.e.d. for subcase 3b

q.e.d. for case 3

q.e.d.

The following theorem refers specifically to a <u>symmetric</u> matrix.

Theorem:

If the n by n symmetric matrix is of rank r, and partitions as:

$$A = \left[\begin{array}{c|c} A_{11} & A_{12} \\ \hline A_{21} & A_{22} \end{array}\right] \qquad (5.11.12)$$

where A_{11} is square, and the leading block-column A_1 is of full rank r (and therefore of order n by r) ,

then A_{11} is invertible.

Proof:

We consider the following cases:

Case 0): $A_{11} = [0]$.

Then we extract an r by r invertible block A^*_{21} from the n by r block-column:

$$A_1 = \left[\begin{array}{c} A_{11} \\ \hline A_{21} \end{array}\right] = \left[\begin{array}{c} \\ \hline A_{21} \end{array}\right] \qquad (5.11.13)$$

which is of full rank r, and the transpose of A^*_{12} is a block of the r by n block-row $[A^*_{11} \mid A^*_{12}] = [\ \mid A^*_{12}]$.

Therefore, if A_{11} is indeed a null block, then:

$$A^* = \left[\begin{array}{c|c} & A^*_{12} \\ \hline A^*_{21} & A^*_{22} \end{array}\right]$$

is invertible as may be verified by:

$$A^{*-1} = \left[\begin{array}{c|c} & A^*_{12} \\ \hline A^*_{21} & A^*_{22} \end{array}\right]^{-1} = \left[\begin{array}{c|c} -A^{*-1}_{21}A^*_{22}A^*_{12} & A^{*-1}_{21} \\ \hline A^{*-1}_{12} & \end{array}\right]$$

$$(5.11.14)$$

and no further proof is needed, as A is shown to be of rank 2r or more, not of rank r.

q.e.d. for case 0.

Case 1): $A_{11} \neq [0]$.

If A_{11} is assumed to be invertible, no further proof is needed. Now let us assume instead, that, contrary to what the theorem states, the rank of A_{11} is less than r.

In that case, an assumption that the theorem is true for symmetric matrices of order 1, 2, ... n-1, and therefore for A_{11}, implies (after, if necessary (re-)ordering rows 1 to r simultaneously with columns 1 to r), the existence of a partitioning of A conform:

$$A = \begin{bmatrix} A^*_{11} & A^*_{12} & A^*_{13} \\ A^*_{21} & A^*_{22} & A^*_{23} \\ A^*_{31} & A^*_{32} & A^*_{33} \end{bmatrix}$$

(5.11.15)

where $A^*_{33} = A_{22}$, i.e. the old leading block-row A'_1 and the old leading block-column A_1 have been split in two, while the second block-column and -row have been re-named third.

As the original block A_{11} is not a null block, and not assumed to be of full rank r, but instead of rank r* (0 < r* < r), the theorem, assumed true for A_{11}, is also applicable to it.

Therefore, by the first theorem in this section, applied separately for A as a whole, and for the leading block-column of A, partitioned conform (5.11.14), the blocks:

$$B_{33} = \begin{bmatrix} A^*_{22}-A^*_{12}A^*_{11}{}^{-1}A^*_{21} & A^*_{23}-A^*_{12}A^*_{11}{}^{-1}A^*_{31} \\ A^*_{32}-A^*_{12}A^*_{11}{}^{-1}A^*_{21} & A^*_{33}-A^*_{12}A^*_{11}{}^{-1}A^*_{31} \end{bmatrix}$$

(5.11.16)

and its leading block-column, are both of rank n-r*.

Therefore, by the theorem, applied for B_{33}, $[A^*_{22}-A_{12}A_{11}{}^{-1}A_{21}]$ is invertible, and the invertability of A_{11} conform (5.11.12) is implied by recursive induction.

q.e.d.

Exercise on rank:

For $\mathbf{A}$ =
$$
\begin{bmatrix}
1 & 1 & 1 & 1 & - & - & - \\
1 & 1 & 2 & 2 & - & - & - \\
1 & 1 & 2 & 3 & 1 & 2 & 2 \\
1 & 1 & 2 & 3 & 1 & 2 & 2 \\
3 & 3 & 6 & 9 & 3 & 6 & 6 \\
1 & 1 & 2 & 3 & 1 & 2 & 3
\end{bmatrix}
$$

you are asked:

(1) Establish the rank r, by part-inversion, re-ordering an r by r invertible block $\mathbf{A}_{11}$ to the top lefthand corner, and:

(2) Find a 7 by (7-r) matrix $\mathbf{C}$, of full rank 7-r, satisfying:

$$\mathbf{A}\,\mathbf{C} = [0].$$

There is an answer-sheet at the end of the chapter.

5.12 **The rank of a product expression**.

Theorem:

The rank of $\mathbf{AB}$ does not exceed the rank of $\mathbf{B}$.

Proof:

Let $\mathbf{A}$ be of order m by p, and $\mathbf{B}$ of order p by n, and of rank r.

Therefore, by the rank completion theorem (see section 5.11), there are n-r independent vectors $\mathbf{c}_j \neq [0]$, satisfying:

$$\mathbf{B}\,\mathbf{c}_j = [0] \tag{5.12.1}$$

and therefore by implication:

$$\mathbf{A} \ \mathbf{B} \ \mathbf{c}_j \ = \ [0] \tag{5.12.2}$$

Now suppose that, contrary to what the theorem states, the rank of **AB** was $r^* > r$.

The product expression **AB** is of order m by n, and the rank completion theorem then states that there are only $n{-}r^* < n{-}r$ independent vectors $\mathbf{c}^* \neq [0]$, satisfying:

$$\mathbf{A} \ \mathbf{B} \ \mathbf{c}^*_j \ = \ [0] \tag{5.12.3}$$

The $n{-}r \ \mathbf{c}_j \neq [0]$ which satisfy (5.12.2) meet that condition.

Therefore, $r^* > r$ cannot be true, and $r^* \leq r$ holds instead.

q.e.d.

Application of the theorem as just proved, to $[AB]' = \mathbf{B}'\mathbf{A}'$, leads to the

Corollary:

The rank of the product expression **AB** does not exceed the rank of **A** (and therefore the rank of neither **A** nor **B**).

In general, the rank of a product expression can be less than the rank of either factor.

Example: $[1 \ \ 1] \begin{bmatrix} 1 \\ -1 \end{bmatrix} = [0].$

Both factors are of rank 1, but the product is of zero rank.

However, the following theorem applies to the product of a matrix and its own transpose:

Theorem (Correlation matrix theorem):

If **A** is of order m by n,

then the rank of **A'A** equals the rank r of **A**.

Proof:

Let **A** be of order m by n, and of rank r.

Then the number of <u>independent</u> vectors $c_j \neq [0]$, satisfying:

$$\mathbf{A}\ c_j = [0], \qquad \text{is n-r.}$$

For any vector **v**, for which $\mathbf{A}\ \mathbf{v} \neq [0]$ applies, we also have:

$$\mathbf{v'A'Av} > 0, \text{ and therefore } \mathbf{A'Av} \neq [0].$$

Therefore, there are n-r, and no more than n-r <u>independent</u> vectors $v_j \neq [0]$, satisfying $\mathbf{A'A}v_j = [0]$.

It was shown in section 5.11, that this implies that **A'A** is of rank n-r.

q.e.d.

It was shown in section 5.11, that if we extract an n by r block-column which is of full rank r:

$$\mathbf{A}_1 \;=\; \left[\begin{array}{c} \mathbf{A}_{11} \\ \hline \mathbf{A}_{21} \end{array}\right]$$

from a symmetric matrix **A**, where $\mathbf{A}_{11}$ is a symmetric diagonal block, then $\mathbf{A}_{11}$ is invertible.

Application of this theorem leads to the following:

Corollary:

If **A** is of rank r, then at least one r by r diagonal block of **A'A** is invertible.

Application of the correlation matrix theorem to a full-rank block-column and its transpose, the corresponding block-row, leads to the same result:

If $[\mathbf{A}_{11} \mid \mathbf{A}_{12}]$ is of order r by n, and of full rank r, then $[\mathbf{A}_{11} \mid \mathbf{A}_{12}] [\mathbf{A}_{11} \mid \mathbf{A}_{12}]' = [\mathbf{A}_{11} \mid \mathbf{A}_{12}] \begin{bmatrix} \mathbf{A}_{11} \\ \hline \mathbf{A}_{21} \end{bmatrix}$

is also of rank r, and, being of order r by r, invertible.

Answersheet on the calculation of the determinant

$$\mathbf{A} = \begin{bmatrix} - & 1 & 2 \\ 3 & -4 & 5 \\ -6 & 7 & -8 \end{bmatrix}$$

The calculation by row operations is as follows:

	x_1	x_2	x_3	=	y_1	y_2	y_3	Σ
	-	1	2		1	-	-	4
	3	-4	5		-	1	-	5
	-6	7	-8		-	-	1	-6
	$\|L\|$	=	$\|A\|$			$\|R\|$	= 1	

we first of all swap rows 1 and 2, to obtain the modified tableau:

	x_1	x_2	x_3	=	y_1	y_2	y_3	Σ
	$\underline{|3|}$	-4	5		-	1	-	5
	-	1	2		1	-	-	4
	-6	7	-8		-	-	1	-6
	$\|L\|$	=	$-\|A\|$			$\|R\|$	= -1	

From which, after the first inversion step:

	x_1	x_2	x_3	=	y_1	y_2	y_3	Σ
	1	-1.33	1.67		-	.33	-	1.67
	-	$\underline{|1|}$	2		1	-	-	4
	-	-1	2		-	2	1	4
	$\|L\|$	=	$-\|A\|/3$			$\|R\|$	= -1/3	

From which, after the second inversion step:

$$
\begin{array}{ccc}
x_1 & x_2 & x_3 \\
\end{array} \quad = \quad
\begin{array}{ccc}
y_1 & y_2 & y_3 \\
\end{array} \quad \Sigma
$$

$$
\begin{bmatrix}
1 & - & 4.33 \\
- & 1 & 2 \\
- & - & |\underline{4}|
\end{bmatrix}
\qquad
\begin{bmatrix}
1.33 & .33 & - \\
1 & - & - \\
1 & 2 & 1
\end{bmatrix}
\qquad
\begin{array}{c}
7 \\
4 \\
8
\end{array}
$$

$$|L| = - |A| / 3 \qquad\qquad |R| = -1/3$$

$$
\begin{array}{ccc}
x_1 & x_2 & x_3 \\
\end{array} \quad = \quad
\begin{array}{ccc}
y_1 & y_2 & y_3 \\
\end{array} \quad \Sigma
$$

$$
\begin{bmatrix}
1 & - & - \\
- & 1 & - \\
- & - & 1
\end{bmatrix}
\qquad
\begin{bmatrix}
.25 & -1.83 & -1.08 \\
.50 & -1 & -.50 \\
.25 & .50 & .25
\end{bmatrix}
\qquad
\begin{array}{c}
1.67 \\
- \\
2
\end{array}
$$

$$|L| = - |A|/12 = 1 \qquad\qquad |R| = -1/12$$

Therefore $|A| = -12$.

Ditto, conform its definition:

$$
|A| = 1 \times -1 \times \begin{vmatrix} 3 & 5 \\ -6 & -8 \end{vmatrix} + 2 \times 1 \times \begin{vmatrix} 3 & -4 \\ -6 & 7 \end{vmatrix} =
$$

$$
- \Big(3 \times (-8) - (-6) \times 5\Big) + 2 \Big(3 \times 7 - (-6) \times (-4)\Big) =
$$

$$
- \Big(-24 + 30\Big) + 2 \Big(21 - 24\Big) = -6 + 2 \times (-3) =
$$

$$
-6 - 6 = -12.
$$

Answersheet on rank

The following operations are performed on the $\mathbf{Ac} = [0]$ system:

First, we transform the leading equation into a c_1 equation:

c_1	c_2	c_3	c_4	c_5	c_6	c_7		Σ
$\lvert\underline{1}\rvert$	1	1	1	–	–	–		4
1	1	2	2	–	–	–		6
1	1	2	3	1	2	2		12
1	1	2	3	1	2	2		12
3	3	6	9	3	6	6		36
1	1	2	3	1	2	3		13

The successor tableau $\mathbf{T_2}$ is:

c_1	c_2	c_3	c_4	c_5	c_6	c_7		Σ
1	1	1	1	–	–	–		4
–	–	1	1	–	–	–		2
–	–	1	2	1	2	2		8
–	–	1	2	1	2	2		8
–	–	3	6	3	6	6		24
–	–	1	2	1	2	3		9

The zero column in the $[\mathbf{A_{22}} - \mathbf{A_{21}}\mathbf{A_{11}}^{-1}\mathbf{A_{21}}]$ block proves the c_2 column to be a linear combination of the columns already allocated to the $\mathbf{A_1}$ block-column (that was in any case obvious in this example).

We therefore interchange columns 2 and 3, and the next inversion step is:

c_1	c_3	c_2	c_4	c_5	c_6	c_7	Σ
1	1	1	1	-	-	-	4
-	|1|	-	1	-	-	-	2
-	1	-	2	1	2	2	8
-	1	-	2	1	2	2	8
-	3	-	6	3	6	6	24
-	1	-	2	1	2	3	9

The resulting $\mathbf{T}_3$ tableau is:

c_1	c_3	c_2	c_4	c_5	c_6	c_7	Σ
1	-	1	-	-	-	-	2
-	1	-	1	-	-	-	2
-	-	-	1	1	2	2	6
-	-	-	1	1	2	2	6
-	-	-	3	3	6	6	18
-	-	-	1	1	2	3	7

The c_2 column still cannot be added to $\mathbf{A}_{11}$ (obviously not), and we interchange the c_4 column with it.

The next inversion step is:

c_1	c_3	c_4	c_2	c_5	c_6	c_7	Σ
1	-	-	1	-	-	-	2
-	1	1	-	-	-	-	2
-	-	|1|	-	1	2	2	6
-	-	1	-	1	2	2	6
-	-	3	-	3	6	6	18
-	-	1	-	1	2	3	7

The resulting $\mathbf{T}_4$ successor tableau is:

$$
\begin{array}{ccccccc}
c_1 & c_3 & c_4 & c_2 & c_5 & c_6 & c_7 & \Sigma \\
\left[\begin{array}{ccccccc}
1 & - & - & 1 & - & - & - \\
- & 1 & - & - & -1 & -2 & -2 \\
- & - & 1 & - & 1 & 2 & 2 \\
- & - & - & - & - & - & - \\
- & - & - & - & - & - & - \\
- & - & - & - & - & - & 1
\end{array}\right] &
\begin{array}{c}
2 \\ -4 \\ 6 \\ - \\ - \\ 1
\end{array}
\end{array}
$$

We now interchange the c_7 and c_2 columns, as well as equations (rows) 4 an 6, and make the last possible inversion step:

$$
\begin{array}{ccccccc}
c_1 & c_3 & c_4 & c_7 & c_5 & c_6 & c_2 & \Sigma \\
\left[\begin{array}{ccccccc}
1 & - & - & - & - & - & 1 \\
- & 1 & - & -2 & -1 & -2 & - \\
- & - & 1 & 2 & 1 & 2 & - \\
- & - & - & |\underline{1}| & - & - & - \\
- & - & - & - & - & - & - \\
- & - & - & - & - & - & -
\end{array}\right] &
\begin{array}{c}
2 \\ -4 \\ 6 \\ 1 \\ - \\ -
\end{array}
\end{array}
\qquad
\begin{array}{l}
\text{equation 1} \\
\text{equation 2} \\
\text{equation 3} \\
\text{equation 6} \\
\text{equation 4} \\
\text{equation 5}
\end{array}
$$

The $\mathbf{T}_5$ successor tableau now is:

$$
\begin{array}{ccccccc}
c_1 & c_3 & c_4 & c_7 & c_5 & c_6 & c_2 & \Sigma \\
\left[\begin{array}{ccccccc}
1 & - & - & - & - & - & 1 \\
- & 1 & - & - & -1 & -2 & - \\
- & - & 1 & - & 1 & 2 & - \\
- & - & - & 1 & - & - & - \\
- & - & - & - & - & - & - \\
- & - & - & - & - & - & -
\end{array}\right] &
\begin{array}{c}
2 \\ -2 \\ 4 \\ 1 \\ - \\ -
\end{array}
\end{array}
\qquad
\begin{array}{l}
\text{equation 1} \\
\text{equation 2} \\
\text{equation 3} \\
\text{equation 6} \\
\text{equation 4} \\
\text{equation 5}
\end{array}
$$

We have now reduced the remaining 2 by 3 block $[\mathbf{A}_{22} - \mathbf{A}_{21}\mathbf{A}_{11}{}^{-1}\mathbf{A}_{12}]$ to a null block, and have established that the rank is r=4.

The corresponding re-ordered matrix **A** is:

$$\mathbf{A}^* = \begin{array}{c} \\ \\ \end{array} \begin{array}{ccccccc} c_1 & c_3 & c_4 & c_7 & c_5 & c_6 & c_2 \\ 1 & 1 & 1 & - & - & - & 1 \\ 1 & 2 & 2 & - & - & - & 1 \\ 1 & 2 & 3 & 2 & 1 & 2 & 1 \\ 1 & 2 & 3 & 3 & 1 & 2 & 1 \\ 3 & 6 & 9 & 6 & 3 & 6 & 3 \\ 1 & 2 & 3 & 2 & 1 & 2 & 1 \end{array} \qquad \begin{array}{l} \text{equation} \ 1 \\ \text{equation} \ 2 \\ \text{equation} \ 3 \\ \text{equation} \ 6 \\ \text{equation} \ 5 \\ \text{equation} \ 4 \end{array}$$

It is in relation to this re-ordered matrix, that we have also
calculated a matrix:

$$\mathbf{C} = \begin{bmatrix} -\mathbf{A}_{11}^{-1}\mathbf{A}_{12} \\ \mathbf{I} \end{bmatrix} = \begin{array}{c} c_1 \\ c_3 \\ c_4 \\ c_7 \\ c_5 \\ c_6 \\ c_2 \end{array} \begin{bmatrix} - & - & -1 \\ 1 & 2 & - \\ -1 & -2 & - \\ - & - & - \\ 1 & - & - \\ - & 1 & - \\ - & - & 1 \end{bmatrix}$$

To conform with the original matrix **A**, we need to re-order the
rows of **C**, to become:

$$\mathbf{C} = \begin{array}{c} c_1 \\ c_2 \\ c_3 \\ c_4 \\ c_5 \\ c_6 \\ c_7 \end{array} \begin{bmatrix} - & - & -1 \\ - & - & 1 \\ 1 & 2 & - \\ -1 & -2 & - \\ 1 & - & - \\ - & 1 & - \\ - & - & - \end{bmatrix}$$

(As all three columns of **C** obey the condition $\mathbf{Ac}_j = [0]$, a re-
ordering of the columns of **C** is superfluous.

CHAPTER VI

DEFINITENESS AND SYMMETRY

6.1 Quadratic forms and symmetric matrices

An expression of the form $\phi = \mathbf{x'Ay}$ is called a <u>form</u> or <u>bilinear form</u>.

We are more specifically concerned with the scalar expression:

$$\phi = \mathbf{x'Ax} \tag{6.1.1}$$

which is is called a <u>quadratic form</u>.

For the expression given in (6.1.1) to be legitimate, it is necessary that **A** is a square matrix. As the value of a quadratic form is just a scalar number, the righthand-side expression in (6.1.1) is its own transpose.

$$\phi = \mathbf{x'Ax} = [\mathbf{x'Ax}]' = \mathbf{x'A'x}$$

$$= 0.5\ \mathbf{x'Ax} + 0.5\ \mathbf{x'A'x} = \mathbf{x'}[0.5\ \mathbf{Ax} + 0.5\ \mathbf{A'x}]$$

$$= \mathbf{x'}[0.5\ \mathbf{A} + 0.5\ \mathbf{A'}]\mathbf{x} \tag{6.1.2}$$

It is therefore firm convention, to always give a quadratic form in symmetric form.

Thus, for $\phi = x_1{}^2 + 2x_1x_2 + x_2{}^2$,

we might write: $\phi = [x_1,\ x_2] \begin{bmatrix} 1 & 2 \\ - & 1 \end{bmatrix} \begin{bmatrix} x_1 \\ x_2 \end{bmatrix}$,

but we conventionally write: $\phi = [x_1,\ x_2] \begin{bmatrix} 1 & 1 \\ 1 & 1 \end{bmatrix} \begin{bmatrix} x_1 \\ x_2 \end{bmatrix}$

$$= x_1{}^2 + x_1x_2 + x_2x_1 + x_2{}^2 .$$

This convention establishes an association between quadratic forms and symmetric matrices.

142

6.2 **Reordering and partitioning of symmetric matrices**

On account of the association with quadratic forms, any partitioning of a symmetric matrix is usually itself symmetric. Thus if we say that the symmetric matrix **A** is partitioned as

$$\mathbf{A} = \begin{bmatrix} \mathbf{A}_{11} & \mathbf{A}_{12} \\ \mathbf{A}_{21} & \mathbf{A}_{22} \end{bmatrix}$$

the understanding that $\mathbf{A}_{11}$ and $\mathbf{A}_{22}$ are also square and symmetric, is implicit: Unless we wish it to be otherwise and clearly state so, that information is implied by simply saying that the symmetric matrix **A** is partitioned. The rows and columns of $\mathbf{A}_{11}$ will be associated with variables 1 to n_1, the rows of $\mathbf{A}_{12}$ with variables 1 to n_1, and its columns with variables $n_1 + 1$ to n, etc.

In that context, the term <u>simultaneous reordering</u> (of the rows and columns of the matrix), should be mentioned. If, for one reason or another, for example to have the terms with non-zero quadratic coefficients as leading terms, we re-name the variables, this is associated, not only with a re-ordering of the columns of **A**, but of the rows as well. For example, we might initially write:

$$\varphi(x_1, x_2, x_3) = x_1 + 2x_2 + 3x_3 + x_2{}^2 + x_2 x_3 + x_3 x_2.$$

The notation involved in a number of theorems and formulae is simplified, if we re-order the reference to the purely linear variable x_1 to the end, and re-name that variable as $x^*{}_3$, x_1 and x_2 then becoming $x^*{}_2$ and $x^*{}_3$. The corresponding re-ordering of the symmetric matrix **B** is from

$$\mathbf{B} = \begin{bmatrix} - & - & - \\ - & 1 & 1 \\ - & 1 & - \end{bmatrix} \quad \text{to} \quad \mathbf{B}^* = \begin{bmatrix} 1 & 1 & - \\ 1 & - & - \\ - & - & - \end{bmatrix}.$$

In the interest of streamlining the formulation of theorems, we will also introduce the convention, that the mere statement of a partitioning, may in fact involve a simultaneous reordering of the rows and columns of the matrix (and of the associated elements of any relevant vectors).

Thus, **B** as given above, is said to permit a partitioning, in which a 2 by 2 block $\mathbf{B}_{11}$ is invertible, whereas we are in fact referring to:

$$\mathbf{B^*}_{11} = \begin{bmatrix} 1 & 1 \\ 1 & - \end{bmatrix}.$$

6.3 Definiteness and extrema of quadratic functions

The term <u>definiteness</u> formally applies to a symmetric matrix, but its relevance relates to the associated quadratic form. The following terms apply:

A square and symmetric matrix $\mathbf{A}$ is:

<u>positive definite</u>   if $\mathbf{x'Ax} > 0$ is true for all $\mathbf{x} \neq 0$,

<u>positive semi-definite</u>   if $\mathbf{x'Ax} \geq 0$ is true for all $\mathbf{x}$,

<u>indefinite</u>   if $\mathbf{x'Ax} > 0$ is true for some $\mathbf{x}$,
                        but also $\mathbf{v'Av} < 0$, for some $\mathbf{v}$,

<u>negative semidefinite</u>   if $\mathbf{x'Ax} \leq 0$, for all $\mathbf{x}$, and

<u>negative definite</u>   if $\mathbf{x'Ax} < 0$ applies for all $\mathbf{x} \neq 0$.

These terms are useful in analyzing the existence of extrema of quadratic functions.

A quadratic function of the vector $\mathbf{x}$ is specified with the help of matrix notation, as follows:

$$\emptyset(\mathbf{x}) = \emptyset_0 + \mathbf{a'x} + \mathbf{x'Bx} \qquad\qquad (6.3.1)$$

The constant $\emptyset_0$ indicates the function-value at $\mathbf{x} = [0]$, the row-vector $\mathbf{a'}$ indicates the linear component, and the quadratic form $\mathbf{x'Bx}$ indicates the quadratic component.

For example,

$$\emptyset(x_1,x_2) = (x_1-3)^2 + (x_2-5)^2 + (2x_1-x_2)^2$$

$$= x_1^2 - 6x_1 + 9 \quad + x_2^2 - 10x_2 + 25 \quad + 4x_1^2 - 4x_1x_2 + x_2^2$$

$$= 36 \quad -6x_1 - 10x_2 \quad + 5x_1^2 - 4x_1x_2 + 2x_2^2$$

is in first instance expressed in matrix notation form as

$$\phi = 36 \quad + [-6 \quad -10] \begin{bmatrix} x_1 \\ x_2 \end{bmatrix} + [x_1 \quad x_2] \begin{bmatrix} 5 & -4 \\ - & 2 \end{bmatrix} \begin{bmatrix} x_1 \\ x_2 \end{bmatrix}$$

In the interest of being able to make use of the properties of definite matrices, we use the symmetric presentation, i.e.:

$$\phi = 36 \quad + [-6 \quad -10] \begin{bmatrix} x_1 \\ x_2 \end{bmatrix} + [x_1 \quad x_2] \begin{bmatrix} 5 & -2 \\ -2 & 2 \end{bmatrix} \begin{bmatrix} x_1 \\ x_1 \end{bmatrix}$$

$$= 36 \quad -6x_1 - 10x_1 \quad + 5x_1^2 - 2x_1x_2 - 2x_2x_1 + x_2^2$$

We now survey the conditions for extreme values of $\phi(\mathbf{x})$. These are, none too surprising, closely related to the type of definiteness of the matrix **B**.

For any $\mathbf{x} = \mathbf{x}^*$, we compare $\phi(\mathbf{x}^*)$, and $\phi(\mathbf{x}^* + \Delta\mathbf{x})$, the old and the new level of $\phi(\mathbf{x})$.

$$\phi(\mathbf{x}^* + \Delta\mathbf{x}) = \phi_0 + \mathbf{a}'[\mathbf{x}^* + \Delta\mathbf{x}] + [\mathbf{x}^* + \Delta\mathbf{x}]'\mathbf{B}[\mathbf{x}^* + \Delta\mathbf{x}]$$

$$= \phi_0 + \mathbf{a}'\mathbf{x}^* + \mathbf{a}'\Delta\mathbf{x} + \mathbf{x}^{*'}\mathbf{B}[\mathbf{x}^* + \Delta\mathbf{x}] + \Delta\mathbf{x}'\mathbf{B}[\mathbf{x}^* + \Delta\mathbf{x}]$$

$$= \phi(\mathbf{x}^*) + \mathbf{a}'\Delta\mathbf{x} + \mathbf{x}^{*'}\mathbf{B}\Delta\mathbf{x} + \Delta\mathbf{x}'\mathbf{B}\mathbf{x}^* + \Delta\mathbf{x}'\mathbf{B}\Delta\mathbf{x}$$

$$= \phi(\mathbf{x}^*) + \mathbf{a}'\Delta\mathbf{x} + 2\mathbf{x}^{*'}\mathbf{B}\Delta\mathbf{x} + \Delta\mathbf{x}'\mathbf{B}\Delta\mathbf{x} \qquad (6.3.2)$$

Therefore:

$$\Delta\phi = \phi(\mathbf{x}^* + \Delta\mathbf{x}) - \phi(\mathbf{x}^*) =$$

$$= \mathbf{a}'\Delta\mathbf{x} + 2\mathbf{x}^{*'}\mathbf{B}\Delta\mathbf{x} + \Delta\mathbf{x}'\mathbf{B}\Delta\mathbf{x} \qquad (6.3.3)$$

For $\mathbf{x} = \mathbf{x}^*$ to be associated with an extreme value of $\phi(\mathbf{x})$, we must first of all require (as usual in differential calculus), that there is no change at all in relation to a small differential variation in $\mathbf{x}$, neglecting the quadratic term, which consists of products of small variations.

The first-order condition is then presented in column form as:

$$\partial\phi/\partial\mathbf{x}(\mathbf{x}^*) \;=\; 2\,\mathbf{B}\,\mathbf{x}^* \;+\; \mathbf{a} \;=\; [0] \qquad\qquad (6.3.4)$$

If the block-equation (6.3.4) is resolvable, i.e. one and only one vector $\mathbf{x}^*$ can be solved from it, by inversion of $\mathbf{B}$, it does not follow from that fact alone, that $\mathbf{x} = \mathbf{x}^*$ is associated with an extreme value of ϕ.

Example:

$$\phi(x_1,x_2) = 2x_1 + 2x_2 \qquad + \quad x_1{}^2 - 4x_1 x_2 + x_2{}^2$$

$$= [2\ \ 2]\begin{bmatrix} x_1 \\ x_2 \end{bmatrix} + [x_1\ \ x_2]\begin{bmatrix} 1 & -2 \\ -2 & 1 \end{bmatrix}\begin{bmatrix} x_1 \\ x_2 \end{bmatrix}$$

The first-order conditions specified by (6.3.4) are:

$$2x_1 \ -4x_2 \ + 2 = 0 \ \rightarrow \ 2x_1 \ -4x_2 = -2$$

$$-4x_1 \ +2x_2 \ + 2 = 0 \ \rightarrow \ -4x_1 \ +2x_2 = -2$$

These conditions resolve uniquely by inversion of

$$\mathbf{B} = \begin{bmatrix} 2 & -4 \\ -4 & 2 \end{bmatrix}$$

to give rise to the solution $x_1 = 1$, $x_2 = 1$.

The first-order derivatives:

$$\partial\phi/\partial x_1 = 2 + 2x_1 \ -4x_2, \text{ and } \partial\phi/\partial x_2 = 2 + 2x_2 \ -4x_1,$$

do indeed vanish at the point $x_1 = 1$, $x_2 = 1$.

But this point is not associated with an extreme value of ϕ. If x_1 and x_2 are both very big and positive or both have a very negative value, the term $-4x_1 x_2$ dominates with a large positive value, and if x_1 and x_2 have opposite signs, the same term can dominate everything else with a large negative value.

The function has no finite maximum or minimum.

This is revealed by the indefiniteness of $\mathbf{B}$:

$$[1 \quad -] \begin{bmatrix} 1 & -2 \\ -2 & 1 \end{bmatrix} \begin{bmatrix} 1 \\ - \end{bmatrix} \qquad [1 \quad -] \begin{bmatrix} 1 \\ -2 \end{bmatrix} = 1 > 0$$

but

$$[1 \quad 1] \begin{bmatrix} 1 & -2 \\ -2 & 1 \end{bmatrix} \begin{bmatrix} 1 \\ 1 \end{bmatrix} \qquad [1 \quad 1] \begin{bmatrix} -1 \\ -1 \end{bmatrix} = -2 < 0$$

For a vector $\mathbf{x} = \mathbf{x}^*$, which obeys the first-order conditions (6.3.4), the formula for a change in the function value, (6.3.3), reduces to:

$$\Delta\phi = \Delta\mathbf{x}'\mathbf{B}\,\Delta\mathbf{x} \tag{6.3.5}$$

For $\phi(\mathbf{x})$ to have a finite extreme value, it is therefore necessary, that $\mathbf{B}$ is semidefinite. If $\mathbf{B}$ is positive semidefinite, then $\phi(\mathbf{x})$ may, or may not have a finite minimum, and if $\mathbf{B}$ is negative semidefinite, there can be a maximum.

Whether there actually is an extreme value, depends on whether or not there is a vector $\mathbf{x} = \mathbf{x}^*$, which obeys the first-order conditions. For a definite matrix $\mathbf{B}$, this is always the case: If $\mathbf{B}$ is positive definite, there is a unique minimum, if $\mathbf{B}$ is negative definite, there is a unique maximum.

Singularity of $\mathbf{B}$ would imply the existence of a vector $\mathbf{x} \neq [0]$, for which $\mathbf{B}\mathbf{x} = [0]$ was true, and that would contradict both positive definiteness and negative definiteness.

A more complicated situation arises in the case of semi-definiteness. If $\mathbf{B}$ is nontrivially semidefinite, i.e. not also definite, then $\mathbf{B}$ is singular, and the first-order conditions (6.3.4) may give rise to a dependent system, which is satisfied by an infinite set of vectors $\mathbf{x}$, or to a contradictory system. In the first case, there is a non-unique maximum or minimum, in the second case there is no finite extreme value of $\phi(\mathbf{x})$.

For example, for $\phi(\mathbf{x}) = 2x_1 + 2x_2 - (x_1 + x_2)^2$, there is a maximum, whenever x_1 and x_2 obey the condition $x_1 + x_2 = 1$. The matrix notation equivalent is in that case:

$$\phi(\mathbf{x}) = \begin{bmatrix} 2 & 2 \end{bmatrix} \begin{bmatrix} x_1 \\ x_2 \end{bmatrix} + \begin{bmatrix} x_1 & x_2 \end{bmatrix} \begin{bmatrix} -1 & -1 \\ -1 & -1 \end{bmatrix} \begin{bmatrix} x_1 \\ x_2 \end{bmatrix},$$

and the first-order conditions (6.3.4) are:

$$\partial\phi/\partial x_1 = -2x_1 \quad -2x_2 \quad + 2 \quad = \quad 0 \; \Big\}$$
$$\partial\phi/\partial x_2 = -2x_1 \quad -2x_2 \quad + 2 \quad = \quad 0$$

Obviously, these conditions are satisfied, whenever x_1 and x_2 obey the condition $x_1 + x_2 = 2$.

But for $\phi(\mathbf{x}) = x_1 - (x_1 + x_2)^2$, there is no finite maximum, nor a finite minimum, as (6.3.4) gives rise to a contradictory system.

6.4 Definiteness and diagonal inversion

An obvious condition for a matrix to be positive definite, is that all its diagonal elements are positive non-zero:

$$\mathbf{e'}_i A \mathbf{e}_i = a_{ii}$$

and the similar requirement $a_{ii} < 0$ will apply for negative definiteness.

The relationship with the existence of an extreme value of $\phi(\mathbf{x})$, as discussed in the previous section, should be clear: For $\phi(\mathbf{x})$ to have a minimum, no $x_i{}^2$ must occur with a negative coefficient, as otherwise the second order condition for a partial minimum relative to x_i cannot be satisfied.

Later on in this book, in Chapter IX, we will explore the relationship between definiteness and functions which are known as the 'latent roots' of a symmetric matrix. Definiteness can, however, also be tested by inversion or attempt to invert along the main diagonal, and this avoids the need to evaluate the latent roots. The following theorems are useful in this context:

Theorem:

If:

the n by n symmetric matrix

$$A = \begin{bmatrix} A_{11} & A_{12} \\ A_{21} & A_{22} \end{bmatrix}$$

(where A_{11} is of order n_1 by n_1)

is positive definite,

then:

(1) A is invertible, and:

(2) A_{11} and A_{22} are both positive definite and invertible, and:

(3) $[A_{22} - A_{21}A_{11}^{-1}A_{12}]$ is positive definite and invertible.

Proof:

$A\,x = [0]$, $x \neq [0]$ implies $x'Ax = 0$, $x \neq [0]$, hence contradicts the assumed positive definiteness of A.

q.e.d. ad (1).

$x'Ax > 0$, for all $x \neq [0]$, implies, for all $x_1 \neq [0]$, $x_2 = [0]$:

$$x'Ax = [x_1', x_2'] \begin{bmatrix} A_{11} & A_{12} \\ A_{21} & A_{22} \end{bmatrix} \begin{bmatrix} x_1 \\ x_2 \end{bmatrix}$$

$$= x_1'A_{11}x_1 + x_1'A_{12}x_2 + x_2'A_{21}x_1 + x_2'A_{22}x_2$$

$$= x_1'A_{11}x_1 > 0 \tag{6.4.1}$$

Hence A_{11} is positive definite, and therefore, by (1), as already proved, invertible. Proof for A_{22} by analogy.

q.e.d. ad (2).

The set of vectors $\mathbf{x}$, to which the property of positive definiteness refers, include the vectors which obey the side-condition:

$$\mathbf{A}_{11}\,\mathbf{x}_1 \;+\; \mathbf{A}_{12}\,\mathbf{x}_2 \;=\; [0] \tag{6.4.2}$$

and therefore, for any arbitrary $\mathbf{x}_2$:

$$\mathbf{x}_1 \;=\; -\mathbf{A}_{11}{}^{-1}\mathbf{A}_{12}\,\mathbf{x}_2 \tag{6.4.3}$$

Under that side-condition, the quadratic form $\mathbf{x}'\mathbf{Ax}$ is evaluated as:

$$\mathbf{x}'\mathbf{Ax} = \mathbf{x}_1{}'\mathbf{A}_{11}\mathbf{x}_1 \;+\; \mathbf{x}_1{}'\mathbf{A}_{12}\mathbf{x}_2 \;+\; \mathbf{x}_2{}'\mathbf{A}_{21}\mathbf{x}_1 \;+\; \mathbf{x}_2{}'\mathbf{A}_{22}\mathbf{x}_2 \;=$$

$$-\mathbf{x}_1{}'[\mathbf{A}_{11}\mathbf{x}_1 + \mathbf{A}_{12}\mathbf{x}_2] \;-\mathbf{x}_2{}'\mathbf{A}_{21}\mathbf{A}_{11}{}^{-1}\mathbf{A}_{12}\mathbf{x}_2 \;+\mathbf{x}_2{}'\mathbf{A}_{22}\mathbf{x}_2$$

$$=\; -\mathbf{x}_2\mathbf{A}_{21}\mathbf{A}_{11}{}^{-1}\mathbf{A}_{12}\mathbf{x}_2 \;+\; \mathbf{x}_2{}'\mathbf{A}_{22}\mathbf{x}_2$$

$$=\; \mathbf{x}_2{}'[\mathbf{A}_{22} - \mathbf{A}_{21}\mathbf{A}_{11}{}^{-1}\mathbf{A}_{12}]\mathbf{x}_2 \;>\; 0 \tag{6.4.4}$$

q.e.d. ad (3).

q.e.d.

Furthermore, from the proof and its method we have the following corollaries:

Corollary 1:
If $\mathbf{A}_{11}$ is positive definite, and $[\mathbf{A}_{22} - \mathbf{A}_{21}\mathbf{A}_{11}{}^{-1}\mathbf{A}_{21}]$ is indefinite, then $\mathbf{A}$ is indefinite.

Corollary 2:
If $\mathbf{A}_{11}$ is positive definite, and $[\mathbf{A}_{22} - \mathbf{A}_{21}\mathbf{A}_{11}{}^{-1}\mathbf{A}_{21}]$ is positive semidefinite, then $\mathbf{A}$ is positive semidefinite.

Corollary 3:
If $\mathbf{A}_{11}$ is positive definite, and $[\mathbf{A}_{22} - \mathbf{A}_{21}\mathbf{A}_{11}{}^{-1}\mathbf{A}_{21}]$ is negative definite, then $\mathbf{A}$ is indefinite.

Recursive application of the theorem to $[\mathbf{A}_{22} - \mathbf{A}_{21}\mathbf{A}_{11}^{-1}\mathbf{A}_{12}]$ and its top lefthand element now leads to:

Corollary 4:
If **A** is positive definite, then **A** can be inverted, by taking positive pivots, going from top left to bottom right, along the main diagonal, without ever interchanging rows. (Finding a a non-positive element in that position would, for a positive definite matrix, contradict the theorem, sub 3.)

Therefore, since the determinant is the recursive product of the pivots:

Corollary 5:
If **A** is positive definite, then $|\mathbf{A}| > 0$ applies.

Theorem:

If the n by n symmetric matrix **A** permits inversion, by positive pivots along the main diagonal, then **A** is positive definite.

Proof:

For n=1, no further proof is needed, Otherwise, we partition **A** into an n-1 by n-1 top lefthand block and a remaining border:

$$\mathbf{A} = \begin{bmatrix} \mathbf{A}_{11} & \mathbf{a}_{1n} \\ \mathbf{a}_{n1}' & a_{nn} \end{bmatrix} \qquad (6.4.5)$$

We now express the quadratic form **x'Ax** as:

$$\phi(\mathbf{x}) = \mathbf{x}_1'\mathbf{A}_{11}\mathbf{x}_1 + \mathbf{x}_1'\mathbf{a}_{1n} + x_n\mathbf{a}'_{n1}\mathbf{x}_1 + a_{nn}x_n^2$$

$$= \mathbf{x}_1'\mathbf{A}_{11}\mathbf{x}_1 + 2x_n\mathbf{a}'_{n1} + a_{nn}x_n^2 \qquad (6.4.6)$$

For $x_n=0$, $\phi(\mathbf{x}) = \mathbf{x'Ax}$ reduces to $\mathbf{x}_1'\mathbf{A}_{11}\mathbf{x}$.

An assumption, that the theorem is true for matrices of order n-1 by n-1, and therefore for $\mathbf{A}_{11}$, therefore implies, for $x_n=0$, $\mathbf{x}\neq[0]$:

$$\mathbf{x'Ax} > 0. \qquad (\mathbf{A}_{11} \text{ is assumed to be pos. def.})$$

Therefore, assuming that the theorem is true for matrices of order n-1 by n-1, and therefore for A_{11}, no further proof for the $x_n=0$ case is needed.

For $x_n \neq 0$, we may, without loss of generality, impose the condition:

$x_n = 1$.

(Neither the sign, nor the absolute value of any particular element of x affects the sign of the quadratic form:

$-x'A[-x] = x'Ax$)

For $x_n=1$, the expression for the quadratic form $x'Ax$ reduces to:

$$\emptyset(x) = x'Ax = x_1'A_{11}x_1 + 2a_{n1}'x1 + a_{nn} \qquad (6.4.7)$$

From which we derive the necessary first order conditions for a minimum value of $\emptyset(x)$, given $x_n=1$:

$$\partial\emptyset/\partial x_1 = 2A_{11}x_1 + 2a_{1n} = [0] \qquad (6.4.8)$$

In (6.4.8) a_{1n} has been substituted for the transpose of a_{n1}', and an equivalent formulation is:

$$x_1'A_{11} + a_{n1}' = [0]' \qquad (6.4.9)$$

Therefore, by (6.4.7) and (6.4.9):

$$\emptyset(x) = x_1'[A_{11}x_1 + a_{n1}] + a_{n1}'x_1 + a_{nn}$$

$$= a_{n1}'x_1 + a_{nn} \qquad (6.4.10)$$

As A_{11} is by assumption positive definite, and therefore invertible, (6.4.8) allows us to express x_1 as:

$$x_1 = -A_{11}^{-1}a_{1n} \qquad (6.4.11)$$

Therefore, by (6.4.10) and (6.4.11):

$$\emptyset(x) = a_{nn} - a_{n1}'A_{11}^{-1}a_{1n} \qquad (6.4.12)$$

The righthand side of (6.4.12) is the nth pivot used in the inversion of A, and is by assumption positive non-zero.

The proof of the theorem now follows by recursive induction.

q.e.d.

6.5 Definiteness and principal minors

In this section we discuss and re-state the relationship between positive diagonal pivots an principal minors, outside the context of quadratic forms and symmetric matrices.

The results on this point, can be proved in that general context, without invoking symmetry. We shall also find occasion to state the rules for recognizing definiteness in symmetric matrices more precisely.

Theorem

If the n by n matrix A and all its principal minors A_{ii} obey the conditions $|A| > 0$ and $|A_{ii}| > 0$

then

- 1) A can be inverted by row-operations, taking positive pivots on the main diagonal only,

- 2) $|A_{22} - A_{21}A_{11}^{-1}A_{12}| > 0$ applies throughout the inversion as stated under 1).

- 3) The properties stated under 1) and 2) apply for all simultaneous reorderings of A.

- 4) After each inversion step, $|[A_{22}-A_{21}A_{11}^{-1}A_{12}]| > 0$ applies, and all principal minors of $|[A_{22}-A_{21}A_{11}^{-1}A_{12}]|$ are also positive.

Proof

Concerning 1 and 2):
An assumption, that the theorem is true for the n-1 by n-1 top
lefthand block $\mathbf{A}_{11}$, completes by recursive induction to the
general validity of 1 and 2), as soon as

$$a_{nn} - a'_{n1}\mathbf{A}_{11}^{-1}a_{1n} > 0 \qquad\qquad\qquad (6.5.1)$$

is proved.

Since, given $|\mathbf{A}_{11}| > 0$, (6.5.1) is indeed a necessary condition
for $|\mathbf{A}| > 0$, we conclude to:

q.e.d. ad 1) and 2).

Concerning 3):
The proof, as supplied for 1) and 2), applies for any simultaneous reordering.

q.e.d. ad 3).

Concerning 4):

If $|[\mathbf{A}_{22}-\mathbf{A}_{21}\mathbf{A}_{11}^{-1}\mathbf{A}_{12}]| \leq 0$ applied after application of a series
of positive pivots, this would contradict $|\mathbf{A}| > 0$. If, following
the application of k positive pivots, the determinant of a p by p
principal minor block of $[\mathbf{A}_{22}-\mathbf{A}_{21}\mathbf{A}_{11}^{-1}\mathbf{A}_{12}]$ were non-positive,
this would imply that the determinant of a k+p by k+p principal
minor block of $\mathbf{A}$ was non-positive.

q.e.d. ad 4.

q.e.d.

Theorem

If the n by n matrix **A** obeys the following conditions:

a) can be inverted as stated for the previous theorem under 1)

and

b) in the process of those calculations, the diagonal elements of **A** at the beginning, of $[\mathbf{A}_{22}-\mathbf{A}_{21}\mathbf{A}_{11}{}^{-1}\mathbf{A}_{12}]$ at the end of each step, and of $\mathbf{A}^{-1}$ at the end are positive,

then

$|\mathbf{A}| > 0$ and $|\mathbf{A}_{ii}| > 0$, as assumed by the previous theorem, holds.

(while otherwise at least one of these properties does not hold).

Proof

We first deal with the special, but relatively simple cases n=1 and n=2.

For n=1 the theorem simply states that the on element is either positive-non-zero or not.

For n=2, we note that $a_{11} > 0$, $a_{22} > 0$, $a_{22} - (a_{12}\, a_{12}) / a_{11} > 0$ does indeed imply $|\mathbf{A}| > 0$ and if one of the diagonal elements of the inverse were not positive, one of the other assumptions must be false. q.e.d. for n=2.

We now consider the non-trivial n > 2 case.

If a) is false, no further proof is needed: by the previous theorem a) is a necessary condition for $|\mathbf{A}| > 0$, all $|\mathbf{A}_{ii}| > 0$.

If b) is seen to be false by way of $a_{ii} < 0$, no further proof is needed: these elements are themselves minor-matrices of order 1 by 1.

If b) is seen to be false by way of the appearance of a negative element of $[A_{22}-A_{21}A_{11}^{-1}A_{12}]$ after k inversion steps in which positive pivots were applied,

then for k < n-1 $|A_{11}| \leq 0$ is shown for a k+1 by k+1 block A_{11}.

If b) is seen to be false by way of the appearance of a negative element $a_{nn} - a'_{n1}A_{11}^{-1}a_{n1}$, after n-1 inversion steps in which positive pivots were applied, then $|A| \leq 0$ is shown.

If b) is seen to be false by way of the appearance of an element $[A^{-1}]_{ii} \leq 0$ in $[A^{-1}]$, following the application of n positive pivots, then $|A_{ii}| \leq 0$ is shown.

This completes the proof, as far as the necessity of a) and b) is concerned, the remaining burden of the proof therefore rests on their sufficiency.

For $|A|$ and the determinants of all leading blocks A_{11}, as actually inverted in the process of inverting A, the proof is already complete: the determinants of these blocks are the recursive products of the pivots as applied.

We now introduce the provisional assumption that the theorem holds for matrices of orders 1 by 1, 2 by 2, ... n-1 by n-1.

Then, by the theorem, assumed true for the n-1 by n-1 leading block A_{11}, $|A_{11}|$ and all its principal minors are shown to be $|A_{ii}| > 0$, as soon as n-1 inversion steps are completed without the appearance of negative diagonal elements.

If A_{11} does indeed conform to $|A_{ii}| > 0$ throughout, then, by property 3) of the previous theorem, the inversion of A_{11} can be done by positive pivots only, in any simultaneous reordering, and since both a) and b) were already proved to be necessary conditions, such an inversion of a re-ordered A_{11} will always conform to b) as well as to a).

The first-order principal minors are related to the diagonal elements of the inverse by:

$$[A^{-1}]_{ii} = |A_{ii}| / |A| \qquad\qquad (6.5.2)$$

therefore:

$$|A_{ii}| = |[A^{-1}]_{ii} \times |A| \qquad\qquad (6.5.3)$$

and $|A_{ii}| > 0$ needs to be assumed (given $[A^{-1}]_{ii} > 0$), irrespective of recursive induction.

The remaining burden of proof therefore rests on the determinants of minor-blocks of n-2 by n-2 or less, not included in the n-1 by n-1 block A_{11} as actually inverted.

For n > 2, an assumption that the theorem applies for the n-1 by n-1 block $[A_{22}-a_{21}a_{11}^{-1}a_{12}]$, as actually inverted during steps 2 to n, implies that:

either:

$[A_{22}-a_{21}a_{11}^{-1}a_{12}]$ has a positive determinant and positive principal minors, and if so, this block can be inverted by positive diagonal pivots in any simultaneous re-ordering, without any negative diagonal elements appearing.

If that is the case then all blocks of orders n-2 by n-2 or less, including those which include a_{nn} as this element occurs in A in its natural ordering, can be inverted by positive pivots, proving the positivity of their determinants.

or alternatively $[A_{22}-a_{21}a_{11}^{-1}a_{12}]$ does not meet that condition, in which case the original assumption that A was inverted by positive pivots without any negative diagonal elements appearing, is contradicted.

q.e.d. for n > 2

q.e.d.

Theorem

If the n by n matrix A and all its principal minors A_{ii} obey the conditions $|A| > 0$ and $|A_{ii}| > 0$

then the inverse A^{-1} shares all properties stated for A.

Proof

By the second theorem in this section $\mathbf{A}$ can be inverted in any simultaneous re-ordering, and by property 4) of the first theorem in this section, the assumed properties of $\mathbf{A}$ then generalize to every block $[\mathbf{A}_{22}-\mathbf{A}_{21}\mathbf{A}_{11}{}^{-1}\mathbf{A}_{12}]$.

An assumption that this theorem is true for $[\mathbf{A}_{22}-\mathbf{A}_{21}\mathbf{A}_{11}{}^{-1}\mathbf{A}_{12}]$ then implies that $[\mathbf{A}^{-1}]_{22} = [\mathbf{A}_{22}-\mathbf{A}_{21}\mathbf{A}_{11}{}^{-1}\mathbf{A}_{12}]^{-1}$ shares the properties assumed for $\mathbf{A}$ (in any simultaneous reordering).

Since $|\mathbf{A}^{-1}| = 1 \ / \ |\mathbf{A}|$, the general proof follows by recursive induction.

q.e.d.

Of the three theorems stated and proved above, the first two and their proofs immediately generalize to non-negative minors and attempting to invert.

The n by n matrix $\mathbf{A}$ is characterized by $|\mathbf{A}| \geq 0$ and $|\mathbf{A}_{ii}| \geq 0$ (all principal minors), if and only if a diagonal block of $\mathbf{A}$, of the same rank as $\mathbf{A}$ itself, can be inverted by application of positive pivots along the main diagonal, without any negative diagonal elements appearing. (As such an inversion reduces the remaining block $[\mathbf{A}_{22}-\mathbf{A}_{21}\mathbf{A}_{11}{}^{-1}\mathbf{A}_{12}]$ to a null block, its proof is implied by the corresponding theorem for an invertible matrix.)

In other words, if we define definiteness in terms of principal minors, classifying $|\mathbf{A}|$ itself as a principal minor as well:

positive   definite: all $|\mathbf{A}_{ii}| > 0$
positive semi-definite: all $|\mathbf{A}_{ii}| \geq 0$
indefinite some $|\mathbf{A}_{ii}| > 0$, some $|\mathbf{A}_{jj}| < 0$
negative   definite: all $|\mathbf{A}_{ii}| < 0$
negative semi-definite: all $|\mathbf{A}_{ii}| \leq 0$

then recognition of definiteness by diagonal inversion is the same in symmetric and non-symmetric matrices alike. For a symmetric matrix this 'definition' of definiteness is equivalent to the usual one.

However, if we define definiteness in the usual way in terms of the sign of the quadratic form $\mathbf{x'A\,x}$, then the theorem from section 6.4:

The n by n symmetric matrix is positive definite, if and only if the determinant $|A|$ and all its principal minors are positive, needs to be modified if symmetry cannot be invoked:

Theorem

The n by n matrix $\mathbf{A}$ obeys the condition $\mathbf{x'A\,x} > 0$ for all $\mathbf{x} \neq [0]$ only if $|A|$ and all its principal minors are positive.

Proof

We partition $\mathbf{A}$ as

$$\mathbf{A} = \begin{bmatrix} a_{11} & \mathbf{a'}_{12} \\ a_{21} & \mathbf{A}_{22} \end{bmatrix} \tag{6.5.4}$$

For $a_{11} \leq 0$, $\mathbf{x'A\,x} > 0$, $\mathbf{x} \neq [0]$ is false: $\mathbf{e'}_1 \mathbf{A}\, \mathbf{e}_1 = a_{11} \leq 0$.

Therefore $a_{11} > 0$.

The set of vectors $\mathbf{x} \neq [0]$, for which $\mathbf{x'A\,x} > 0$ needs to apply, includes those vectors for which $\mathbf{a'}_1\mathbf{x} = a_{11}x_1 + \mathbf{a'}_{12}\mathbf{x}_2 = 0$, and hence $x_1 = a_{11}^{-1}\mathbf{a}_{12}\mathbf{x}_2$ applies.

For that subset of all vectors $\mathbf{x}$, for which $\mathbf{x'A\,x} > 0$ applies, the quadratic form is:

$$\mathbf{x'A\,x} = x_1(a_{11}x_1 + a_{12}'\mathbf{x}_2) + \mathbf{x'}_2\,[a_{21}x_1 + \mathbf{A}_{22}]\,\mathbf{x}_{22} =$$

$$\mathbf{x'}_2[-a_{21}a_{11}^{-1}\mathbf{a'}_{12} + \mathbf{A}_{22}]\,\mathbf{x}_2$$

$$= \mathbf{x'}_2\,[\mathbf{A}_{22} - a_{21}a^{-1}\mathbf{a'}_{12}]\,\mathbf{x}_2 \tag{6.5.5}$$

Therefore $\mathbf{x'A\,x} > 0$, can only be true for all $\mathbf{x} \neq [0]$, if $a_{11} > 0$ applies, and if

$$\mathbf{x'}_2\,[\mathbf{A}_{22} - a_{21}a_{11}^{-1}\mathbf{a'}_{12}]\,\mathbf{x}_2 \text{ is true for all } \mathbf{x}_{22} \neq [0].$$

An assumption that the theorem is true for the n-1 by n-1 matrix $[A_{22}-a_{21}a_{11}{}^{-1}a'_{12}]$ (in all simultaneous reorderings), now gives rise top a proof by recursive induction.

q.e.d.

Note the one difference between the above proof, and the similar proof in the symmetric case where the existence of $A_{11}{}^{-1}$ immediately followed from $x'_1 a x_1 > 0$ for all $x_1 \neq [0]$. In the non-symmetric case a proof on that point is needed, hence the recursive induction from n-1 to n, rather than form $n-n_1$ to n.

However, the stronger statement 'if and only if' is false, at least for some non-symmetric matrices.

The positive minors and the positive determinant are a necessary condition for the consistent positivity of the quadratic form, but that condition is sufficient, only if **A** is known to be symmetric.

The proof of the second theorem in section 6.4 invokes symmetry, in a more essential way, and is false for non-symmetric matrices, as may be illustrated for

$$\mathbf{A} = \begin{bmatrix} 1 & -2 \\ 8 & 1 \end{bmatrix} \; ; \quad |A| = 18 \; ; \quad [9 \quad -1] \begin{bmatrix} 1 & -2 \\ 8 & 1 \end{bmatrix} \begin{bmatrix} 9 \\ -1 \end{bmatrix} = -10$$

(with the two principal minors being $a_{11} = a_{22} = 1$.)

Nor is the opposite sign of the off-diagonal elements the only reason for this theorem to be restricted to symmetric matrices, as may be illustrated by

$$\mathbf{A} = \begin{bmatrix} 5 & 2 \\ 8 & 4 \end{bmatrix} \; ; \quad |A| = 4 \; ; \quad [1 \quad -1] \begin{bmatrix} 5 & 2 \\ 8 & 4 \end{bmatrix} \begin{bmatrix} 1 \\ -1 \end{bmatrix} = -1$$

Corollary:
If $|A + A'|$ and all its principal minors are positive non-zero, (non-negative) then the same applies for $|A|$. (But the similar property of $|A|$ itself does always carry over for $|A + A'|$.)

In applying the results of this and the previous section in testing for definiteness -now again conventionally defined in relation to a symmetric matrix-, it is useful to observe that a 2 by 2 matrix

$$A = \begin{bmatrix} - & \alpha \\ \alpha & \beta \end{bmatrix}$$

is indefinite, whenever $\alpha \neq 0$ applies, irrespective of the value of β, as may be shown by evaluating the quadratic form $\mathbf{x'Ax}$ in this case as:

$$\mathbf{x'Ax} = 2 \alpha x_1 x_2 + \beta x^2$$

Therefore $\mathbf{x'Ax} > 0$ is true if x_1 and x_2 are both non-zero and of the same sign, and $\mathbf{x'Ax} < 0$ is true if x_1 and x_2 are both non-zero and of opposite sign, and the absolute value of x_1 is more than half the absolute value of βx_2.

Example:

$$\begin{bmatrix} 1 & 1 \end{bmatrix} \begin{bmatrix} - & 1 \\ 1 & 10 \end{bmatrix} \begin{bmatrix} 1 \\ 1 \end{bmatrix} = 2 + 10 = 12$$

but

$$\begin{bmatrix} 10 & -1 \end{bmatrix} \begin{bmatrix} - & 1 \\ 1 & 10 \end{bmatrix} \begin{bmatrix} 10 \\ -1 \end{bmatrix} = -20 + 10 = -10$$

To test for definiteness, we now proceed as follows:

If $a_{ii} > 0$ (and therefore $\mathbf{e'}_i \mathbf{Ae}i = a_{ii} > 0$) applies for some i, and also $a_{jj} < 0$ for some j, we immediately find the matrix indefinite. If $a_{ii} = 0$, $a_{ij} = a_{jj} \neq 0$ applies for some i,j then (see above) we have identified a 2 by 2 indefinite block, and therefore find the whole matrix indefinite.

If $a_{ii} > 0$ applies for all i, we begin by presuming positive definiteness.

We then begin by inverting or attempting to invert along the main diagonal. If a zero appears on the diagonal, while there are still non-zero elements in A_{22} elsewhere in the relevant row and column, we have identified an indefinite 2 by 2 block, and A is indefinite.

If a negative element appears on the diagonal, the same conclusion follows.

Otherwise, we continue enlarging A_{11} (if necessary after simultaneous reordering, to bring a positive non-zero element in position n_1,n_1), until inversion of A by positive pivots only, is complete, and A is shown to be positive definite, or until $[A_{22}-A_{21}A_{11}{}^{-1}A_{12}]$ is reduced to a zero block, proving A to be positive semi-definite, by corollary 2 of the previous section.

If all the diagonal elements of the initial matrix A are negative, we presume negative definiteness, and proceed as before on minus A.

Exercises on definiteness:

Establish the type of definiteness of:

$$A = \begin{bmatrix} 4 & 2 & 2 & 4 \\ 2 & 2 & 2 & 3 \\ 2 & 2 & 1 & 3 \\ 4 & 3 & 3 & 6 \end{bmatrix}$$

$$B = \begin{bmatrix} 1 & 2 & 3 \\ 2 & 1 & 2 \\ 3 & 2 & 1 \end{bmatrix}, \quad \begin{bmatrix} 3 & 2 & -1 \\ 2 & 4 & 2 \\ -1 & 2 & 3 \end{bmatrix}, \quad \begin{bmatrix} 1 & 1 & 1 \\ 1 & 1 & 2 \\ 1 & 2 & 1 \end{bmatrix}$$

Part-answersheet on definiteness

The relevant part-inversion of **A**, i.e. on the **Ax** = **y** system, is
as follows:

$$
\begin{array}{cccc}
x_1 & x_2 & x_3 & x_4
\end{array}
\quad = \quad
\begin{array}{cccc}
y_1 & y_2 & y_3 & y_4
\end{array}
\quad \Sigma
$$

$$
\begin{bmatrix}
\underline{|4|} & 2 & 2 & 4 \\
2 & 2 & 2 & 3 \\
2 & 2 & 1 & 3 \\
4 & 3 & 3 & 6
\end{bmatrix}
\begin{bmatrix}
1 & - & - & - \\
- & 1 & - & - \\
- & - & 1 & - \\
- & - & - & 1
\end{bmatrix}
\begin{matrix}
13 \\
10 \\
9 \\
17
\end{matrix}
$$

$$
\begin{array}{cccc}
x_1 & x_2 & x_3 & x_4
\end{array}
\quad = \quad
\begin{array}{cccc}
y_1 & y_2 & y_3 & y_4
\end{array}
\quad \Sigma
$$

$$
\begin{bmatrix}
1 & .5 & .5 & 1 \\
- & 1 & 1 & 1 \\
- & 1 & - & 1 \\
- & 1 & 1 & 2
\end{bmatrix}
\begin{bmatrix}
.25 & - & - & - \\
-.5 & 1 & - & - \\
-.5 & - & 1 & - \\
-1 & - & - & 1
\end{bmatrix}
\begin{matrix}
3.25 \\
3.5 \\
2.5 \\
4
\end{matrix}
$$

At this point indefiniteness is revealed:

$$
[\mathbf{A}_{22} - \mathbf{A}_{21}\mathbf{A}_{11}^{-1}\mathbf{A}_{12}] =
\begin{bmatrix}
1 & 1 & 1 \\
1 & - & 1 \\
1 & 1 & 2
\end{bmatrix}
$$

is indefinite, as may be illustrated by:

$$
\begin{bmatrix} - & 1 & 1 \end{bmatrix}
\begin{bmatrix}
1 & 1 & 1 \\
1 & - & 1 \\
1 & 1 & 2
\end{bmatrix}
\begin{bmatrix} - \\ 1 \\ 1 \end{bmatrix} =
\begin{bmatrix} - & 1 & 1 \end{bmatrix}
\begin{bmatrix} 1 \\ 1 \\ 3 \end{bmatrix} = 4 > 0,
$$

but

$$
\begin{bmatrix} - & -2 & 1 \end{bmatrix}
\begin{bmatrix}
1 & 1 & 1 \\
1 & - & 1 \\
1 & 1 & 2
\end{bmatrix}
\begin{bmatrix} - \\ -2 \\ 1 \end{bmatrix} =
\begin{bmatrix} - & -2 & 1 \end{bmatrix}
\begin{bmatrix} -1 \\ 1 \\ - \end{bmatrix} = -2 < 0.
$$

(It was shown in section 6.5, that a zero on the diagonal,
generally implies indefiniteness, except in the case of a null
column and corresponding null row.)

For the other three matrices, the results are: B_1 and B_3 are indefinite, B_2 is positive semidefinite.

CHAPTER VII

LATENT ROOTS AND CHARACTERISTIC VECTORS

7.1 The characteristic equation

A vector $\mathbf{x} \neq [0]$ may be chosen in such a way, that, on multiplying a square matrix by the vector, the result is a vector which is proportional to the original one. We then say that such a vector is a <u>characteristic vector</u> (of the matrix), and, as far as columns (to be discussed in this section) are concerned, we may add, for the sake of completeness, that it is a <u>characteristic column vector</u>.

The relation between the vector, the matrix, and the ratio is therefore:

$$\mathbf{A}\,\mathbf{x} = \mathbf{x}\,\lambda \qquad\qquad (7.1.1)$$

For (7.1.1) to be a legitimate relationship, $\mathbf{A}$ needs to have the same number of columns as the order of $\mathbf{x}$, and also the same number of rows, otherwise the product does not even have the same order as $\mathbf{x}\,\lambda$. The number λ is called the <u>latent root</u>.

There are a number of alternative names for the same concept. Of the equivalents for the term <u>latent root</u>, the German word <u>Eigenvalue</u> is the most commonly used, sometimes also the literal translation of 'Eigenvalue' <u>own value</u> or <u>proper value</u>.

Example:

$$\mathbf{A} = \begin{bmatrix} 1 & 2 & 3 \\ 3 & 2 & 1 \\ 1 & 4 & 1 \end{bmatrix} ; \qquad \begin{bmatrix} 1 & 2 & 3 \\ 3 & 2 & 1 \\ 1 & 4 & 1 \end{bmatrix} \begin{bmatrix} 1 \\ 1 \\ 1 \end{bmatrix} = \begin{bmatrix} 6 \\ 6 \\ 6 \end{bmatrix} = \begin{bmatrix} 1 \\ 1 \\ 1 \end{bmatrix} \times 6$$

As each row of **A** adds to 6, we find **A s** = **s** × 6, i.e. the summation-vector is a characteristic vector, with λ = 6 as root, and (7.1.1) is met, for this matrix **A**, for λ = 6 as root.

We must require **x** to be a non-zero vector, as otherwise no requirement on λ is contained in (7.1.1) at all.

To find the relevant requirement on λ, we re-order (7.1.1) as:

$$\mathbf{A}\,\mathbf{x} - \mathbf{x}\,\lambda = [\mathbf{A} - \mathbf{I}\,\lambda]\,\mathbf{x} = [0] \qquad\qquad (7.1.2)$$

For (7.1.2) to hold for $\mathbf{x} \neq [0]$, it is necessary, that $[\mathbf{A} - \mathbf{I}\,\lambda]$ is singular.

Therefore:

$$|\mathbf{A} - \mathbf{I}\,\lambda| = 0 \qquad\qquad (7.1.3)$$

This relationship is known as the <u>characteristic equation</u>

Example:

$$\mathbf{A} = \begin{bmatrix} 2 & 1 & 1 \\ 1 & 2 & 1 \\ 1 & 1 & 2 \end{bmatrix}$$

$$|\mathbf{A} - \mathbf{I}\lambda| = \begin{vmatrix} 2-\lambda & 1 & 1 \\ 1 & 2-\lambda & 1 \\ 1 & 1 & 2-\lambda \end{vmatrix} =$$

$$(2-\lambda) \times \begin{vmatrix} 2-\lambda & 1 \\ 1 & 2-\lambda \end{vmatrix} - 1 \times \begin{vmatrix} 1 & 1 \\ 1 & 2-\lambda \end{vmatrix} + 1 \times \begin{vmatrix} 1 & 2-\lambda \\ 1 & 1 \end{vmatrix} =$$

$$(2-\lambda)\Big((2-\lambda)(2-\lambda) - 1\Big) - \Big(1\times(2-\lambda) - 1\Big) + \Big(1 - 1 \times (2-\lambda)\Big) =$$

$$(2-\lambda)\left(3 - 4\lambda + \lambda^2\right) - \left(1 - \lambda\right) + \left(-1 + \lambda\right) =$$

$$\left((6 - 8\lambda + 2\lambda^2) - (3\lambda - 4\lambda^2 + \lambda^3)\right) - \left(1 - \lambda\right) + \left(-1 + \lambda\right) =$$

$$6 - 8\lambda + 2\lambda^2 \quad - 3\lambda + 4\lambda^2 \quad - \lambda^3 \quad - 1 + \lambda \quad - 1 + \lambda =$$

$$4 - 9\lambda + 6\lambda^2 - \lambda^3 = 0.$$

This equation solves as $\lambda_1 = 4$, and $\lambda_2 = \lambda_3 = 1$, and we confirm these roots, as follows:

For $\lambda_1 = 4$:

$$\begin{bmatrix} 2 & 1 & 1 \\ 1 & 2 & 1 \\ 1 & 1 & 2 \end{bmatrix} \begin{bmatrix} 1 \\ 1 \\ 1 \end{bmatrix} = \begin{bmatrix} 4 \\ 4 \\ 4 \end{bmatrix} = \begin{bmatrix} 1 \\ 1 \\ 1 \end{bmatrix} \times 4$$

For $\lambda_2 = 1$:

$$\begin{bmatrix} 2 & 1 & 1 \\ 1 & 2 & 1 \\ 1 & 1 & 2 \end{bmatrix} \begin{bmatrix} 1 \\ -1 \\ - \end{bmatrix} = \begin{bmatrix} 1 \\ -1 \\ - \end{bmatrix} = \begin{bmatrix} 1 \\ -1 \\ - \end{bmatrix} \times 1$$

For $\lambda_3 = 1$:

$$\begin{bmatrix} 2 & 1 & 1 \\ 1 & 2 & 1 \\ 1 & 1 & 2 \end{bmatrix} \begin{bmatrix} - \\ 1 \\ -1 \end{bmatrix} = \begin{bmatrix} - \\ 1 \\ -1 \end{bmatrix} = \begin{bmatrix} - \\ 1 \\ -1 \end{bmatrix} \times 1$$

7.2 The minors form of the characteristic equation

The example given in the previous section, develops the characteristic equation according to the definition of a determinant, as given in Chapter V. There is, however, a different way of developing and presenting the characteristic equation, according to the separate terms of a polynomial expression in λ.

To this purpose, it is first of all useful to rewrite (7.1.3) as:

$$|I\lambda - A| = 0 \qquad\qquad\qquad (7.2.1)$$

The characteristic equation is now written, with separate indication of its coefficient for any p^{th} power of λ as:

$$\phi(\lambda) = |I\lambda - A| = \sum_{p=0}^{n} c_p \lambda^p = 0 \qquad\qquad (7.2.2)$$

To develop (7.2.2) further, we need to introduce a term which we have not used so far.

A <u>principal minor</u> is minor, whose minor-matrix is obtained, by removing a number of rows and <u>corresponding</u> columns from the parent-matrix. We speak of first-order, second-order, third-order principal minors, depending on whether the minor-matrix is of order n-1, n-2, n-3, etc.

Thus, $\mathbf{A} = \begin{bmatrix} 1 & 2 & 3 \\ 4 & 5 & 6 \\ 7 & 8 & 9 \end{bmatrix}$

has three first-order principal minors:

$$\begin{vmatrix} 5 & 6 \\ 8 & 9 \end{vmatrix}, \qquad \begin{vmatrix} 1 & 3 \\ 7 & 9 \end{vmatrix}, \qquad \text{and} \qquad \begin{vmatrix} 1 & 2 \\ 4 & 5 \end{vmatrix},$$

which arise, on removing the first, second, and third rows and <u>corresponding</u> columns.

And (for a matrix of order 3 by 3), also 3 second order principal minors, viz:

$$\begin{vmatrix} 1 \end{vmatrix}, \qquad \begin{vmatrix} 5 \end{vmatrix}, \qquad \text{and} \qquad \begin{vmatrix} 9 \end{vmatrix}.$$

The coefficient c_p of λ^p, as figuring in (7.2.2), is now obtained as the summation of all p^{th} order principal minors, which are the determinants of n-p by n-p minor-matrices, of $-\mathbf{A}$. The constant is $|-\mathbf{A}|$ itself, and the leading term λ^p has a coefficient of 1 by definition.

To see why this is so, we first of all express $|I\lambda-\mathbf{A}|$ as if there were n different numbers λ_j. For reasons which will become clear when we derive some results, we indicate that form of $|I\lambda - A|$ as a separate function:

$$f(\lambda_1, \lambda_2, \ldots\ldots \lambda_n) =$$

$$\begin{vmatrix} \lambda_1 - a_{12} & -a_{12} & \ldots\ldots\ldots\ldots\ldots\ldots & -a_{1n} \\ -a_{21} & \lambda_2 - a_{22} & \ldots\ldots\ldots\ldots\ldots\ldots & -a_{2n} \\ \ldots\ldots\ldots\ldots\ldots\ldots\ldots\ldots \\ -a_{n1} & -a_{n2} & \ldots\ldots\ldots\ldots\ldots\ldots & \lambda_n - a_{nn} \end{vmatrix} \qquad (7.2.3)$$

where the various number λ_j, are in fact all the same, and should not be confused with the roots of the characteristic equation, as discussed in the previous section.

From (7.2.3), replacing the separate λ_i ($i \neq j$) in the result by their value λ:

$$\partial f / \partial \lambda_j = |I_{n-1}\lambda - A_{jj}| \qquad (7.2.4)$$

We shall refer to $f(\lambda_1, \lambda_1, \ldots\ldots \lambda_n)$, when it actually meets the condition:

$$\lambda_1 = \lambda_2 = \ldots\ldots\ldots \lambda_n = \lambda \qquad (7.2.5)$$

as $\phi(\lambda)$.

Therefore, by (7.2.4) and (7.2.5):

$$d\phi/d\lambda = \sum_{j=1}^{n} \partial\phi/\partial\lambda_j = \sum_{j=1}^{n} |I_{n-1}\lambda - A_{jj}| \qquad (7.2.6)$$

By (7.2.2), we find for $\lambda = 0$:

$$d\phi/d\lambda(0) = c_1 \qquad (7.2.7)$$

Therefore, by (7.2.6) and (7.2.7):

$$c_1 = \sum_{j=1}^{n} |-A_{jj}| \qquad (7.2.8)$$

$$c_0 = |-A| \qquad (7.2.9)$$

follows by evaluating $\phi(\lambda)$ as given by the two most lefthand members of (7.2.2), for $\lambda=0$.

Recursive differentiation of (7.2.4), relative to λ, and application of (7.2.4) and its derivatives in evaluating the derivatives of the principal minors of $|I\lambda-A|$, now gives rise to a formula for the pth-order derivative, which is:

$$d^p\phi/d^p\lambda \;=\; p!\; \Sigma\; |-A^*_p| \tag{7.2.10}$$

where the symbol $\Sigma\; |-A^*_p|$ stands for aggregated series of the the principal minors of $-A$, and the multiplicative factor p! arises because the same minor is obtained from different terms.

The similar recursive differentiation of $\phi(\lambda)$, as expressed by (7.2.2), gives rise to:

$$d^p\phi/d^p\lambda(0) \;=\; p!\; c_p \tag{7.2.11}$$

Example:

$$A = \begin{bmatrix} 1 & 2 & - \\ 3 & -4 & 5 \\ - & 6 & 7 \end{bmatrix} \qquad |I\lambda-A| = \begin{vmatrix} \lambda-1 & -2 & - \\ -3 & \lambda+4 & -5 \\ - & -6 & \lambda-7 \end{vmatrix}$$

$$|-A| = -1 \times \begin{vmatrix} 4 & -5 \\ -6 & -7 \end{vmatrix} \;+\; 2 \times \begin{vmatrix} -3 & -5 \\ - & -7 \end{vmatrix} =$$

$$-1 \times (-28 \;\; -30) \;+\; 2 \times (21 - 0) \;=\; 58 + 42 \;=\; 100\;.$$

first-order principal minors:

$$\begin{vmatrix} 4 & -5 \\ -6 & -7 \end{vmatrix} = -58; \qquad \begin{vmatrix} -1 & -2 \\ -3 & 4 \end{vmatrix} = -10; \qquad \begin{vmatrix} -1 & - \\ - & -7 \end{vmatrix} = 7;$$

these three principal minors add up to -61.

second-order principal minors:

$$|-1| = -1; \qquad |4| = 4; \qquad |-7| = -7; \quad \text{these add up to } -4.$$

The characteristic equation therefore is:

$$|I\lambda - A| = \lambda^3 - 4\;\lambda^2 - 61\;\lambda + 100 = 0.$$

Check:

$$|I\lambda - A| = \begin{vmatrix} \lambda-1 & -2 & - \\ -3 & \lambda+4 & -5 \\ - & -6 & \lambda-7 \end{vmatrix} =$$

$$(\lambda-1) \begin{vmatrix} \lambda+4 & -5 \\ -6 & \lambda-7 \end{vmatrix} + 2 \begin{vmatrix} -3 & -5 \\ - & \lambda-7 \end{vmatrix} =$$

$(\lambda-1) \left((\lambda+4)(\lambda-7) - 30 \right) + 2 \left(-3(\lambda-7) \right) =$

$(\lambda-1) \left(\lambda^2 - 3\lambda - 28 - 30 \right) + 2 \left(-3\lambda + 7 \right) =$

$\left(\lambda^3 - 3\lambda^2 - 58\lambda \right) - \left(\lambda^2 - 3\lambda - 58 \right) + 2 \left(-3\lambda + 21 \right) =$

$\lambda^3 - 3\lambda^2 - 58\lambda \quad -\lambda^2 + 3\lambda + 58 \quad - 6\lambda + 42 \quad =$

$\lambda^3 - 4\lambda^2 - 61\lambda + 100$.

Application of (7.2.10), for p=2:

$$d^p\phi/d^p\lambda = \Sigma\, d|I\lambda - A_{jj}|/d\lambda =$$

$$d \begin{vmatrix} \lambda+4 & -5 \\ -6 & \lambda-7 \end{vmatrix} / d\lambda + d \begin{vmatrix} \lambda-1 & -2 \\ -3 & \lambda+4 \end{vmatrix} / d\lambda + d \begin{vmatrix} \lambda-1 & - \\ - & \lambda-7 \end{vmatrix} / d\lambda$$

$= |\lambda-7| + |\lambda+4| \quad + \quad |\lambda+4| + |\lambda-1| \quad + \quad |\lambda-7| + |\lambda-1|$

$= 2 \times \left(|\lambda-1| + |\lambda+4| + |\lambda-7| \right)$.

Therefore: $\qquad\qquad d\phi/d\lambda(0) = 2 \times \left(-1 + 4 - 7 \right)$

Hence, given $d\phi/d\lambda(0) = 2\,c_p$: $\qquad c_p = -1 + 4 - 7 = 4$.

The sum of the diagonal elements of a square matrix is known as the <u>trace</u>. (Notation: tr (**A**), where it has to be assumed that **A** is square.)

Traces obviously obey the following definitional identity:

$$\text{tr } (\mathbf{A} + \mathbf{B}) = \text{tr } (\mathbf{A}) + \text{tr } (\mathbf{B}) =$$

$$\sum_{i=1}^{n} (a_{ii} + b_{ii}) = \sum_{i=1}^{n} a_{ii} + \sum_{i=1}^{n} b_{ii} \qquad (7.2.12)$$

where it has to be assumed that both $\mathbf{A}$ and $\mathbf{B}$ are of order n by n.

Theorem:

If $\mathbf{A}$ is of order n by n, than the roots λ_1 to λ_n of $\mathbf{A}$ obey:

$$\sum \lambda_i = \sum a_{ii} = \text{tr } (\mathbf{A}) \qquad (7.2.13)$$

Proof:

We develop the characteristic function:

$$\emptyset (\lambda) = (\lambda-\lambda_1)(\lambda-\lambda_2) \ldots (\lambda-\lambda_n) = \sum_{p=0}^{n} c_p \lambda^{n-p} \qquad (7.2.14)$$

of which the second term is:

$$c_1 \lambda^{n-1} = \sum_{i=1}^{n} \lambda_i \lambda^{n-1} \qquad (7.2.15)$$

Application of (7.2.10) shows that this term is:

$$c_1 \lambda^{n-1} = \sum a_{ii} \lambda^{n-1} = \text{tr } (\mathbf{A}) \lambda^{n-1} \qquad (7.2.16)$$

q.e.d.

Example:

$$\mathbf{A} = \begin{bmatrix} 2 & 3 \\ 1 & 4 \end{bmatrix} \begin{bmatrix} 1 \\ 1 \end{bmatrix} = \begin{bmatrix} 5 \\ 5 \end{bmatrix} = \begin{bmatrix} 1 \\ 1 \end{bmatrix} \times 5$$

As the trace is $2 + 4 = 6$, the other root must be $\lambda_2 = 6 - 5 = 1$, as is confirmed by:

$$\mathbf{A} = \begin{bmatrix} 2 & 3 \\ 1 & 4 \end{bmatrix} \begin{bmatrix} 3 \\ -1 \end{bmatrix} = \begin{bmatrix} 3 \\ -1 \end{bmatrix}$$

Exercise:

For $\mathbf{A} = \begin{bmatrix} 1 & 2 & 3 \\ 3 & 2 & 1 \\ 1 & 4 & 1 \end{bmatrix}$

develop the characteristic equation, both according to the formula given in section 7.1, and by the rules given in this section. Check that the resulting equation fits the root associated with $\mathbf{A}\,\mathbf{s} = \mathbf{s} \times 6$. (Should these consistency checks fail, you may find the answersheet at the end of the chapter helpful in spotting errors.)

7.3 Polynomial functions and their factorization

A general survey of the problems which arise in resolving the characteristic equation, i.e. finding numbers λ_j for which (7.1.3) is met, goes beyond the scope of this book.

However, at the level of theory, we still need to say something about the existence of roots, i.e. whether and under what conditions there <u>are</u> such numbers, and their relationship to each other.

A function of the form:

$$f(x) = c_0 + c_1 x + c_2 x^2 + \ldots\ldots c_n x^n = \sum_{p=0}^{n} c_p x^p \qquad (7.3.1)$$

is known as a <u>polynomial</u> function of x, and, of the nth <u>order</u> or <u>degree</u>. If the coefficients c_p are all ordinary real numbers (as different from complex numbers, to be discussed below), we speak of a <u>real</u> polynomial function.

The corresponding equation:

$$f(x) = \sum_{p=0}^{n} c_p x^p = 0 \qquad\qquad (7.3.2)$$

is then named a <u>polynomial equation</u> (of the nth order).

When writing a polynomial equation explicitly, it is usual to write the terms of (7.3.2) in the reverse order, and to divide all coefficients by c_p. The usual form of writing a polynomial equation explicitly is therefore:

$$x^n + c_{n-1}x^{n-1} + c_{n-2}x^{n-2} + \ldots\ldots\ldots + c_0 = 0$$

e.g.:

$$x^5 + 3x^4 - 6x^3 + 12x^2 - 3x + 15 = 0$$

The values of x, for which (7.3.2) is true, are called the <u>roots</u> of the polynomial equation.

Polynomial functions can be subjected to an operation which is not unlike 'long division' as it applies to integer numbers.

The following definitions apply:

A polynomial function $f(x)$ of order n may be <u>divided</u> by a another polynomial function $d(x)$, the <u>division function</u>, of order k $\leq$n. The division consists of establishing a <u>quotient function</u> $q(x)$, and a <u>remainder function</u> $r(x)$, satisfying:

$$f(x) \equiv d(x) \times q(x) + r(x) \qquad\qquad (7.3.3)$$

where $r(x)$ is of the lowest possible order.

The operation is done by recursively re-defining $q^*(x)$ and $r^*(x)$, as they occur in:

$$f(x) \equiv d(x) \times q^*(x) + r^*(x) \qquad\qquad (7.3.4)$$

where $q^*(x)$ is:

$$q^*(x) = \sum_{p=1}^{n-k} q_p\, x^p \qquad\qquad (7.3.5)$$

Initially, l is set at $l=n-k+1$, in which case $q^*(x)$ is identically equal to zero, therefore initially, $r^*(x)=f(x)$. At each divisionstep, l is reduced by one, and q_1 is set at the ratio between the coefficient associated with the highest powers of x in $r^*(x)$ and $d(x)$, thereby causing the vanishment of the coefficient associated with the highest power of x in the next $r^*(x)$.

Example:
$$f(x) = x^5 + 2x^4 - 3x^3 + 7x^2 + 3x - 4$$

$$d(x) = x^2 - x + 2$$

The division is performed as follows:

```
                    f(x)      =   x⁵  + 2x⁴  - 3x³  + 7x²  + 3x  - 4
q*₃ =  x³   →    x³ d(x)  =   x⁵   - x⁴  + 2x³
                                 ___________________

r*(x) =                          3x⁴  - 5x³  + 7x²  + 3x  - 4
q*₂ = 3x²   →   3x² d(x)  =      3x⁴  - 3x³  + 6x²
                                      ________________

r*(x) =                               - 2x³  +  x²  + 3x  - 4
q*₁ = -2x   →   -2x d(x)  =           - 2x³  + 2x²  - 4x
                                           ______________

                                            - x²  + 7x  - 4
q*₀ = -1    →    -d(x)    =                  - x²  +  x  - 2
                                                ___________

r*(x) =          r(x)     =                          6x  - 2
```

Therefore:

$$q(x) = x^3 + 3x^2 - 2x - 1$$

$$f(x) = x^5 + 2x^4 - 3x^3 + 7x^2 + 3x - 4$$

$$= (x^2 - x + 2)(x^3 + 3x^2 - 2x - 1) + 6x - 2$$

The mechanics of this operation recursively determine the next coefficient of q(x), starting with the one associated with the highest power of x, and hence also uniquely determine the remainder function. If f(x) is of order n, and d(x) is of order k, then the order of r(x) will at most be k-1, as otherwise a further division step could be made. If the division function is of the first order, the order of the remainder function is zero, simply a number, and we speak of the remainder.

Theorem (Remainder theorem):

If f(x) is divided by d(x) = x-a, the remainder is r = f(a).

Proof:

Apply (7.3.3) for d(x) = x-a, as follows:

$$f(x) \equiv (x-a) \times q(x) + r \qquad (7.3.6)$$

and evaluate for x = a

q.e.d.

Corollary:

If x=a is a root of f(x) = 0, then r = 0 (there is no remainder).

This corollary is often of practical help, if we wish to solve the other roots of a polynomial equation of degree 3, of which one root is known.

Example:

$$\mathbf{A} = \begin{bmatrix} 1 & 2 & 3 \\ 3 & 2 & 1 \\ 1 & 4 & 1 \end{bmatrix} \qquad \text{(from section 7.1)}$$

As each row of **A** adds to 6, we already know one root, $\lambda = 6$:

$$\begin{bmatrix} 1 & 2 & 3 \\ 3 & 2 & 1 \\ 1 & 4 & 1 \end{bmatrix} \begin{bmatrix} 1 \\ 1 \\ 1 \end{bmatrix} = \begin{bmatrix} 6 \\ 6 \\ 6 \end{bmatrix} = \begin{bmatrix} 1 \\ 1 \\ 1 \end{bmatrix} \times 6$$

The characteristic equation of this matrix is:

$$\lambda^3 - 4\lambda^2 - 8\lambda - 24 = 0$$

(For details of its calculation, see the answersheet at the end of the chapter.)

We now divide $f(\lambda) = \lambda^3 - 4\lambda^2 - 8\lambda - 24$ by $d(\lambda) = \lambda - 6$, as follows:

$$
\begin{array}{llll}
q^*_2 = \lambda^2 \rightarrow \quad \lambda^2 d(\lambda) = & \lambda^3 & -4\lambda^2 & -8\lambda & -24 \\
 & \lambda^3 & -6\lambda^2 & & \\
\hline
r^*(\lambda) = & & 2\lambda^2 & -8\lambda & -24 \\
q^*_1 = 2\lambda \rightarrow \quad 2 1 d(\lambda) = & & 2\lambda^2 & -12\lambda & \\
\hline
r^*(\lambda) = & & & 4\lambda & -24 \\
q^*_0 = 4 \rightarrow \quad 4 d(\lambda) = & & & 4\lambda & -24 \\
\hline
\end{array}
$$

The theory of polynomial functions and polynomial equations becomes more uniform, and therefore easier to develop further, if we are able to give content to the following statement:

A polynomial equation of the nth order has n roots.

As far as ordinary real numbers is concerned, this clearly is not so, as may be illustrated by the second degree polynomial equation:

$$x^2 + 1 = 0$$

For ordinary real numbers, $x^2 + 1 \geq 1 > 0$ will apply.

However, we are able to maintain this statement as true, if we accept <u>complex</u> numbers as roots.

The expression $r = a + b\, i$, where a and b are 'ordinary' real numbers, and

$$i = \sqrt{-1} \qquad\qquad (7.3.7)$$

i.e. the root of $x^2 + 1 = 0$, is a complex number.

This extension of the set of numbers also is the significance of the term <u>real</u> polynomial function. A real polynomial function is a polynomial function of which the coefficients c_p are real numbers. If some of the c_p are complex, then we speak of a complex function.

We now first of all note the following

Theorem (Fundamental Theorem of Algebra):

A polynomial equation of degree n (real or complex), has at least one root (real or complex).

We shall refrain from providing a proof of this theorem. For reference see, among others, Hawkins and Hawkins [21], Chapter 5, and Spiegel [42], p. 125.

The combination of the Fundamental Theorem of Algebra, with (recursive application of) the Remainder Theorem, gives rise to the following

Corollary:

A polynomial function of degree n, can be written as the product of n polynomial functions of degree 1.

Hence, if the leading coefficient is 1, i.e. the function $f(x)$ has the form:

$$f(x) = x^n + c_{n-1}x^{n-1} + c_{n-2}x^{n-2} + \ldots\ldots\ldots c_0 \qquad (7.3.8)$$

then it can be written as:

$$f(x) = (x - r_1)(x - r_2) \ldots\ldots\ldots (x - r_n)$$

and the numbers r_1 to r_n are all roots of $f(x) = 0$.

This corollary on its own, is not as useful as it appears at first sight, because all roots could be complex, such as the roots of $x^8 + 1 = 0$.

However, we have the following further

Theorem:

If
$x = \alpha + \beta i$ is a root of the real polynomial equation $f(x) = 0$,

then
$x = \alpha - \beta i$ is also a root of $f(x) = 0$.

Proof:

We first of all state, and will proceed to prove, the following

Lemma:

$$(\alpha + \beta i)^p = \gamma + \delta i$$

implies

$$(\alpha - b_i)^p = \gamma - \delta i$$

Proof (of the lemma):

For p=1, the lemma merely states the trivial identity between $\alpha + \beta i$ and itself.

Otherwise, we introduce the provisional assumption, that the lemma is true, for some k, for p = 2, 3, ... k, in addition to p=1.

Therefore, by assumption, for k = p+1:

$$(\alpha + \beta i)^k = (\alpha + \beta i)(\alpha + \beta i)^p = (\alpha + \beta i)(\gamma + \delta i)$$

$$= (\alpha\gamma - \beta\delta) + (\alpha\delta + \beta\gamma)i$$

as well as:

$$(\alpha - \beta i)^k = (\alpha - \beta i)(\alpha - \beta i)^p = (\alpha - \beta i)(\gamma - \delta i)$$

$$= (\alpha\gamma - \beta\delta) - (\alpha\delta + \beta\gamma)i$$

The general proof of the lemma now follows by recursive induction.

q.e.d. for the lemma.

Therefore, <u>denoting</u> the pth power of $\alpha + \beta i$ as $\gamma_p + \delta_p i$:

$$f(\alpha + \beta i) = \sum_{p=0}^{n} c_p (\alpha + \beta i)^p = \sum_{p=0}^{n} c_p (\gamma_p + \delta_p i) = 0$$

$$(7.3.9)$$

and hence, taking the real and the irrational parts separately:

$$\sum_{p=0}^{n} c_p \gamma_p = 0 \qquad\qquad (7.3.10)$$

as well as:

$$\sum_{p=0}^{n} c_p \delta_p = 0 \qquad\qquad (7.3.11)$$

implies

$$f(\alpha - \beta i) = \sum_{p=0}^{n} c_p (\alpha - \beta i)^p = \sum_{p=0}^{n} c_p (\gamma_p - \delta_p) \qquad (7.3.12)$$

Upon adding i times (7.3.11) to (7.3.10), we re-confirm (7.3.9). Upon subtracting i times (7.3.11) from (7.3.10), the righthand side of (7.3.12) is seen to be:

$$f(\alpha - \beta i) = 0 \qquad\qquad (7.3.13)$$

q.e.d.

We therefore speak of a pair of complex roots, not only for a second degree polynomial equation, but whenever a complex number is a root of a polynomial equation. The usual form of indicating this is:

$$x = \alpha \pm \beta i$$

Note the following corollary of this theorem, in relation to the expression of a polynomial function conform (7.3.8):

$$(x - r_1)(x - r_2) = (x - \alpha - \beta i)(x - \alpha + \beta i)$$
$$= (x - \alpha)^2 - (\beta i)^2 = (x - \alpha)^2 + \beta^2$$
$$= x^2 - 2\alpha x + \alpha^2 + \beta^2 \qquad (7.3.14)$$

Therefore:

A real polynomial function, of degree $n > 2$, can always be factorised into a number of real polynomial functions of degree 1 or 2.

Corollary: If n is odd, at least one root is real.

The reference to complex roots goes a long way in justifying the statement that a polynomial equation of the nth order has n roots, but we must mention another complication.

Consider the polynomial equation:

$$x^5 - 2x^4 + x^3 = x^3(x-1)^2 = 0$$

It is immediately obvious that $x = 0$, and $x = 1$, are both roots of this equation. Are there any other roots ? No, $x^3 = 0$ implies $x = 0$, and cannot be met by any (real or complex) number $x \neq 0$.

We therefore say that $x=0$ is a <u>repeated</u> (in this case: thrice repeated) root of $x^5 - 2x^4 + x^3 = 0$, and $x=1$ is a (twice) repeated root.

If $f(x)$ factorises in such a way that there is only one factor $(x-r)$ in (7.3.8), we say that $x=r$ is a <u>distinct</u> root.

Note, that the concept of distinct versus repeated roots applies strictly in relation to f(x) = 0, not with any other righthand side:

f(x) = x⁴ -1 factorizes as:

$$x^4 - 1 = (x^2 - 1)\,(x^2 + 1) = (x - 1)\,(x + 1)\,(x + i)\,(x - i)$$

and x⁴ = 1 has four distinct roots: x = 1, -1, i and -i.

7.4 Column vectors, row vectors, and their calculation

Once we have established, for some value of λ, $|A-I\lambda| = 0$, we note that the singularity of $[A-I\lambda]$ also implies row dependence, i.e.:

$$\mathbf{w'}\,[\mathbf{A-I}\lambda] = [0]'$$

and therefore:

$$\mathbf{w'A} = \lambda\,\mathbf{w'} \tag{7.4.1}$$

for some $\mathbf{w'} \neq [0]'$.

Such a vector is known as a <u>characteristic row vector</u>, or, if the context is clear, row-vector for short.

Except in the case of a symmetric matrix, characteristic row-vectors normally bear no immediately visible relationship to the corresponding column vectors. In particular, they are not simply the transposed column-vectors.

For example, for
$$\mathbf{A} = \begin{bmatrix} 1 & 2 & 3 \\ 3 & 2 & 1 \\ 1 & 4 & 1 \end{bmatrix}$$

(from section 7.1), we had no difficulty in establishing at least one column vector:

$$\begin{bmatrix} 1 & 2 & 3 \\ 3 & 2 & 1 \\ 1 & 4 & 1 \end{bmatrix} \begin{bmatrix} 1 \\ 1 \\ 1 \end{bmatrix} = \begin{bmatrix} 6 \\ 6 \\ 6 \end{bmatrix} = \begin{bmatrix} 1 \\ 1 \\ 1 \end{bmatrix} \times 6$$

However, in general, the resolution of a characteristic vector is best done by part-resolution of the $[A-I\lambda]x = [0]$ system, or, in the case of a row-vector, of the $[A'-I\lambda]w = [0]$ system.

In this case, (7.4.1) is first transposed as:

$$A' w = w \lambda \qquad\qquad (7.4.2)$$

i.e. the row-vector is to be found as the column-vector of A'.

This system is then re-ordered as:

$$[A' - I\lambda] w = [0] \qquad\qquad (7.4.3)$$

i.e. in this example:

$$w_1 + 3w_2 + w_3 = 6w_1 \quad\rightarrow\quad -5w_1 + 3w_2 + w_3 = 0$$

$$2w_1 + 2w_2 + 4w_3 = 6w_2 \quad\rightarrow\quad 2w_1 - 4w_2 + 4w_3 = 0$$

$$3w_1 + w_2 + w_3 = 6w_3 \quad\rightarrow\quad 3w_1 + w_2 - 5w_3 = 0$$

We set up something like the standard row-operations tableau. In the interest of avoiding fractional vectors, we apply a modification of the all-integer elimination procedure from section 5.9.

We proceed as follows:

$$\begin{array}{ccccc} w_1 & w_2 & w_3 & = 0 & \Sigma \\ \boxed{-5} & 3 & 1 & & -1 \\ 2 & -4 & 4 & & 2 \\ 3 & 1 & -5 & & -1 \end{array} \qquad \begin{array}{l} -5\,w_1 + 3w_2 + w_3 = 0 \\ 2\,w_1 - 4w_2 + 4w_3 = 0 \\ 3\,w_1 + w_2 - 5w_3 = 0 \end{array}$$

Multiply the second and third rows by -5:

$$
\begin{array}{ccc}
w_1 & w_2 & w_3
\end{array} \; = 0 \qquad \Sigma
$$

$$
\begin{bmatrix}
|\underline{-5}| & 3 & 1 \\
-10 & 20 & -20 \\
-15 & -5 & 25
\end{bmatrix}
\qquad
\begin{array}{c}
-1 \\
-10 \\
5
\end{array}
\qquad
\begin{array}{rrrl}
-5\,w_1 & + 3w_2 & + w_3 & = 0 \\
-10\,w_1 & 20w_2 & -20w_3 & = 0 \\
-15\,w_1 & -5w_2 & +25w_3 & = 0
\end{array}
$$

Subtract the leading row twice from row 2, and three times from row 3:

$$
\begin{array}{ccc}
w_1 & w_2 & w_3
\end{array} \; = 0 \qquad \Sigma
$$

$$
\begin{bmatrix}
-5 & 3 & 1 \\
- & |\underline{14}| & -22 \\
- & -14 & 22
\end{bmatrix}
\qquad
\begin{array}{c}
-1 \\
-8 \\
8
\end{array}
\qquad
\begin{array}{rrrl}
-5\,w_1 & + 3w_2 & + w_3 & = 0 \\
 & +14w_2 & -22w_3 & = 0 \\
 & -14w_2 & +22w_3 & = 0
\end{array}
$$

Multiply the first row by 14:

$$
\begin{array}{ccc}
w_1 & w_2 & w_3
\end{array} \; = 0 \qquad \Sigma
$$

$$
\begin{bmatrix}
-70 & 42 & 14 \\
- & |\underline{14}| & -22 \\
- & -14 & 22
\end{bmatrix}
\qquad
\begin{array}{c}
-14 \\
-8 \\
8
\end{array}
\qquad
\begin{array}{rrrl}
-70\,w_1 & +42w_2 & +14w_3 & = 0 \\
 & 14w_2 & -22w_3 & = 0 \\
 & -14w_2 & +22w_3 & = 0
\end{array}
$$

Subtract 3 times the second row from the top row, and add the second row to the third row::

$$
\begin{array}{ccc}
w_1 & w_2 & w_3
\end{array} \; = 0 \qquad \Sigma
$$

$$
\begin{bmatrix}
-70 & - & 80 \\
- & 14 & -22 \\
- & - & -
\end{bmatrix}
\qquad
\begin{array}{c}
10 \\
-8 \\
-
\end{array}
\qquad
\begin{array}{rrrl}
-70\,w_1 & & +80w_3 & = 0 \\
 & 14w_2 & 22w_3 & = 0
\end{array}
$$

Divide row 1 (and 3, but that is trivial), by the first pivot, i.e. by -5:

$$
\begin{array}{ccccc}
w_1 & w_2 & w_3 & = 0 & \Sigma
\end{array}
$$

$$
\left[\begin{array}{ccc}
14 & - & -16 \\
- & 14 & -22 \\
- & - & -
\end{array}\right]
\quad
\begin{array}{c}
-2 \\
-8 \\
-
\end{array}
\qquad
\begin{array}{l}
14\,w_1 \qquad\qquad -16w_3 = 0 \\
\qquad\quad 14w_2 \quad -22w_3 = 0
\end{array}
$$

Therefore, for $w_3 = 14$:

$$14w_1 \quad -16w_3 = 0 \quad \rightarrow \quad 14w_1 = 16w_3 = 16 \times 14 \quad \rightarrow \quad w_1 = 16$$

$$14w_2 \quad -22w_3 = 0 \quad \rightarrow \quad 14w_2 = 16w_3 = 22 \times 14 \quad \rightarrow \quad w_2 = 22$$

From which: $\mathbf{w}' = [16 \ \ 22 \ \ 14]$

There is in this case an incidental common factor 2, and we find:

$$\mathbf{w}' = [8 \ \ 11 \ \ 7]$$

to be the vector with the smallest integer numbers, which will meet (7.4.1).

Check:

$$
[8 \ \ 11 \ \ 7]
\left[\begin{array}{ccc}
1 & 2 & 3 \\
3 & 2 & 1 \\
1 & 4 & 1
\end{array}\right]
= [48 \ \ 66 \ \ 42] = [8 \ \ 11 \ \ 7] \times 6
$$

A number of comments on this calculation are appropriate at this point.

First of all, it should be clear, that, if it were not for our desire to avoid fractions, the similar generalization of the standard method of row-operations, would have done just as well.

The corresponding calculation tableaus would in that case have been:

$$
\begin{array}{ccccc}
w_1 & w_2 & w_3 & = 0 & \Sigma
\end{array}
$$

$$
\left[\begin{array}{ccc}
|{-5}| & 3 & 1 \\
2 & -4 & 4 \\
3 & 1 & -5
\end{array}\right]
\quad
\begin{array}{c}
-1 \\
2 \\
-1
\end{array}
\qquad
\begin{array}{l}
-5\,w_1 \ + 3w_2 \ \ + w_3 \ = 0 \\
2\,w_1 \ \ - 4w_2 \ + 4w_3 = 0 \\
3\,w_1 \ \ + w_2 \ \ - 5w_3 = 0
\end{array}
$$

$$
\begin{array}{cccc}
w_1 & w_2 & w_3 & = 0 \qquad \Sigma \\
\end{array}
$$

$$
\left[\begin{array}{ccc}
1 & -0.6 & -0.2 \\
- & |-2.8| & 4.4 \\
- & 2.8 & -4.4
\end{array}\right]
\qquad
\begin{array}{c}
0.2 \\
2.2 \\
-1.6
\end{array}
\qquad
\begin{array}{llll}
w_1 & -0.6w_2 & -0.2w_3 & = 0 \\
    & -2.8w_2 & +4.4w_3 & = 0 \\
    & 2.8w_2 & -4.4w_3 & = 0
\end{array}
$$

$$
\begin{array}{cccc}
w_1 & w_2 & w_3 & = 0 \qquad \Sigma \\
\end{array}
$$

$$
\left[\begin{array}{ccc}
1 & - & -1.14 \\
- & 1 & -1.57 \\
- & - & -
\end{array}\right]
\qquad
\begin{array}{c}
0.14 \\
-0.57 \\
-
\end{array}
\qquad
\begin{array}{lll}
w_1 &   & -1.14w_3 & = 0 \\
    & w_2 & -1.57w_3 & = 0
\end{array}
$$

Therefore, setting w_3 at $w_3 = 1$:

$$
\begin{array}{lllll}
w_1 & - 1.14w_3 = 0 & \rightarrow & w_1 = 1.14w_3 & \rightarrow & w_1 = 1.14 \\
w_2 & - 1.57w_3 = 0 & \rightarrow & w_2 = 1.57w_3 & \rightarrow & w_2 = 1.57
\end{array}
$$

Secondly, there are matrices, for which some elements of the
characteristic vector can only be a zero. In that case, an
attempt to express other elements of the vector in them, is bound
to be unsuccessful, and we may have to interchange columns as
well as rows. As far as the method of elimination after division
by the pivot is concerned, the relevant algorithm is exactly the
same as the one which we discussed in section 5.10 for the
evaluation of the rank of a matrix.

Example:

$$
\mathbf{A} = \left[\begin{array}{cccc}
1 & 1 & 1 & 1 \\
2 & 2 & 2 & - \\
- & 2 & 3 & 3 \\
4 & 4 & 4 & 2
\end{array}\right]
$$

This is a singular matrix, used in section 5.7 to illustrate that
failure to invert by row operations proves singularity. We now
wish to part-resolve the $\mathbf{A}x = [0]$ system, and proceed as follows:

$$
\begin{array}{cccc}
\mathbf{x_1} & \mathbf{x_2} & \mathbf{x_3} & \mathbf{x_4} \\
\end{array}
\qquad \Sigma
$$

$\mathbf{x_1}$	$\mathbf{x_2}$	$\mathbf{x_3}$	$\mathbf{x_4}$	Σ
$\underline{|1|}$	1	1	1	4
2	2	2	–	6
–	2	3	3	8
4	4	4	2	14

$\mathbf{x_1}$	$\mathbf{x_2}$	$\mathbf{x_3}$	$\mathbf{x_4}$	Σ
1	1	1	1	4
–	–	–	-2	-2
–	2	3	3	8
–	–	–	-2	-2

Interchange the second and third rows, to obtain the following tableaus:

$\mathbf{x_1}$	$\mathbf{x_2}$	$\mathbf{x_3}$	$\mathbf{x_4}$	Σ
1	1	1	1	4
–	$\underline{|2|}$	3	3	8
–	–	–	-2	-2
–	–	–	-2	-2

$$
\begin{aligned}
x_1 + x_2 + x_3 + x_4 &= 0 \\
2x_2 + 3x_3 + 3x_4 &= 0 \\
-2x_4 &= 0 \\
-2x_4 &= 0
\end{aligned}
$$

$\mathbf{x_1}$	$\mathbf{x_2}$	$\mathbf{x_3}$	$\mathbf{x_4}$	Σ
1	–	-0.5	-0.5	–
–	1	1.5	1.5	2
–	–	–	-2	-2
–	–	–	-2	-2

$$
\begin{aligned}
x_1 \qquad -0.5x_3 - 0.5x_4 &= 0 \\
x_2 + 1.5x_3 + 1.5x_4 &= 0 \\
-2\ x_4 &= 0 \\
-2\ x_4 &= 0
\end{aligned}
$$

Now interchange columns 3 and 4:

$\mathbf{x_1}$	$\mathbf{x_2}$	$\mathbf{x_4}$	$\mathbf{x_3}$	Σ
1	–	-0.5	-0.5	–
–	1	1.5	1.5	2
–	–	$\underline{|-2|}$	–	-2
–	–	-2	–	-2

$$
\begin{aligned}
x_1 \qquad -0.5x_4 - 0.5x_3 &= 0 \\
x_2 + 1.5x_4 + 1.5x_3 &= 0 \\
-2\ x_4 \qquad\qquad &= 0 \\
-2\ x_4 \qquad\qquad &= 0
\end{aligned}
$$

$$
\begin{array}{cccc}
x_1 & x_2 & x_4 & x_3
\end{array}
\qquad \Sigma
$$

$$
\begin{bmatrix}
1 & - & - & -0.5 \\
- & 1 & - & 1.5 \\
- & - & 1 & - \\
- & - & - & -
\end{bmatrix}
\quad
\begin{matrix}
0.5 \\
2 \\
1 \\
-
\end{matrix}
\qquad
\begin{matrix}
x_1 & & & -0.5\,x_3 = 0 \\
& x_2 & & +1.5\,x_3 = 0 \\
& & x_4 & = 0
\end{matrix}
$$

Therefore, setting x_3 (<u>not</u> x_4) at the arbitrary value $x_3 = 1$:

$$x_1 \;\; - 0.5\,x_3 = 0 \qquad \rightarrow \qquad x_1 = 0.5\,x_3 = 0.5$$

$$x_2 \;\; + 1.5\,x_3 = 0 \qquad \rightarrow \qquad x_2 = -1.5\,x_3 = -1.5$$

$x_4 = 0$ is given as such, and the vector associated with the
zero root of **A** is:

$$
\mathbf{x} =
\begin{bmatrix}
0.5 \\
-1.5 \\
1 \\
-
\end{bmatrix}
\; ; \quad
\begin{bmatrix}
1 & 1 & 1 & 1 \\
2 & 2 & 2 & - \\
- & 2 & 3 & 3 \\
4 & 4 & 4 & 2
\end{bmatrix}
\begin{bmatrix}
0.5 \\
-1.5 \\
1 \\
-
\end{bmatrix}
=
\begin{bmatrix}
- \\
- \\
- \\
-
\end{bmatrix}
$$

Exercise:

For $\mathbf{A} =$
$$
\begin{bmatrix}
2 & 4 & 4 & 4 & - \\
1 & 2 & 2 & 2 & - \\
1 & 2 & 2 & 2 & - \\
1 & 2 & 2 & 2 & - \\
- & - & - & - & -
\end{bmatrix}
$$

develop the characteristic equation, find all the roots (which
are all real), and the corresponding vectors.

(There is an answersheet at the end of the chapter).

7.5 **Complex vectors**

If some or all of the roots are complex, the corresponding vectors will also be complex. We shall be applying a number of results, which were in first instance obtained in the context of real numbers, to complex roots as well.

The normal arithmetic of real numbers, i.e. addition, subtraction, multiplication, division, permit generalization to complex numbers. We review these operations, as follows:

- addition: $\qquad$ $(\alpha + \beta i) + (\gamma + \delta i) = \alpha + \gamma + (\beta + \delta)i$

- subtraction: $\qquad$ $(\alpha + \beta i) - (\gamma + \delta i) = \alpha - \gamma + (\beta - \delta)i$

- multiplication: $(\alpha + \beta i) \times (\gamma + \delta i) = \alpha\gamma - \beta\delta + (\alpha\delta + \beta\gamma)i$

- division: $\qquad$ $1 / (\gamma + \delta i) = (\gamma - \delta i) / (\gamma^2 + \delta^2)$

as may be verified by multiplying both sides by $\gamma + \delta i$.

Therefore:

$$(\alpha + \beta i) / (\gamma + \delta i) = \left((\alpha\delta + \beta\gamma) + (\alpha\gamma - \beta\delta)i\right) / (\gamma^2 + \delta^2)$$

It follows that, although the arithmetic gets a bit more complicated, the vectors associated with complex roots can be calculated in the same way as discussed in the previous section for real roots.

Example:

$$A = \begin{bmatrix} 2 & 2 & -2 \\ - & - & 2 \\ 1 & -3 & 4 \end{bmatrix}$$

We develop the characteristic equation:

$$|A - I\lambda| = \begin{vmatrix} 2-\lambda & 2 & -2 \\ - & -\lambda & 2 \\ 1 & -3 & 4-\lambda \end{vmatrix} =$$

$$(2-\lambda) \begin{vmatrix} -\lambda & 2 \\ -3 & 4-\lambda \end{vmatrix} \quad -2 \begin{vmatrix} - & 2 \\ 1 & 4-\lambda \end{vmatrix} \quad -2 \begin{vmatrix} - & -\lambda \\ 1 & -3 \end{vmatrix} =$$

$$(2-\lambda)\left((-\lambda)(4-\lambda)+6\right) \quad -2\left(0-2\right) \qquad\qquad -2\left(0+\lambda\right) \quad =$$

$$(2-\lambda)\left(-4\lambda+\lambda^2+6\right) \qquad\qquad +4 \qquad\qquad\qquad -2\lambda \qquad =$$

$$(2-\lambda)\left(6-4\lambda+\lambda^2\right) \qquad\qquad +4 \qquad\qquad\qquad -2\lambda \qquad =$$

$$12 \quad -8\lambda \quad +2\lambda^2 \quad -6\lambda \quad +4\lambda \quad -\lambda^3 \qquad +4 \qquad -2\lambda \qquad =$$

$$-\lambda^3 + 6\lambda^2 - 16\lambda + 16 = 0$$

or equivalently:

$$\lambda^3 - 6\lambda^2 + 16\lambda - 16 = 0$$

One root of **A** is identifiable by its vector: the rows of **A** all
add to 2, i.e: **A s** = **s** × 2.

Applying the factorization technique from section 7.3, we express
the characteristic equation as:

$$(\lambda-2)(\lambda^2 -4\lambda +8) = 0;$$

The remaining roots are now solved from: $\lambda^2 -4\lambda +8 = 0$,

with the help of the formula:

$$\lambda = \left(4 \pm \sqrt{(16-32)}\right) / 2 = 2 \pm 2i$$

Therefore:

$$\lambda_1 = 2, \quad \lambda_2 = 2+2i, \quad \lambda_3 = 2-2i.$$

We now apply the method of part-inversion, as set out in section
7.4, to the complex matrix $[\mathbf{A} - \mathbf{I}\lambda_3]$ ($\lambda = 2 -2i$). In the
interest of avoiding the complication of developing complex
fractional numbers, we shall use a generalization of the adjoint
all integer method of inversion:

$$
\begin{array}{ccc}
x_1 & x_2 & x_3 \\
\end{array} \quad = 0 \qquad \Sigma
$$

$$
\left[
\begin{array}{ccc}
\underline{|2i|} & 2 & -2 \\[4pt]
- & -2 +2i & 2 \\[4pt]
1 & -3 & 2 +2i
\end{array}
\right]
\qquad
\begin{array}{c}
2i \\[4pt]
2i \\[4pt]
2i
\end{array}
$$

Multiply the third row by 2i, to obtain the intermediate tableau:

$$
\begin{array}{ccc}
x_1 & x_2 & x_3 \\
\end{array} \quad = 0 \qquad \Sigma
$$

$$
\left[
\begin{array}{ccc}
2i & 2 & -2 \\[4pt]
- & -2 +2i & 2 \\[4pt]
2i & -6i & -4 +4i
\end{array}
\right]
\qquad
\begin{array}{c}
2i \\[4pt]
2i \\[4pt]
-4
\end{array}
$$

Subtract the leading row from the third row, to obtain the successor tableau:

$$
\begin{array}{ccc}
x_1 & x_2 & x_3 \\
\end{array} \quad = 0 \qquad \Sigma
$$

$$
\left[
\begin{array}{ccc}
2i & 2 & -2 \\[4pt]
- & -2 +2i & 2 \\[4pt]
- & -2 -6i & -2 +4i
\end{array}
\right]
\qquad
\begin{array}{c}
2i \\[4pt]
2i \\[4pt]
-4 -2i
\end{array}
$$

Now multiply rows 1 and 3 by $(-2 +2i)$, and obtain the intermediate tableau:

$$
\begin{array}{ccc}
x_1 & x_2 & x_3 \\
\end{array} \quad = 0 \qquad \Sigma
$$

$$
\left[
\begin{array}{ccc}
-4 -4i & -4 +4i & 4 -4i \\[4pt]
- & \underline{|1-2+2i|} & 2 \\[4pt]
- & 16 +8i & -4 -12i
\end{array}
\right]
\qquad
\begin{array}{c}
-4 -4i \\[4pt]
2i \\[4pt]
12 -4i
\end{array}
$$

Subtract 2 times the second row from the leading row, and subtract it $(2 +6i)$ times from the third row, to obtain the successor tableau:

$$
\begin{array}{ccc}
x_1 & x_2 & x_3 \\
\end{array} \quad = 0 \qquad \Sigma
$$

$$
\left[
\begin{array}{ccc}
-4 -4i & - & -4i \\[4pt]
- & -2 +2i & 2 \\[4pt]
- & - & -
\end{array}
\right]
\qquad
\begin{array}{c}
-4 -8i \\[4pt]
2i \\[4pt]
-
\end{array}
$$

Therefore, setting x_3 at the arbitrary value $x_3 = 1$:

$(-4\ -4i)x_1\quad -4i\ x_3\ = 0\qquad \rightarrow\qquad (-4\ -4i)x_1\ =\ 4i\ x_3\ =\ 4i$

Therefore:

$x_1\ =\ 4i/(-4\ -4i)\ =\ i/(-1\ -i)\ =\ 1/(i\ -\ 1)\ =\ -0.5\ -0.5\ i$

Similarly, for x_2:

$(-2\ +\ 2i)\ x_2\ +\ 2\ x_3\ = 0\qquad \rightarrow\qquad (-2\ +\ 2i)\ x_3\ =\ -\ 2x_3\ =\ -2$

Therefore: $x_2\ =\ -2/(-2\ +\ 2i)\ =\ 1/(1\ -\ i)\ =\ (1\ +\ i)/2$

Check:

$$\begin{bmatrix} 2 & 2 & -2 \\ - & - & 2 \\ 1 & -3 & 4 \end{bmatrix} \begin{bmatrix} -0.5 & -0.5i \\ 0.5 & +0.5i \\ 1 & \end{bmatrix} = \begin{bmatrix} -2 \\ 2 \\ 2 -2i \end{bmatrix} = \begin{bmatrix} -0.5 & -0.5i \\ 0.5 & +0.5i \\ 1 & \end{bmatrix} (2\ -2i)$$

Theorem:

If

$x_1\ =\ \mathbf{r} + \mathbf{c}i$ is a characteristic vector of the n by n matrix $\mathbf{A}$ (with $\lambda_1\ =\ \alpha + \beta i$ as root),

then

$x_2\ =\ \mathbf{r} - \mathbf{c}i$ is also a vector of $\mathbf{A}$, (with $\lambda_2\ =\ \alpha - \beta i$ as root).

Proof:

By assumption:

$\mathbf{Ax}_1\ =\ \mathbf{A}[\mathbf{r} +\mathbf{c}\ i]$

$\qquad =\ [\mathbf{r} + \mathbf{c}\ i](\alpha + \beta\ i)\ =\ \mathbf{r}\ \alpha\ -\mathbf{c}\ \beta\ +\ [\mathbf{r}\ \beta\ +\mathbf{c}\ \alpha]i\qquad (7.5.1)$

Therefore, taking real and irrational parts separately:

$$\mathbf{A}\,\mathbf{r} = \mathbf{r}\,\alpha - \mathbf{c}\,\beta \qquad\qquad (7.5.2)$$

and

$$\mathbf{A}\,\mathbf{c}\,i = [\mathbf{r}\,\beta + \mathbf{c}\,\alpha]\,i \qquad\qquad (7.5.3)$$

and by implication:

$$\mathbf{A}\,\mathbf{c} = \mathbf{r}\,\beta + \mathbf{c}\,\alpha \qquad\qquad (7.5.4)$$

On taking the difference rather than the sum of (7.5.2) and (7.5.3), we obtain:

$$\mathbf{A}\,[\mathbf{r} - \mathbf{c}\,i] = \mathbf{r}\,\alpha - \mathbf{c}\,\beta - [\mathbf{r}\,\beta + \mathbf{c}\,\alpha]\,i \qquad (7.5.5)$$

and therefore:

$$\mathbf{A}\mathbf{x}_2 = \mathbf{A}[\mathbf{r} - \mathbf{c}i]$$

$$= [\mathbf{r} - \mathbf{c}i](\alpha - \beta i) = \mathbf{r}\alpha - \mathbf{c}\beta - [\mathbf{r}\beta + \mathbf{c}\alpha]i \qquad (7.5.6)$$

q.e.d.

We shall refer to this property (which appears to apply to roots as well as vectors) as the <u>complementary pair property</u>.

Example:

$$\begin{bmatrix} 2 & 2 & -2 \\ - & - & 2 \\ 1 & -3 & 4 \end{bmatrix} \begin{bmatrix} -0.5 & -0.5i \\ 0.5 & +0.5i \\ 1 & \end{bmatrix} = \begin{bmatrix} -2 \\ 2 \\ 2 - 2i \end{bmatrix} = \begin{bmatrix} -0.5 & -0.5i \\ 0.5 & +0.5i \\ 1 & \end{bmatrix} (2 - 2i)$$

was established earlier in this section.

$$\begin{bmatrix} 2 & 2 & -2 \\ - & - & 2 \\ 1 & -3 & 4 \end{bmatrix} \begin{bmatrix} -0.5 & +0.5i \\ 0.5 & -0.5i \\ 1 & \end{bmatrix} = \begin{bmatrix} -2 \\ 2 \\ 2 + 2i \end{bmatrix} = \begin{bmatrix} -0.5 & +0.5i \\ 0.5 & -0.5i \\ 1 & \end{bmatrix} (2 + 2i)$$

is therefore implied.

Exercise:

For **A** = $\begin{bmatrix} 1 & 1 \\ -1 & 1 \end{bmatrix}$

calculate the (complex) roots, as well as the corresponding column vectors.

(There is an answersheet at the end of the chapter.)

7.6 Distinct roots and orthogonality

The 1 by n row-vector **w'** and the n by 1 column vector **x** are <u>orthogonal</u> to each other, if they obey the condition:

$$\mathbf{w'}\,\mathbf{x} = 0 \qquad\qquad (7.6.1)$$

Theorem:

$$\left.\begin{array}{l} \mathbf{A}\,\mathbf{x} = \mathbf{x}\,\lambda_1 \\[2mm] \mathbf{w'}\,\mathbf{A} = \lambda_2\mathbf{w'} \end{array}\right\} \quad \lambda_1 \neq \lambda_2 \qquad (7.6.2)$$

implies (7.6.1)

Proof:

Form the expression:

$$\begin{aligned} \mathbf{w'A}\,\mathbf{x} &= [\mathbf{w'A}]\,\mathbf{x} = \mathbf{w'}[\mathbf{A}\,\mathbf{x}] \\[2mm] &= \lambda_2\,\mathbf{w'}\,\mathbf{x} = \mathbf{w'}\,\mathbf{x}\,\lambda_1 \qquad (7.6.3) \end{aligned}$$

Therefore either **w'x** = 0, or $\lambda_1 = \lambda_2$. Since $\lambda_1 \neq \lambda_2$ was assumed, **w'x** = 0 follow.

q.e.d.

Theorem:

If:

X_1 is an n by k matrix of column vectors of **A**, obeying

$$A X_1 = X_1 \hat{L}_1 \qquad (7.6.4)$$

where $\hat{L}$ is a k by k diagonal matrix (therefore all its diagonal elements λ_j are roots of **A**, x_j being the corresponding column vectors), obeying the condition:

$$\lambda_j \neq \lambda_i \text{ for all } i \neq j$$

while

$$A x_{k+1} = x_{k+1} \lambda_{k+1} \qquad (7.6.5)$$

$$\lambda_{k+1} \neq \lambda_j \ (j = 1, 2, \ldots..k)$$

holds for another k+1th root as well.

then:

x_k cannot be expressed as a combination of the columns of X_1.

Proof:

We introduce the provisional assumption, that the theorem is true for k* = k-1. Therefore, by assumption, X_1 is of full rank k.

Now let us assume, by contra-assumption to what the theorem states:

$$X_1 w = x_{k+1} \qquad (7.6.7)$$

If that were so,

$$A X_1 w = X_1 \hat{L} w = A x_{k+1} = x_{k+1} \lambda_{k+1} \qquad (7.6.8)$$

would also be true.

This would show that:

$$\mathbf{X}^* = [\mathbf{X}_1 \mid -\mathbf{x}_{k+1}] \tag{7.6.9}$$

was of rank k-1 or less, as both $\begin{bmatrix} \mathbf{w} \\ \hline 1 \end{bmatrix}$ and $\begin{bmatrix} \hat{\mathbf{L}}\mathbf{w} \\ \hline \lambda_{k+1} \end{bmatrix}$

could express the null vector as a combination of its columns.

This would contradict the assumed full rank of $\mathbf{X}_1$, and the provisional assumption is shown to imply the theorem. Since the theorem is true by not having content for k=1, the general proof follows by recursive induction.

q.e.d.

Corollary:

If all roots are distinct, the (column) vectors permit the formation of an invertible matrix of characteristic vectors.

7.7 Repeated roots and vanishing minors

A sufficient condition for λ to be a repeated root of $\mathbf{A}$, is the vanishment of $|\mathbf{I}\lambda-\mathbf{A}|$ and the <u>sum</u> of its first-order principal minors. This is clear from section 7.2, where we saw that these two expressions are, respectively the constant and the coefficient of the first power of γ, if we treat γ as a root of $|\mathbf{I}\lambda-\mathbf{A}|$, given any λ.

There are two types of matrices, for which that condition only arises, if the stronger condition of the separate vanishment of <u>all</u> first-order principal minors of $|\mathbf{I}\lambda-\mathbf{A}|$ is also satisfied.

If $\mathbf{A}$ is a <u>triangular</u> matrix, the roots are its diagonal elements, and the first-order principal minors are the determinants of n-1 by n-1 diagonal blocks of $\mathbf{A}$, which are themselves triangular.

If two or more of these diagonal elements of $\mathbf{A}$ are $\lambda=\lambda_1=\lambda_2$, all n-1 by n-1 diagonal blocks of $[I\lambda-A]$ contain at least one zero and are singular.

If **A** is <u>symmetric</u>, a similar result applies for quite different reasons. As this is a slightly more complicated case, we formally state it as a theorem, and provide a proof.

Theorem:

If λ is a repeated root of the n by n symmetric matrix **A**,

then all principal minors of $|I\lambda-A|$ vanish.

Proof:

Assume by contra-assumption, the existence of an invertible n-1 by n-1 block $[I\lambda-A_{11}]$. We then express $|I\lambda-A|$, conform sections 5.6 and 5.7 as:

$$|I\lambda-A| =$$

$$|I\lambda-A_{11}| \; (\lambda - a_{nn} - a'_{n1} \, [I\lambda-A_{11}]^{-1} \, a_{1n}) \qquad (7.7.1)$$

From which:

$$d|I\lambda-A|/d\lambda =$$

$$\eth|I\lambda-A_{11}|/\eth\lambda \; (\lambda - a_{nn} - a'_{n1} \, [I\lambda-A_{11}]^{-1} \, a_{1n}) \; +$$

$$|I\lambda-A_{11}| \; \eth(\lambda - a_{nn} - a'_{n1} \, [I\lambda-A_{11}]^{-1} \, a_{1n})/\eth\lambda \qquad (7.7.2)$$

We note from (7.7.1) that, if $[I\lambda-A_{11}]$ is indeed invertible, then $|I\lambda-A| = 0$ requires:

$$\lambda - a_{nn} - a'_{n1} \, [I\lambda-A_{11}]^{-1} \, a_{1n} = 0 \qquad (7.7.3)$$

This consideration implies the vanishment of the first term on the righthand side of (7.7.2), and a necessary condition for the vanishment of $d|I\lambda-A|/d\lambda$ is seen to be:

$$\partial(\lambda - a_{nn} - \mathbf{a'}_{n1} [\mathbf{I}\lambda - \mathbf{A}_{11}]^{-1} \mathbf{a}_{1n})/\partial\lambda = 0 \qquad (7.7.4)$$

Conform section 4.8 we are able to differentiate $[\mathbf{I}\lambda - \mathbf{A}_{11}]$ relative to λ as:

$$\partial[\mathbf{I}\lambda - \mathbf{A}_{11}]^{-1}/\partial\lambda = - [\mathbf{I}\lambda - \mathbf{A}_{11}]^{-1} [\mathbf{I}\lambda - \mathbf{A}_{11}]^{-1} \qquad (7.7.5)$$

Therefore:

$$\partial(\lambda - a_{nn} - \mathbf{a'}_{n1} [\mathbf{I}\lambda - \mathbf{A}_{11}]^{-1} \mathbf{a}_{1n})/\partial\lambda =$$

$$1 + \mathbf{a'}_{n1} [\mathbf{I}\lambda - \mathbf{A}_{11}]^{-1} [\mathbf{I}\lambda - \mathbf{A}_{11}]^{-1} \mathbf{a}_{1n} \qquad (7.7.6)$$

For a <u>symmetric</u> matrix, the last term on the righthand side of (7.7.6) is a quadratic form and positive definite.

Therefore:

$$\partial(\lambda - a_{nn} - \mathbf{a'}_{n1} [\mathbf{I}\lambda - \mathbf{A}_{11}]^{-1} \mathbf{a}_{1n})/\partial\lambda \geq 1 \qquad (7.7.7)$$

(where the borderline case of = 1 will only apply if $\mathbf{A}$ is block-diagonal by way of $\mathbf{a}_{1n} = [0]$.)

and (7.7.4) is contradicted.

We must therefore assume instead, that no n-1 by n-1 invertible block $[\mathbf{I}\lambda - \mathbf{A}_{11}]$ exists.

q.e.d.

We now come to discuss multiple repeated roots.

Here the general necessary condition is, that for a k-fold repeated root all <u>sums</u> of determinants of the <u>groups</u> of minor-matrices of orders n-1 by n-1, n-2 by n-2, ... down to n-k by n-k of $[\mathbf{I}\lambda - \mathbf{A}]$ are zero.

For the triangular matrix, it is immediately obvious that, if λ is a k-fold repeated root, then $[\mathbf{I}\lambda - \mathbf{A}]$ contains k zero diagonal elements, and as a result all separate minor-matrices of orders n-1 by n-1 down to n-k by n-k will have a zero determinant.

The similar generalization does in fact also apply for the symmetric matrix, but a more sophisticated analysis is required

to prove it: we need to invoke recursive induction. We proved,
(for the triangular matrix as well as for the symmetric matrix),
that $[I\lambda-A]$ has a repeated zero root, only if each n-1 by n-1
block $[I\lambda-A_{11}]$ has a zero root. For λ to be not just a twice-
repeated root, but a thrice or more repeated root, it is
necessary that a differential variation in λ does not breach the
condition $|I\lambda-A_{11}| = 0$, in other words, that λ is a repeated root
of (each separate) n-1 by n-1 diagonal block $[I\lambda-A_{11}]$.

Therefore any theorems concerning repeated roots also apply to
these minor-matrices. It follows that the second-order principal
minors of $|I\lambda-A|$, which are all first-order principal minor
matrices of an n-1 by n-1 diagonal block $[I\lambda-A_{11}]$, all need to
vanish separately, if λ is a thrice or more repeated root of A.
Clearly this argument applies recursively and the stricter
condition that all k^{th} order principal minors of $[I\lambda-A]$ need to
vanish separately is proved both for the triangular matrix and
for the symmetric matrix.

Note however, that this theorem does <u>not</u> apply for <u>all</u> types of
matrices.

Example:

$$A = \begin{bmatrix} 1 & -1 \\ 1 & -1 \end{bmatrix}$$

has the characteristic equation:

$$|I\lambda-A| = (\lambda-1)(\lambda+1) + 1 = \lambda^2 - 1 + 1 = \lambda^2 = 0.$$

Answersheet on the characteristic equation

The development of $|A - I\lambda|$ by the definition of a determinant is:

$$\begin{vmatrix} 1-\lambda & 2 & 3 \\ 3 & 2-\lambda & 1 \\ 1 & 4 & 1-\lambda \end{vmatrix} =$$

$$(1-\lambda) \begin{vmatrix} 2-\lambda & 1 \\ 4 & 1-\lambda \end{vmatrix} - 2 \begin{vmatrix} 3 & 1 \\ 1 & 1-\lambda \end{vmatrix} + 3 \begin{vmatrix} 3 & 2-\lambda \\ 1 & 4 \end{vmatrix} =$$

$$(1-\lambda) \left((2-\lambda)(1-\lambda) - 4 \right) - 2 \left(3(1-\lambda) - 1 \right) + 3 \left(12 - (2-\lambda) \right) =$$

$$(1-\lambda) \left(2 - 3\lambda + \lambda^2 - 4 \right) - 2 \left(3 - 3\lambda - 1 \right) + 3 \left(12 - 2 + \lambda \right) =$$

$$(1-\lambda) \left(-2 - 3\lambda + \lambda^2 \right) \qquad - 2 \left(2 - 3\lambda \right) \qquad + 3 \left(10 + \lambda \right) =$$

$$-2 - 3\lambda + \lambda^2 + 2\lambda + 3\lambda^2 - \lambda^3 \qquad -4 + 6\lambda \qquad +30 + 3\lambda \quad =$$

$$24 + 8\lambda + 4\lambda^2 - \lambda^3$$

The conventional formulation of the characteristic equation as a polynomial equation therefore is:

$$\lambda^3 - 4\lambda^2 - 8\lambda - 24 = 0$$

Check on the known root $\lambda = 6$:

$$6^3 - 4 \times 6^2 - 8 \times 6 - 24 = 216 - 4 \times 36 - 8 \times 6 - 24 =$$

$$216 - 144 - 48 - 24 = 0$$

The corresponding development by the principal minors is:

$$c_0 = |-A| = \begin{vmatrix} -1 & -2 & -3 \\ -3 & -2 & -1 \\ -1 & -4 & -1 \end{vmatrix} =$$

$$-1 \begin{vmatrix} -2 & -1 \\ -4 & -1 \end{vmatrix} \quad + 2 \begin{vmatrix} -3 & -1 \\ -1 & -1 \end{vmatrix} \quad - 3 \begin{vmatrix} -3 & -2 \\ -1 & -4 \end{vmatrix} =$$

$$-1 \times (2 - 4) \quad\quad + 2 (3 - 1) \quad\quad - 3 (12 - 2) \quad =$$

$$-1 \times -2 \quad\quad\quad + 2 \times 2 \quad\quad\quad - 3 \times 10 \quad\quad =$$

$$2 \quad + \quad 4 \quad - \quad 30 \quad = \quad -24$$

$$c_1 = \quad |A_{11}| \quad\quad\quad + |A_{22}| \quad\quad\quad + |A_{33}| \quad =$$

$$\begin{vmatrix} -2 & -1 \\ -4 & -1 \end{vmatrix} \quad + \quad \begin{vmatrix} -1 & -3 \\ -1 & -1 \end{vmatrix} \quad + \quad \begin{vmatrix} -1 & -2 \\ -3 & -2 \end{vmatrix} \quad =$$

$$(2 - 4) \quad\quad + \quad (1 - 3) \quad\quad + \quad (2 - 6) \quad =$$

$$-2 \quad\quad\quad - 2 \quad\quad\quad - 4 \quad = -8$$

$$c_2 = - a_{11} - a_{22} - a_{33} = -1 - 2 - 1 = -4$$

confirming $\lambda^3 - 8\lambda^2 - 4\lambda - 24 = 0$

Answersheet on the characteristic equation, roots and vectors

of:

$$\mathbf{A} = \begin{bmatrix} 2 & 4 & 4 & 4 & - \\ 1 & 2 & 2 & 2 & - \\ 1 & 2 & 2 & 2 & - \\ 1 & 2 & 2 & 2 & - \\ - & - & - & - & - \end{bmatrix}$$

The easiest way of developing the characteristic equation of this matrix, is to use the minors form of the characteristic equation. A and all the minor-matrices of greater order than 1 by 1 are singular.

As this form of the characteristic equation relates to sums of minors of $-\mathbf{A}$, we find:

$$\lambda^5 - 8\,\lambda^4 = (\lambda - 8)\,\lambda^4 = 0.$$

Therefore:

$$\lambda_1 = 8; \qquad \lambda_2 = \lambda_3 = \lambda_4 = \lambda_5 = 0.$$

Calculation of vectors (using an adaptation of the all integer elimination method):

$\lambda = 8$

	x_1	x_2	x_3	x_4	x_5	= 0	Σ
	|-6|	4	4	4	–		6
	1	-6	2	2	–		-1
	1	2	-6	2	–		-1
	1	2	2	-6	–		-1
	–	–	–	–	-8		-8

Multiply rows 2, 3, and 4 by -6, and subtract the leading row from them:

	x_1	x_2	x_3	x_4	x_5	= 0	Σ
	-6	4	4	4	–		6
	–	32	-16	-16	–		–
	–	-16	32	-16	–		–
	–	-16	-16	32	–		–
	–	–	–	–	-8		-8

Divide by the (as far as the elimination procedure is concerned, incidental) common factor 2:

x_1	x_2	x_3	x_4	x_5	$= 0$	Σ
-3	2	2	2	-		3
-	16	-8	-8	-		-
-	-8	16	-8	-		-
-	-8	-8	16	-		-
-	-	-	-	-4		-4

Multiply row 1 by 8, and rows 3 and 4 by -2:

x_1	x_2	x_3	x_4	x_5	$= 0$	Σ
-24	16	16	16	-		24
-	|16|	-8	-8	-		-
-	16	-32	16	-		-
-	16	16	-32	-		-
-	-	-	-	-4		-4

Now subtract row 2 from rows 1, 3, and 4:

x_1	x_2	x_3	x_4	x_5	$= 0$	Σ
-24	-	24	24	-		24
-	16	-8	-8	-		-
-	-	-24	24	-		-
-	-	24	-24	-		-
-	-	-	-	-4		-4

Divide rows 1 to 4 by the (systematic) common factor 8, to obtain:

$$
\begin{array}{ccccc}
x_1 & x_2 & x_3 & x_4 & x_5 \\
\end{array}
$$

	x_1	x_2	x_3	x_4	x_5	$= 0$	Σ
	-3	$-$	3	3	$-$		3
	$-$	2	-1	-1	$-$		$-$
	$-$	$-$	-3	3	$-$		$-$
	$-$	$-$	3	-3	$-$		$-$
	$-$	$-$	$-$	$-$	-4		-4

Multiply row 2 by 3, and rows 1, 3, and 4 by -1, to obtain:

	x_1	x_2	x_3	x_4	x_5	$= 0$	Σ
	3	$-$	-3	-3	$-$		-3
	$-$	6	-3	-3	$-$		$-$
	$-$	$-$	$\lvert\underline{3}\rvert$	-3	$-$		$-$
	$-$	$-$	-3	3	$-$		$-$
	$-$	$-$	$-$	$-$	-4		-4

Now subtract row 3 from rows 1, 2, and 4, to obtain:

	x_1	x_2	x_3	x_4	x_5	$= 0$	Σ
	3	$-$	$-$	-6	$-$		-3
	$-$	6	$-$	-6	$-$		$-$
	$-$	$-$	3	-3	$-$		$-$
	$-$	$-$	$-$	$-$	$-$		$-$
	$-$	$-$	$-$	$-$	-4		-4

The last, fifth equation implies (as it did right from the beginning -we might as well have left it out-): $x_5 = 0$.

Setting x_4 at $x_4 = 1$, and dividing once more all coefficients by 3, resp. 6, we infer:

$$3x_1 - 6x_4 = 0 \quad \rightarrow \quad x_1 - 2x_4 = 0 \quad \rightarrow \quad x_1 = 2x_4 = 2.$$

Similarly, for x_2:

$$6x_2 - 6x_4 = 0 \quad \rightarrow \quad x_1 - x_4 = 0 \quad \rightarrow \quad x_1 = x_4 = 1.$$

and for x_3:

$$3x_3 - 3x_4 = 0 \quad \rightarrow \quad x_3 - x_4 = 0 \quad \rightarrow \quad x_3 = x_4 = 1.$$

Therefore: $\quad \mathbf{x} = \begin{bmatrix} 2 \\ 1 \\ 1 \\ 1 \end{bmatrix}$

Now for $\lambda = 0$

x_1	x_2	x_3	x_4	x_5	$= 0$	Σ
$|\underline{2}|$	4	4	4	–		14
1	2	2	2	–		7
1	2	2	2	–		7
1	2	2	2	–		7
–	–	–	–	–		–

Multiply rows 2, 3, and 4 by 2, and subtract the leading row from the thus multiplied rows, to obtain:

x_1	x_2	x_3	x_4	x_5	$= 0$	Σ
2	4	4	4	–		14
–	–	–	–	–		–
–	–	–	–	–		–
–	–	–	–	–		–
–	–	–	–	–		–

For $x_5 = 1$, we find $x_1 = x_2 = x_3 = x_4 = 0$.

Otherwise, we can set at choice x_2, x_3, or x_4 at 1, and find $x_1 = -2$, the other variables being zero-valued.

Summary: $\lambda=8$ $\lambda=0$ $\lambda=0$ $\lambda=0$ $\lambda=0$

$$\mathbf{X} = \begin{bmatrix} 2 & -2 & -2 & -2 & - \\ 1 & 1 & - & - & - \\ 1 & - & 1 & - & - \\ 1 & - & - & 1 & - \\ - & - & - & - & 1 \end{bmatrix}$$

Answersheet on complex vectors:

$$\mathbf{A} = \begin{bmatrix} 1 & 1 \\ -1 & 1 \end{bmatrix}$$

Characteristic equation: $\begin{vmatrix} 1-\lambda & 1 \\ -1 & 1-\lambda \end{vmatrix}$

$= (1-\lambda)^2 + 1 = 0 \qquad \rightarrow \qquad (1-\lambda)^2 = (\lambda-1)^2 = -1 \quad \rightarrow$

$\lambda - 1 = \pm\sqrt{-1} = \pm i \quad \rightarrow \quad \lambda = 1 \pm i$

For this two by two example, elimination by row-operations is actually superfluous, but the set-up tableau is given here for the sake of completeness, for $\lambda = 1+i$:

$x_1 \qquad x_2 \qquad = 0$

$$\begin{bmatrix} -i & 1 \\ -1 & -i \end{bmatrix} \qquad \begin{array}{l} (1-\lambda)x_1 \quad + x_2 = -i\,x_1 \quad + x_2 = 0 \\ -x_1 \quad +(1-\lambda)x_2 = -x_1 \quad -ix_2 = 0 \end{array}$$

To avoid complex division, we impose the arbitrary requirement $x_2 = 1$, and use the second equation to calculate x_1:

$-x_1 \quad -ix_2 = 0 \qquad \rightarrow \qquad -x_1 = ix_2 = i \qquad \rightarrow \qquad x_1 = -i$

Check on the first equation:

$$-i\ x_1 + x_2 = -i \times (-i) + 1 = i^2 + 1 = -1 + 1 = 0$$

The other vector is now found by the complementary pair property.

We now verify the correctness of the calculation of both vectors:

$\lambda = 1+i$:

$$\begin{bmatrix} 1 & 1 \\ -1 & 1 \end{bmatrix} \begin{bmatrix} -i \\ 1 \end{bmatrix} = \begin{bmatrix} -i + 1 \\ i + 1 \end{bmatrix} = \begin{bmatrix} -i \\ 1 \end{bmatrix} \times (1 + i)$$

$\lambda = 1-i$:

$$\begin{bmatrix} 1 & 1 \\ -1 & 1 \end{bmatrix} \begin{bmatrix} i \\ 1 \end{bmatrix} = \begin{bmatrix} i + 1 \\ -i + 1 \end{bmatrix} = \begin{bmatrix} i \\ 1 \end{bmatrix} \times (1 - i)$$

CHAPTER VIII

MORE ABOUT ROOTS AND VECTORS

8.1 Transformation and similarity

Theorem

If **x** is a column vector of the n by n matrix **A**, and λ is the corresponding latent root, and **T** is an n by n invertible matrix,

then:

$$\mathbf{B} = \mathbf{T}\,\mathbf{A}\,\mathbf{T}^{-1} \tag{8.1.1}$$

has

$$\mathbf{y} = \mathbf{T}\,\mathbf{x} \tag{8.1.2}$$

as vector, and λ as root.

Proof:

Pre-multiply:

$$\mathbf{A}\,\mathbf{x} = \mathbf{x}\,\lambda \tag{8.1.3}$$

by **T**, and substitute $\mathbf{T}^{-1}\,\mathbf{y}$ for **x**, conform (8.1.2), and obtain:

$$T\,A\,x \;=\; T\,x\,\lambda \;=\; T\,A\,T^{-1}\,y \;=\; y\,\lambda \qquad\qquad (8.1.4)$$

q.e.d.

Example:

$$A = \begin{bmatrix} 1 & 1 \\ 1 & 1 \end{bmatrix} \qquad T = \begin{bmatrix} 0.5 & - \\ - & -1 \end{bmatrix} \qquad B = \begin{bmatrix} 1 & -0.5 \\ -2 & 1 \end{bmatrix}$$

$$\begin{bmatrix} 1 & 1 \\ 1 & 1 \end{bmatrix} \begin{bmatrix} -1 \\ 1 \end{bmatrix} = \begin{bmatrix} - \\ - \end{bmatrix} \qquad \begin{bmatrix} 1 & -0.5 \\ -2 & 1 \end{bmatrix} \begin{bmatrix} -0.5 \\ -1 \end{bmatrix} = \begin{bmatrix} - \\ - \end{bmatrix}$$

An important special case of this theorem relates to the simultaneous reordering of the rows and the columns of **A**: permutation operators are invertible matrices, with the special property that the inverse of a permutation operator is its transpose.

Example (indicating the transformed matrix as **A***):

$$A = \begin{bmatrix} 4 & 5 \\ 1 & - \end{bmatrix} \; ; \qquad \begin{bmatrix} 4 & 5 \\ 1 & - \end{bmatrix} \begin{bmatrix} 5 \\ 1 \end{bmatrix} = \begin{bmatrix} 25 \\ 5 \end{bmatrix} = \begin{bmatrix} 5 \\ 1 \end{bmatrix} \times 5$$

$$A^* = \begin{bmatrix} - & 1 \\ 1 & - \end{bmatrix} \begin{bmatrix} 4 & 5 \\ 1 & - \end{bmatrix} \begin{bmatrix} - & 1 \\ 1 & - \end{bmatrix} = \begin{bmatrix} - & 1 \\ 5 & 4 \end{bmatrix}$$

$$\begin{bmatrix} - & 1 \\ 5 & 4 \end{bmatrix} \begin{bmatrix} 1 \\ 5 \end{bmatrix} = \begin{bmatrix} 5 \\ 25 \end{bmatrix} = \begin{bmatrix} 1 \\ 5 \end{bmatrix} \times 5$$

The concept of transformation can be generalized, from vectors to square matrices.

If the n by n matrices are related by:

$$B \;=\; X\,A\,X^{-1} \qquad\qquad (8.1.5)$$

(and therefore also $A = X^{-1}$),

then **A** and **B** are said to be <u>similar</u> to each other. and **X** is said to be the <u>transformation operator</u>.

Note, that (8.1.5) implies:

$$\mathbf{X}\,[\mathbf{A} - \mathbf{I}\lambda]\,\mathbf{X}^{-1} = [\mathbf{X}\,\mathbf{A} - \mathbf{X}\,\lambda]\,\mathbf{X}^{-1}$$

$$= \mathbf{X}\,\mathbf{A}\,\mathbf{X}^{-1} - \mathbf{X}\,\lambda\,\mathbf{X}^{-1} = \mathbf{B} - \mathbf{I}\lambda \qquad (8.1.6)$$

or in words:

If $\mathbf{B}$ is similar to $\mathbf{A}$, then $[\mathbf{B} - \mathbf{I}\lambda]$ is similar to $[\mathbf{A} - \mathbf{I}\lambda]$.

This will apply, irrespective of whether λ actually is a root of $\mathbf{A}$ or $\mathbf{B}$.

Note also, that if any power of $\mathbf{B}$ is developed, we obtain:

$$\mathbf{B}^p = \mathbf{B}\,\mathbf{B}^{p-1} = \mathbf{X}\,\mathbf{A}\,\mathbf{X}^{-1}\,\mathbf{B}^{p-1} \qquad (8.1.7)$$

From which, by recursion:

$$\mathbf{B}^p = \mathbf{X}\,\mathbf{A}^p\,\mathbf{X}^{-1} \qquad (8.1.8)$$

i.e. if $\mathbf{A}$ and $\mathbf{B}$ are similar to each other, so are all their powers, and the similarity is revealed by the same transformation operator.

8.2 Diagonalization

We now collect the characteristic column vectors in a matrix, to which we refer as $\mathbf{X}$.

Post-multiplication of $\mathbf{A}$ by $\mathbf{X}$ now leads to:

$$\mathbf{A}\,\mathbf{X} = \mathbf{X}\,\hat{\mathbf{L}} \qquad (8.2.1)$$

Here $\hat{\mathbf{L}}$ is a diagonal matrix, each l_{jj} is a root of $\mathbf{A}$, as corresponding to the column-vector $\mathbf{x}_j$.

Depending on the structure and numerical content of $\mathbf{A}$, we may, or we may not, be able to assemble an invertible matrix $\mathbf{X}$.

If all roots are distinct, then (as we saw in the previous section), all we need to do is to assemble $\mathbf{X}$ as containing a column $\mathbf{x}_j$ in association with each of the n distinct roots λ_j.

In that case, post-multiplication of (8.2.1) by $\mathbf{X}^{-1}$ yields:

$$\mathbf{A} = \mathbf{X}\,\hat{\mathbf{L}}\,\mathbf{X}^{-1} \tag{8.2.2}$$

From which, on pre-multiplication by $\mathbf{X}^{-1}$:

$$\mathbf{X}^{-1}\,\mathbf{A} = \hat{\mathbf{L}}\,\mathbf{X}^{-1} \tag{8.2.3}$$

i.e. $\mathbf{X}^{-1}$ is a matrix of row-vectors.

If (8.2.1) is pre-multiplied by $\mathbf{X}^{-1}$, instead of post-multiplied, we obtain:

$$\mathbf{X}^{-1}\,\mathbf{A}\,\mathbf{X} = \hat{\mathbf{L}} \tag{8.2.4}$$

and we say that $\mathbf{A}$ has been <u>diagonalized</u>.

Or, using the term defined in the previous section, we can say that $\mathbf{A}$ is similar to the diagonal matrix $\hat{\mathbf{L}}$.

Exercise:

For $\mathbf{A} = \begin{bmatrix} 2 & 8 & - \\ 2 & 2 & - \\ 1 & 1 & 1 \end{bmatrix}$

you are given the information that an invertible matrix $\mathbf{X}$ of three characteristic column vectors exists.

You are asked:

(1) Find the roots of $\mathbf{A}$.

(2) Find a three by 3 invertible matrix of column vectors.

(3) Hence express **A X** and **A** as:

$$\mathbf{A\,X} = \mathbf{X\,\hat{L}} = [\mathbf{x}_1, \mathbf{x}_2, \mathbf{x}_3] \begin{bmatrix} \lambda_1 & - & - \\ - & \lambda_2 & - \\ - & - & \lambda_3 \end{bmatrix}$$

and therefore:

$$\mathbf{A\,X\,X}^{-1} = \mathbf{X\,\hat{L}\,X}^{-1} = [\mathbf{x}_1, \mathbf{x}_2, \mathbf{x}_3] \begin{bmatrix} \lambda_1 & - & - \\ - & \lambda_2 & - \\ - & - & \lambda_3 \end{bmatrix} \mathbf{X}^{-1}$$

$$= \mathbf{A}.$$

8.3 **Triangularization**

For a general (non-symmetric) matrix, there is no theorem which states that a transformation conform (8.2.4) always exists.

Some examples of matrices which cannot be diagonalized are:

$$\mathbf{A} = \begin{bmatrix} 1 & 1 \\ - & 1 \end{bmatrix} ; \qquad\qquad \mathbf{B} = \begin{bmatrix} 2 & -1 \\ 4 & -2 \end{bmatrix}$$

The characteristic equation of **A** is $(1-\lambda)^2 = 0$, and resolves as a repeated root of $\lambda_1 = \lambda_2 = 1$.

The characteristic equation of **B** is:

$$(2-\lambda)(-2-\lambda) + 4 = \lambda^2 - 4 + 4 = \lambda^2 = 0,$$

and resolves with a repeated root $\lambda_1 = \lambda_2 = 0$.

Yet $[\mathbf{A}-\mathbf{I}\lambda]$, resp. $[\mathbf{B}-\mathbf{I}\lambda]$, are nevertheless both of rank 1, and there is only one independent column vector in both cases.

The apparently suggestive proposition that **A** can always be diagonalized, is untrue, at least for some matrices.

A weaker theorem, which <u>is</u> generally valid, is as follows:

Theorem: (Weak triangularization theorem or Schur's theorem)

For every n by n matrix **A**, there is a (at least one) pair of n by n matrices **X**, and **U**, which permit us to express **A** as:

$$\mathbf{A} = \mathbf{X}\,\mathbf{U}\,\mathbf{X}^{-1} \qquad\qquad (8.3.1)$$

where **U** is upper triangular.

This theorem is known in the literature (for example Pullman [37], p. 204) as Schur's theorem, and will here also be referred to as the triangularization theorem, and because we shall be stating a stronger version of it, the above theorem, as generally referred to as Schur's theorem, is the weak triangularization theorem.

We now first of all illustrate Schur's Theorem, in its weaker form.

$$\mathbf{A} = \mathbf{U} = \begin{bmatrix} 1 & - & 2 & 3 \\ - & 1 & - & 4 \\ - & - & 1 & - \\ - & - & - & 1 \end{bmatrix}$$

already is in the form required, and the theorem states that, regardless of repeated roots, there always is a transformation-operator, which brings **A** in that form. The diagonal elements are all roots of **A**, and there may, or may not be, one or more equal diagonal elements, and hence repeated roots. In this example, there is a 4-fold repeated root $\lambda = 1$.

This matrix has two column vectors, e_1 and e_2, and two row-vectors, e'_3 and e'_4.

As will be shown below, this is sufficient information to know that there is a further transformation, which turns **U** into a blockdiagonal matrix, with two separate triangular diagonal blocks.

Theorem (Triangularization theorem):

For every n by n matrix **A**, there is a (at least one) pair of n by n matrices **X** and **U**, which permit us to express **A** as:

$$\mathbf{A} = \mathbf{X} \mathbf{U} \mathbf{X}^{-1} \tag{8.3.1}$$

where **U** meets the following conditions:

- (1) **U** is block-diagonal, and each of the k n_p by n_p diagonal blocks $\mathbf{U}_{pp}$ of **U** is upper-triangular.

- (2) For any p, all the diagonal elements of $\mathbf{U}_{pp}$ are equal to the same latent root λ_p of **A**.

- (3) $u_{i-1,i} = 1$ or $u_{i-1,i} = 0$
  applies for i = 2, 3, ... n,
  and $u_{ih} = 0$ applies for i < h-1 (and for i > h).

(i.e. the diagonal blocks have unity elements or zeros in the positions immediately above the diagonal, and zeros only further up as well as below the diagonal).

Matrices (blocks) of the structure as indicated are known as <u>Jordan blocks</u>, and we reserve, in the rest of this section the letter **U**, or $\mathbf{U}_p$ (or $\mathbf{U}_{22}$, $\mathbf{U}_{33}$, etc, if a block of a larger matrix has that structure), for a <u>Jordan block</u>, and **B** for a block-diagonal matrix (block), of which the diagonal blocks are Jordan blocks. Despite the use of separate letters to indicate blocks, indices will relate to the whole matrix.

We now proceed to the proof of the triangularization theorem.

Proof:

We consider any vector $\mathbf{x}_j$, satisfying $\mathbf{A} \mathbf{x}_j = \mathbf{x}_j \lambda_j$.

We may (if necessary after simultaneous reordering), assume $x_1 = 1$, and form the transformation-operator:

$$\mathbf{T}_1 = \left[\begin{array}{c|c} 1 & \\ \hline -\mathbf{x}_2 & \mathbf{I}_{n-1} \end{array} \right] \tag{8.3.2}$$

and therefore:

$$T_1^{-1} = \left[\begin{array}{c|c} 1 & \\ \hline x_2 & I_{n-1} \end{array}\right] \qquad (8.3.3)$$

From which:

$$A^*_1 = T_1\, A\, T_1^{-1} = T_1\, [A\, T_1^{-1}] =$$

$$\left[\begin{array}{c|c} 1 & \\ \hline -x_2 & I_{n-1} \end{array}\right] \left[\begin{array}{c|c} \lambda & a'_{12} \\ \hline x_2\lambda & A_{22} \end{array}\right] = \left[\begin{array}{c|c} \lambda & a'_{12} \\ \hline & A^*_{22} \end{array}\right] \qquad (8.3.4)$$

where A^*_{22} is:

$$A^*_{22} = A_{22} - x_2\, a'_{12}\, \lambda \qquad (8.3.5)$$

An assumption that the theorem is true for matrices of order n-1 by n-1, and therefore for A^*_{22} now leads to a proof of Schur's theorem, by recursive induction: We transform A^*_1 again, using a blockdiagonal transformation-operator:

$$T_2 = \left[\begin{array}{c|c} 1 & \\ \hline & T_{22} \end{array}\right] \qquad (8.3.6)$$

where T_{22} is the n-1 by n-1 transformation-operator X, which the theorem, assumed to be true for A^*_{22}, states to exist.

Proofs of the stronger properties are now obtained as follows:

Case 1): $A\, x = x\lambda$, $w'A = \lambda w'$, $w'x \neq 0$
is true for at least one pair of vectors x and w'.

Then we may assume without loss of generality, $w_1 \neq 0$ as well as $x_1 = 1$, and (by suitable scaling of w'), also $w'x = 1$.

After transformation conform (8.3.4) a row-vector of A^*_1 is:

$$w^{*'} = w'T_1^{-1} = [w_1\ ,\ w'_2] \left[\begin{array}{c|c} 1 & \\ \hline x_2 & I \end{array}\right] = [1,\ w'_2] \qquad (8.3.7)$$

(the transformation sets the leading element of $w^{*'}$ at $w^*_1 = 1$.)

We now form a corresponding transformation-operator:

$$\mathbf{T_3} \;=\; \left[\begin{array}{c|c} 1 & \mathbf{w'}_2 \\ \hline & \mathbf{I} \end{array}\right]$$

$$(8.3.8)$$

We now pre-multiply (8.3.4) by $\mathbf{T_3}$ and post-multiply by $\mathbf{T_3}^{-1}$, and we we obtain, on re-naming the transformed matrix as $\mathbf{A^*}_2$:

$$\mathbf{A^*}_2 = \mathbf{T_3}\,\mathbf{T_1}\,\mathbf{A}\,\mathbf{T_1}^{-1}\,\mathbf{T_3}^{-1} \;=\; \left[\begin{array}{c|c} \lambda & \\ \hline & \mathbf{A^*}_{22} \end{array}\right]$$

$$(8.3.9)$$

(The transformation is in effect the transpose of (8.3.4), and sweeps out the leading row, but no further change on $\mathbf{A^*}_{22}$ arises, as the last term in a transposed version of (8.3.5) vanishes on account of $\mathbf{a^*}_{21}$ being a null vector.)

An assumption that the theorem is true for $\mathbf{A^*}_{22}$ then leads to a proof by recursive induction.

q.e.d. for case 1)

Case 2):
λ is a distinct root of the characteristic equation of $\mathbf{A}$.

For a distinct root $\Sigma\,|A_{jj}-I\lambda| \neq 0$, and hence $|A_{jj}-I\lambda| \neq 0$ is required for at least one first-order principal minor. We may then assume, if necessary on simultaneous reordering:

$x_1 \neq 0$, $w_1 \neq 0$, and case 1 as proved above is applicable.

q.e.d. for case 2 (λ is a distinct root).

If none of the above cases is known to apply, we first transform $\mathbf{A}$ according to (8.3.4).

An assumption that the theorem is true for the n-1 by n-1 matrix $\mathbf{A^*}_{22}$ allows us to assume that, if necessary after further transformation, using a block-diagonal transformation-operator, $\mathbf{A^*}_{22} = \mathbf{B}_{22}$, (i.e. $\mathbf{A^*}_{22}$ is brought in the structure stated by the theorem).

Hence, by assumption, we replace (8.3.4) by:

$$A^*_3 = \left[\begin{array}{c|c} \lambda_1 & a^{*\prime}_{12} \\ \hline & B_{22} \end{array}\right] \tag{8.3.10}$$

(where $a^{*\prime}_{12}$ is: $a^{*\prime}_{12} = a^{\prime}_{12} T_{22}^{-1}$)

We now consider the transformation-operator:

$$T_4 = \left[\begin{array}{c|c} 1 & t^{\prime}_{12} \\ \hline & I_{n-1} \end{array}\right] \tag{8.3.11}$$

As with all triangular matrices with unity elements on the diagonal and only one row containing non-zero off-diagonal elements, T_4^{-1} differs from T_4 itself, only by way the off-diagonal vector $t^{\prime}_{12}$ changing its sign.

This transformation-operator T_4 gives rise to a transformed matrix

$$A^*_4 = T_4 \, A^*_3 \, T_4^{-1} \tag{8.3.12}$$

where A^*_4 has the same partitioning as A^*_3 with $a^{*\prime}_{12}$ after transformation relating to the similar vector before transformation (indicated below as $a^{\prime}_{12}$) by:

$$a^{*\prime}_{12} = a^{\prime}_{12} + t^{\prime}_{12} B_{22} - \lambda_1 t^{\prime}_{12} \tag{8.3.13}$$

or in elementswise form:

$$a^*_{1j} = a_{1j} + t_{1,j-1} \times b_{j-1,j} + t_{1j} (\lambda_p - \lambda_1) \tag{8.3.14}$$

We are now in a position to state proofs for some further cases, as follows:

Case 3):
$\lambda_p \neq \lambda_1$ for at least one p (λ is not an n-fold repeated root)

Then we recursively resolve t_{1j}, for $a^*_{1j} = 0$ from (8.3.14) as:

$$t_{1,j} = (a_{1,j} + t_{1,j-1} \times b_{j-1,j}) / (\lambda_1 - \lambda p) \tag{8.3.15}$$

for all j associated with a complete diagonal block U_{pp} of B, which then becomes a complete diagonal block of A^*_4, and a proof by recursive induction arises.

q.e.d. for case 3.

Case 4):
λ is an n-fold repeated root.

We consider a number of subcases (in which asterisks are reserved for elements of A^*_4 as given in (8.3.12), i.e. it is assumed that A itself already has the structure of A^*_3 as given in (8.3.10):

Subcase 4a) $a_{1,j} = b_{j-1,j} = 0$ applies for some j.

Then, despite the impossibility to resolve (8.3.14), for that particular j as stated in (8.3.15), $a^*_{1,j} = 0$ holds for that j in any case, and (8.3.14) -with the last term vanishing-, can then be resolved for the other columns of U_{pp}, as:

$$t_{1,j-1} = -a_{1j} \;/\; b_{j-1,j} = -a_{1j}$$

A proof by recursive induction then arises as well.

q.e.d. for case 4a.

Subcase 4b): $a_{12} = 0$, $B_{22} = U_{22}$ applies.
($a_{j-1,j} = 1$ holds for j = 3, 4, ... n, but not for j=2)

Then we resolve $a'^*_{1j} = [0]'$ from (8.3.14) without the last term:

$$t_{1,j-1} = -a_{1,j} \quad (j = 3, 4, \ldots n)$$

and find $A^*_{22} = U_{22}$ (indices 2 to n), to be a complete diagonal block of $A^*_4 = B$ (even without invoking recursive induction).

q.e.d. for subcase 4b

Subcase 4c) $a_{12} \neq 0$, $\mathbf{a'}_{12} = [0]'$.

Then

$$\mathbf{T}_5 = \begin{bmatrix} \mathbf{I}_{j-1} & & \\ & a_{12} & \\ & & \mathbf{I}_{n-j} \end{bmatrix} \qquad (8.3.16)$$

applied recursively for $j = 2, 3, \ldots\ldots$, until either $j=n$, or $b_{j,j+1} = 0$ is met, will serve as an operator, to divide the jth column by a_{12}, and also multiply the jth row by a_{12}, and hence requiring recursion to ensure $\mathbf{A} = \mathbf{B}$, as long as this multiplication turns $b_{j,j+1} = 1$ into $b^*_{j,j+1} = a_{12}$.

q.e.d. for subcase 4c.

Subcase 4d): $a_{12} \neq 0$, $a_{1,j} \neq 0$, $b_{j-1,j} = 0$ (for some $j > 2$).

Then:

$$\mathbf{T}_6 = \begin{bmatrix} 1 & & & & \\ & 1 & & t_{2,j} & \\ & & \mathbf{I}_{j-3} & & \\ & & & a_{1,j} & \\ & & & & \mathbf{I}_{n-j} \end{bmatrix} \qquad (t_{2,j} = a_{1,j} / a_{1,2})$$

$$(8.3.17)$$

provides the operator, to subtract the second column from the jth, with a multiple suitable ensure $a^*_{1,j} = 0$ on post-multiplication by $\mathbf{T}_6^{-1}$, while the other changes cancel on pre-multiplication by $\mathbf{T}_6$.

Subcase 4c then arises.

q.e.d. for subcase 4d)

q.e.d. for case 4.

q.e.d. for the triangularization theorem.

We now proceed to state some corollaries concerning the vectors and their relationships to the transformation-operator $\mathbf{X}$.

- (4) The leading column of each n by n_p block-column $\mathbf{X}_p$ is a column vector of $\mathbf{A}$.

Similarly, the row-vectors of $\mathbf{U}$ are associated with zero rows in $[\mathbf{U}-\mathbf{I}\lambda_p]$, associated with the bottom righthand elements of the $\mathbf{U}_{pp}$

Therefore:

- (5) The last row of each n_p by n block-row of $\mathbf{X}^{-1}$ is a row-vector

The combination of (4) and (5) implies:

- (6) **A** pair of a row- and column-vector, satisfying $\mathbf{w}'_p\,\mathbf{x}_p = 1$, arises, whenever an associated 1 by 1 block $\mathbf{U}_{pp}$ exists.

By case 1 of the proof, this last property also applies in the other direction:

- (7) We can require $\mathbf{U}_{pp} = [\lambda_p]$ ($n_p = 1$), whenever we find a pair of vectors $\mathbf{w}'_p\,\mathbf{x}_p \neq 0$.

The transformation stated by the stronger form of the triangularization theorem is known as the <u>Weierstrass-Jordan normal form</u> (Bodewig [6], pp. 83-87, referring to Hawkes [19], and to Aitken and Turnbull [3]).

Both publications do indeed contain the theorem, but neither source refers to Weierstrass or Jordan on that particular point. This is the more remarkable, because Hawkes develops it, in aid of providing a proof of another theorem, for which he does refer to Weierstrass. We shall simply speak of triangularization, and, because of the particular requirements of this theorem, of triangularization to unity elements a line above the diagonal.

The reference to Jordan, which appears to be the French mathematician Camille Jordan [27], is, however, fairly generally recognized, and the $\mathbf{U}_{pp}$, with their specific structure:

$$
U_{pp} = \begin{bmatrix} \lambda_p & 1 & - & - & \ldots\ldots & - & - \\ - & \lambda_p & 1 & - & \ldots\ldots & - & - \\ & & & \ldots\ldots\ldots\ldots & & & \\ - & - & & \ldots\ldots\ldots\ldots & \lambda_p & 1 \\ - & - & & \ldots\ldots\ldots\ldots & - & \lambda_p \end{bmatrix}
$$

are indeed known as <u>Jordan blocks</u> (see for example Graham [17] p.63)

We end this section with an obvious comment:

We formulated the theorems on triangularization in terms of upper-triangular matrices. Obviously, the similar requirements could also be stated in terms of lower triangularity: Simultaneous reordering transforms the one into the other.

8.4 Power convergence and the expansion theorem

In this and subsequent sections of this chapter, we shall find it necessary, to develop powers of matrices, including powers of Jordan blocks, the matrices U, of the structure stated to exist by the (strong) triangularization theorem.

The case of an n-fold repeated root without non-trivial blockdiagonalization will come up as a complication in a number of proofs, while proofs for other types of matrices come more readily, as we are in that case in a position to call upon recursive induction. It is therefore useful to discuss the powers of that particular type of matrix here, rather than at a number of separate points further in the text of this and subsequent sections.

It is then first of all useful, to split U into its diagonal and above-diagonal terms:

$$
U = I_n \lambda + M \tag{8.4.1}
$$

where M is (the above-diagonal unity elements):

$$M = \left[\begin{array}{c|c} & I_{n-1} \\ \hline & \end{array}\right] \tag{8.4.2}$$

Owing to the special structure of the two terms on the righthand side of (8.4.1), products (and by implication recursive products) of them, can be developed in any order and permit interchanging of the factors:

$$I_n \lambda\, M = M\, I_n \lambda = M\lambda \tag{8.4.3}$$

This consideration permits us to expand $U^p = [I\lambda + M]^p$, as if the two terms on the righthand side of (8.4.1) were scalar numbers:

$$U^p = [I_n \lambda + M]^p = \sum_{h=0}^{p} \left(\frac{p!}{h!\,(p-h)!}\right) M^{p-h}\, \lambda^h \tag{8.4.4}$$

Since M^n and all higher powers of M are null matrices, (8.4.4) truncates, for $p > n-1$, to become:

$$U^p = [I_n \lambda + M]^p = \sum_{h=0}^{n-1} \left(\frac{p!}{h!\,(p-h)!}\right) M^{p-h}\, \lambda^h \tag{8.4.5}$$

The application of (8.4.4) and (8.4.5) is further simplified, by the fact that M and its powers only contain n-p non-zero (unity) elements in position 1,p+1, 2,p+2, 3,p+3, etc., and otherwise consist of zeros only.

As a result, each U^p has the property:

$$[U^p]_{i,i+h} = [U^p]_{1,1+h}$$

$$(i = 1, 2, \ldots\ldots n, \quad h = 0, 1, 2, \ldots\ldots n-i) \tag{8.4.6}$$

i.e. the numerical content of any element of U^p only depends on the difference between its row- and column indices.

Example:

$$\begin{bmatrix} 1 & 1 & - & - \\ - & 1 & 1 & - \\ - & - & 1 & 1 \\ - & - & - & 1 \end{bmatrix}^4 = \begin{bmatrix} 1 & 4 & 6 & 4 \\ - & 1 & 4 & 6 \\ - & - & 1 & 4 \\ - & - & - & 1 \end{bmatrix}$$

It is at this stage useful to note the following peculiarity of the product of two (upper) triangular matrices $\mathbf{T}$ and $\mathbf{U}$, both of order n by n.

$$\mathbf{T}\,\mathbf{U} = \left[\begin{array}{c|c} \mathbf{T}_{11} & \mathbf{T}_{12} \\ \hline & \mathbf{T}_{22} \end{array}\right] \left[\begin{array}{c|c} \mathbf{U}_{11} & \mathbf{U}_{12} \\ \hline & \mathbf{U}_{22} \end{array}\right] =$$

$$\left[\begin{array}{c|c} \mathbf{T}_{11}\mathbf{U}_{11} & \mathbf{T}_{11}\mathbf{U}_{12} + \mathbf{T}_{12}\mathbf{U}_{22} \\ \hline & \mathbf{T}_{22}\mathbf{U}_{22} \end{array}\right] \tag{8.4.7}$$

We first of all note, that, due to the vanishment of the bottom lefthand block, the diagonal blocks are the product of the diagonal blocks of the two matrices only.

Therefore, for $\mathbf{P} = \mathbf{U}^{p}$, we infer by recursive induction:

$$\mathbf{P} = \left[\begin{array}{c|c} \mathbf{U}_{11} & \mathbf{U}_{12} \\ \hline & \mathbf{U}_{22} \end{array}\right]^{p} = \left[\begin{array}{c|c} \mathbf{U}_{11}{}^{p} & \mathbf{P}_{12} \\ \hline & \mathbf{U}_{22}{}^{p} \end{array}\right] \tag{8.4.8}$$

i.e. we obtain the diagonal blocks of the power of a triangular matrix (the generalization to lower triangularity will be obvious), by raising the corresponding block of the matrix itself to the relevant power. One off-diagonal block is a null block, and only the other off-diagonal block has a more complicated structure.

Note incidentally, that the validity of (8.4.8) applies quite generally to <u>all</u> triangular, and indeed to block-triangular matrices, and to other products of (block)triangular matrices than just the powers of a (block)triangular matrix, provided the partitioning is consistent: No use of triangularity of $\mathbf{U}_{11}$, $\mathbf{U}_{22}$, $\mathbf{T}_{11}$, and $\mathbf{T}_{22}$ and their powers was made in order to arrive at (8.4.7) and (8.4.8).

The product of two or more block-triangular matrices, which have diagonal blocks of orders which make the products of these diagonal blocks legitimate expressions, is a block-triangular matrix which has these products as diagonal blocks.

Example:

$$\left[\begin{array}{cc|c} 1 & 2 & 3 \\ \hline & & 4 \end{array}\right] \left[\begin{array}{c|c} 5 & 6 \\ 7 & 8 \\ \hline & 9 \end{array}\right] = \left[\begin{array}{c|c} 19 & 49 \\ \hline & 36 \end{array}\right] \quad ; \quad \begin{bmatrix} 1 & 2 \end{bmatrix} \begin{bmatrix} 5 \\ 7 \end{bmatrix} = \begin{bmatrix} 19 \end{bmatrix}$$

The reason for developing this result here, was simply that we need to apply it in relation to a particular type of triangular matrix.

In developing the powers of a Jordan block, we can treat any top lefthand or bottom righthand block of **U** as a Jordan block of lower order in its own right.

This means that we can take the truncated form of (8.4.4), i.e. (8.4.5), as the general one, provided we apply it to a block of which the order does not exceed p. Therefore, if we wish to calculate all elements of $\mathbf{U}^p$, it will suffice, to calculate each $[\mathbf{U}^p]_{1j}$, treating this element as the 1,n element of a top lefthand block.

For this one element, only the last term on the righthand side of (8.4.5) is non-zero: all the lower powers of **M** have their non-zero elements further to the left. This one element is:

$$[\mathbf{U}^p]_{1n} = \left(\frac{p!}{(n-1)! \, (p-n+1)!} \right) \lambda^{p-n+1} \tag{8.4.9}$$

We now work out out p!, as figuring in (8.4.9), and divide each factor n+h in it, for n+h = n, ... , p, by one of the factors in (p-n+1)!, and combine with a factor λ, to obtain:

$$[\mathbf{U}^p]_{1n} = \left(\frac{1 \times 2 \times \dots \times (n-1) \ \times \ n \times (n+1) \times \dots \times p}{1 \times 2 \times \dots \times (n-1) \ \times \ 1 \times 2 \times \dots \times (p-n+1)} \right) \lambda^{p-n+1}$$

$$= \ n\lambda \times \left(\frac{n+1}{2} \lambda \right) \times \left(\frac{n+2}{3} \lambda \right) \times \dots \times \left(\frac{p}{(p-n+1)} \lambda \right) \tag{8.4.10}$$

It is now seen, that, if $[\mathbf{U}^p]_{1n}$ is at all non-zero, we obtain recursion from p to p+1, (or from p-1 to p) simply by adding the last factor to (8.4.10).

This gives rise to the recursive formula:

$$[U^p]_{1n} = ((\lambda p)/(p-n+1)) \, [U^{p-1}]_{1n} \qquad (p \geq n) \qquad (8.4.11)$$

For p = n-1, the recursion formula (8.4.11) would give rise to an infinite ratio. This is correct: we need to place a non-zero element in the 1,n cell for the first time.

Reference to (8.4.5), and (8.4.9) in their original forms indicates that this element is:

$$[U^{n-1}]_{1n} = 1 \qquad (8.4.12)$$

For p < n-1 we obviously have a zero element in the 1,n position.

The generalizations of (8.4.11) and (8.4.12) with respect to other elements of U^p are as follows:

$$[U^p]_{ij} = ((\lambda p)/(p-j+i)) \, [U^{p-1}]_{ij} \qquad (p > j-i) \qquad (8.4.13)$$

and:

$$[U^p]_{ij} = 1 \qquad (p = j-i) \qquad (8.4.14)$$

while for j > i+p, there are only zero elements.

Exercise:

Using the recursion formula (8.4.13), calculate:

$$\begin{bmatrix} 2 & 1 & - & - & - \\ - & 2 & 1 & - & - \\ - & - & 2 & 1 & - \\ - & - & - & 2 & 1 \\ - & - & - & - & 2 \end{bmatrix}^4$$

(There is an answersheet at the end of the chapter)

We are now in a position to state, and proceed to prove the following

Theorem (Power-convergence theorem):

If the roots of the n by n matrix **A** are all (in the unit circle):

$$|\lambda_j| < 1 \qquad (j = 1, 2, \ldots\ldots n) \qquad\qquad (8.4.15)$$

then $\mathbf{A}^p$ converges to the null matrix, as p goes to infinity.

Proof:

We deal separately with three separate cases.

Case a): **A** is diagonalizable.

Then $\mathbf{A} = \mathbf{X}\ \hat{\mathbf{L}}\ \mathbf{X}^{-1}$ and $\mathbf{A}^p = \mathbf{X}\ \hat{\mathbf{L}}^p\ \mathbf{X}^{-1}$

q.e.d. for case a).

Case b):
A is not diagonalizable, but triangularizes blockdiagonally, non-trivially into more than one triangular diagonal block, as

$$\mathbf{A} = \mathbf{X}\ \mathbf{U}\ \mathbf{X}^{-1}$$

An assumption, that the theorem is true for matrices of smaller order than n by n, and therefore for the diagonal blocks of **U**, then leads to a proof by recursive induction.

Case c)
A can only be triangularized, not blockdiagonally triangularized into more than one diagonal block.

Then **A** has an n-fold repeated root, and $\mathbf{U} = \mathbf{X}\ \mathbf{A}\ \mathbf{X}^{-1}$ can be required to be a <u>Jordan block</u>.

Reference to the recursion-formula (8.4.13) then shows, that for $|\lambda|p < p-j+i$, and therefore for i = 1, j = n, and $|\lambda| < 1$:

$$p > (n-1)/(1 - |\lambda|) \qquad\qquad (8.4.16)$$

all elements of $\mathbf{U}^p$ start to converge to zero with increasing p.

q.e.d. for case c.

q.e.d. for the power-convergence theorem.

The power convergence theorem now permits us to prove the following theorem, to which the title of this section refers as the expansion theorem, and which is also known in the literature (see for example Pullman [37], p.41), as the Lagrange-Sylvester theorem.

Theorem (expansion theorem):

If the roots of **A** are within the unit circle,

then

$$B = [I-A]^{-1} \qquad (8.4.17)$$

exists, and is evaluated as:

$$B = I + A + A^2 + A^3 + A^4 + \ldots \ldots \qquad (8.4.18)$$

Proof:

By assumption, no root of **A** is equal to one (all being smaller in absolute value).

Therefore $B = [I-A]^{-1}$ exists.

The convergence of the righthand-side of (8.4.18) is implied by the power-convergence theorem. Remains therefore to prove the equality between the two members of (8.4.18).

We denote the righthand-side of (8.4.18) as **E** for expansion, and evaluate [I-A]E, as follows:

$$E = I + A + A^2 + A^3 + A^4 + A^5 + A^6 + \ldots\ldots\ldots\ldots$$

less: $A\ E = \qquad A + A^2 + A^3 + A^4 + A^5 + A^6 + A^7 + \ldots\ldots\ldots$

$$[I-A]\ E = I \qquad\qquad\qquad\qquad\qquad\qquad\qquad \ldots$$

According to the power-convergence theorem, the missing term at the end, (minus) A to the power infinite, is a null matrix.

Therefore $E = [I - A]^{-1}$.

q.e.d.

Example:

$$A = \begin{bmatrix} 0.2 & 0 \\ 2.0 & 0.5 \end{bmatrix}$$

$$[I-A]^{-1} = \begin{bmatrix} 1.00 & 0 \\ 0 & 1.00 \end{bmatrix} + \begin{bmatrix} 0.20 & 0 \\ 2.00 & 0.50 \end{bmatrix} + \begin{bmatrix} 0.04 & 0 \\ 1.40 & 0.25 \end{bmatrix} + \begin{bmatrix} 0.01 & 0 \\ 0.78 & 0.13 \end{bmatrix}$$

$$+ \begin{bmatrix} 0.00 & 0 \\ 0.41 & 0.06 \end{bmatrix} + \begin{bmatrix} 0.00 & 0 \\ 0.21 & 0.03 \end{bmatrix} + \begin{bmatrix} 0.00 & 0 \\ 0.10 & 0.02 \end{bmatrix} + \begin{bmatrix} 0.00 & 0 \\ 0.04 & 0.01 \end{bmatrix}$$

$$+ \begin{bmatrix} 0.00 & 0 \\ 0.03 & 0.00 \end{bmatrix} + \begin{bmatrix} 0.00 & 0 \\ 0.02 & 0.00 \end{bmatrix} + \begin{bmatrix} 0.00 & 0 \\ 0.00 & 0.00 \end{bmatrix} = \begin{bmatrix} 1.25 & 0 \\ 4.99 & 2.00 \end{bmatrix}$$

For a non-negative A, the result of this calculation is in general downwardly biased: the omitted remaining terms are non-negative.

In fact, due to the rounding of $0.2^3 = 0.008$ to 0.01, and the rounding of $0.5^3 = 0.125$ to 0.13, the diagonal elements happen to be exact.

The true inverse is:

$$\begin{bmatrix} 0.2 & 0 \\ 2.0 & 0.5 \end{bmatrix}^{-1} = \begin{bmatrix} 1.25 & 0 \\ 5.00 & 2.00 \end{bmatrix}$$

8.5 **Idempotent and nilpotent matrices**

A matrix, for which, for some q > 1,

$$\mathbf{A}^q = \mathbf{A} \tag{8.5.1}$$

applies, is called a <u>periodic</u> of <u>period</u> q-1.

If (8.5.1) applies for q=2, than **A** is said to be <u>idempotent</u>. It may be the case that the <u>word</u> idempotent, i.e. 'capable of itself', suggest the more <u>wider</u> definition for which (following Graham [16], p. 30), the wider them 'periodic' is used here, but it is clear that the narrower definition:

$$\mathbf{A}^2 = \mathbf{A}$$

is the definition used for 'idempotent' in most modern textbooks.

If

$$\mathbf{A}^q = [0] \tag{8.5.2}$$

is true for some q, we say that **A** is <u>nilpotent</u>. There is in this case no need to exclude **A** itself, as for a periodic, where **A** = **A** is true for every matrix.

Theorem:

If **A** has an n-fold repeated zero root,

then **A** is nilpotent.

Proof:

Consider the triangularization of **A**.

If **A** diagonalizes completely (**A** = [0]), no further proof is needed.

If **A** triangularizes blockdiagonally into two or more diagonal (upper) triangular blocks, some of which may not be null blocks, an assumption that the theorem is true for these separate blocks, immediately leads to a proof by recursive induction.

If **A** is of rank n-1 then (see section 8.3), **A** may be triangularized as:

$$\mathbf{A}^* = \mathbf{X}\,\mathbf{A}\,\mathbf{X}^{-1} =
\begin{bmatrix}
- & 1 & - & - & \cdots\cdots\cdots\cdots & - & - & - \\
- & - & 1 & - & \cdots\cdots\cdots\cdots & - & - & - \\
\cdots & \cdots\cdots\cdots\cdots\cdots\cdots\cdots\cdots\cdots & & & & & & \\
- & - & & \cdots\cdots\cdots\cdots\cdots & & - & 1 & - \\
- & - & & \cdots\cdots\cdots\cdots\cdots & & - & - & 1 \\
- & - & & \cdots\cdots\cdots\cdots\cdots & & - & - & -
\end{bmatrix}$$

$$(8.5.3)$$

and $\mathbf{Q} = \mathbf{A}^{*q}$ has the property:

$$\left. \begin{aligned} \mathbf{q}_j &= [0], \quad \text{for } j \leq q, \\[2mm] \mathbf{q}_j &= \mathbf{e}_{j-1}, \quad \text{for } j > q, \end{aligned} \right\} \qquad (8.5.4)$$

as may be verified by recursive induction.

Therefore:

$$\mathbf{A}^{*n} = [0]$$

q.e.d. for an **A** of rank n-1.

If the rank of **A** is less than n, then every n_i by n_i diagonal block of **A*** is of rank n_i-1 by n_i-1, and the theorem will apply with q being the greatest of the separate n_i.

q.e.d.

A null matrix has an n-fold repeated zero root, and the roots of $\mathbf{A}^q$ are λ^q. Hence a nilpotent matrix cannot have non-zero roots.

Therefore:

A is nilpotent, if and only if **A** has an n-fold repeated zero root.

Theorem:

If **A** is a periodic,

then:

- (1) **A** is diagonalizable, and

- (2) all its roots are either $\lambda = 0$,

$$\text{or of absolute value } |\lambda| = 1.$$

Proof:

From the assumed:

$$\mathbf{A}^q = \mathbf{A} \tag{8.5.5}$$

we infer, on applying the triangularization theorem:

$$\mathbf{X} \, \mathbf{A}^q \, \mathbf{X}^{-1} = \mathbf{X} \mathbf{A} \mathbf{X}^{-1} = \mathbf{A}^* \tag{8.5.6}$$

where **A*** is triangular and blockdiagonal with all the diagonal blocks **A***$_{jj}$ being <u>Jordan blocks</u>, (see section 8.3), and **A*** is also a periodic.

If **A*** blockdiagonalizes non-trivially into several diagonal blocks **A***$_{jj}$, an assumption that the theorem is true for the separate diagonal blocks of **A***, leads immediately to a proof by recursive induction.

If **A** has an n-fold repeated root, (2) also follows from the requirement:

$$|a^*{}_{jj}{}^p| = |a^*{}_{jj}| = |\lambda_j{}^p| = |\lambda| \tag{8.5.7}$$

A proof concerning the diagonalizability of **A** is however, needed for the case of an n-fold repeated root.

If **A** has an n-fold repeated zero root and triangularizes conform (8.5.3), then (8.5.4) contradicts **A** being a periodic.

(Except in the trivial case of **A** = [0], no matrix with an n-fold repeated zero root is a periodic.)

Now suppose **A** had an n-fold repeated root of an absolute value of 1, while **A***q was of rank n-1 > 1.

Then (8.4.13) contradicts periodicity: the absolute value of the above-diagonal elements of **U** increases consistently with increasing p.

q.e.d. for the diagonalizability of **A**.

In all other cases, proof of the theorem now follows by recursive induction.

q.e.d.

Note that the theorem does <u>not</u> state that periodics have no complex roots. On the contrary: In Chapter 10, we shall be discussing rotation operators. To the extent that the angle of rotation is a rational fraction of the full circle, a rotation-operator is a periodic matrix, and the roots normally are complex. But that stronger property <u>does</u> apply to idempotent matrices.

Theorem:

If **A** is idempotent, then

- 1) **A** has n independent vectors, and

- 2) the corresponding roots are either zero or one.

Proof:

If **A** is of rank r, then n-r independent vectors give rise to zero roots, whereas the columns of **A** themselves (and by implication r independent combinations of them), are vectors associated with a unity root.

q.e.d.

The existence of r independent vectors satisfying

$$\mathbf{A} \mathbf{x}_j = \mathbf{x}_j \neq [0] \qquad (j = 1, 2, \ldots . r) \qquad (8.5.8)$$

and n-r independent vectors satisfying

$$\mathbf{A} \mathbf{x}_j = [0] \ (\mathbf{x}_j \neq [0]) \qquad (j = r+1, \ldots n) \qquad (8.5.9)$$

also is a sufficient condition to confirm that **A** is idempotent.

Theorem

If r independent vectors satisfy (8.5.8), and n-r vectors satisfy (8.5.9), then:

● 1) **A** is idempotent, and

● 2) **A** is similar to a symmetric matrix.

Proof:

By assumption there are n independent vectors. Therefore **A** diagonalizes as:

$$\mathbf{A} = \mathbf{X} \left[\begin{array}{c|c} I_r & \\ \hline & \end{array}\right] \mathbf{X}^{-1} \qquad (8.5.10)$$

and 1) is immediately seen to be implied: $\mathbf{A}^2 = \mathbf{A}$.

Concerning 2) we note:

$$\mathbf{X}^{-1} \mathbf{A} \mathbf{X} = \left[\begin{array}{c|c} I_r & \\ \hline & \end{array}\right] \qquad (8.5.11)$$

confirming 2)

q.e.d.

The following may now be observed concerning recognizing idempotent matrices, periodics, and nilpotent matrices.

First of all, $A^2 = A$ unambigiously separates idempotent matrices from other types of matrices.

A nilpotent matrix is either a null matrix, or similar to a matrix which blockdiagonalizes into one or more (non-zero) Jordan blocks. It is therefore sufficient to develop the nth power.

If A^n is not a null matrix, then A is not nilpotent.

Concerning periodics, we state (at this stage without proof), that symmetric matrices have real roots only, and that a symmetric periodic therefore has roots of 0, 1, and -1, and therefore A^2 has roots of 0 and 1.

Therefore if A is a symmetric periodic, then A has an idempotent second power, i.e. obeys the condition:

$$A^4 = A^2$$

Exercise:

Establish which of the following matrices is idempotent, nilpotent or a periodic:

$$\begin{bmatrix} 0.5 & -0.5 \\ -0.5 & 0.5 \end{bmatrix} \quad \begin{bmatrix} 0.5 & 0.5 \\ 0.5 & 0.5 \end{bmatrix} \quad \begin{bmatrix} 0.5 & 1 \\ 0.25 & 0.5 \end{bmatrix}$$

$$\begin{bmatrix} 0.25 & 0.25 \\ 0.25 & 0.25 \end{bmatrix} \quad \begin{bmatrix} 1 & - \\ - & - \end{bmatrix} \quad \begin{bmatrix} - & - \\ 1 & - \end{bmatrix}$$

$$\begin{bmatrix} - & 1 \\ 1 & - \end{bmatrix} \quad \begin{bmatrix} - & - \\ - & - \end{bmatrix}$$

(There is an answersheet at the end of the chapter)

8.6 **The roots and vectors of permutation operators**

An important class of periodic matrices, which is used in matrix algebra itself, are <u>permutation operators</u>. These normally have complex roots (of absolute value $|\lambda| = 1$.)

Throughout this section we reserve the letter **P**, for indicating an n by n permutation operator. Recall (from section 3.3), the definition of a permutation operator, which is here re-stated in the notation used in this section.

A permutation operator is a square matrix, of which the columns, as well as the rows, are all unit vectors.

$$\mathbf{p}_j = \mathbf{e}_i \qquad (j = 1, 2, \ldots n) \qquad (8.6.1)$$

as well as

$$\mathbf{p'}_i = \mathbf{e'}_j \qquad (i = 1, 2, \ldots n) \qquad (8.6.2)$$

Theorem (permutation operator decomposition theorem):

If a permutation operator is decomposable, then the (square) diagonal blocks are themselves permutation operators, and the off-diagonal blocks are zero blocks.

Proof:

By assumption, **P** is decomposable, and may therefore be partitioned (if necessary after simultaneous reordering), as:

$$\mathbf{P} = \left[\begin{array}{c|c} \mathbf{P}_{11} & \\ \hline \mathbf{P}_{21} & \mathbf{P}_{22} \end{array} \right] \qquad (8.6.3)$$

where $\mathbf{P}_{11}$ is square and of order as n_1 by n_1 $(1 \leq n_1 < n)$.

Given this assumed partitioning, each row of $\mathbf{P}_{11}$ contains a unity element by (8.6.2). Similarly, each column of $\mathbf{P}_{22}$ contains a unity element by (8.6.1). Therefore, since **P** as a whole contains only n unity elements, none are present in $\mathbf{P}_{21}$.

q.e.d.

The partitioning conform (8.6.3) therefore implies:

$$\mathbf{P} = \left[\begin{array}{c|c} \mathbf{P}_{11} & \\ \hline & \mathbf{P}_{22} \end{array}\right] \tag{8.6.4}$$

A special, but by no means trivial case of a decomposable permutation-operator arises if there are unity elements on the main diagonal. In that case $\mathbf{P}$ permits simultaneous (re-)ordering as:

$$\mathbf{P} = \left[\begin{array}{c|c} p_{11} & \\ \hline & \mathbf{P}_{22} \end{array}\right] = \left[\begin{array}{c|c} 1 & \\ \hline & \mathbf{P}_{22} \end{array}\right] \tag{8.6.5}$$

Obviously, there can be several diagonal permutation-operator blocks, some of which may, or may not be, unity elements.

Example:

$$\mathbf{P} = \left[\begin{array}{cc|cc|cc} 1 & - & - & - & - & - \\ - & 1 & - & - & - & - \\ \hline - & - & 1 & 1 & - & - \\ - & - & 1 & - & - & - \\ \hline - & - & - & - & 1 & 1 \\ - & - & - & - & 1 & - \end{array}\right]$$

Theorems concerning indecomposable permutation-operators and their proofs are much simplified, if we may assume a particular ordering. We shall refer to a permutation-operator, which satisfies:

$$\left. \begin{array}{l} p^*_{j+1,j} = 1 \ (j = 1, 2, \ldots\ldots n-1) \\[1.5em] p^*_{1,n} = 1 \end{array} \right\} \tag{8.6.6}$$

as a permutation-operator in <u>lower semidiagonal form</u>.

One peculiarity of such matrices, and of the corresponding systems of equations, is that they permit, what might be called 'circulary reordering', without actually changing. We will speak of circulary reordering to the right, if indices 1 to n-1 are renamed as 2 to n, while index n is renamed as 1. The meaning of the term circulary reordering to the left will then be obvious.

We illustrate circular reordering to the right, as follows:

$$A = \begin{bmatrix} - & - & - & 4 \\ 1 & - & - & - \\ - & 2 & - & - \\ - & - & 3 & - \end{bmatrix} \qquad A^* = \begin{bmatrix} - & - & - & 3 \\ 4 & - & - & - \\ - & 1 & - & - \\ - & - & 2 & - \end{bmatrix}$$

The corresponding permutation-operator is, however, not affected by such a transformation.

For the sake of streamlining the formulation of theorems and proofs, we will use the term 'permitting (re-)ordering', and it should be understood, that this includes the trivial case, where the ordering in question is already present.

Theorem:

If P is indecomposable,

then P permits a simultaneous reordering, into lower semi-diagonal form.

Proof:

If $p_{11} = 1$ applies, P is immediately seen to be decomposable, conform (8.6.5).

The same consideration applies, for at some stage finding, for some $j > 1$, $p_{jj} = 1$.

Otherwise, we recursively assign, for $j = 1, 2, \dots n-1$, the index $i^* = j+1$, to any i^{th} row and column of P, for which $p_{ij} = 1$ applies for $i > j$, placing the old $(j+1)^{th}$ index in the old position i.

Since, for $2 \leq i < j$, we already have $p_{i,i-1} = 1$, we always find $p_{i,j} = 1$, either for $i > j$, or else for $i=j$, or $i=1$.

If $p_{1,j} = 1$ is found for $j < n$, then a j by j block P_{11} is found to be a complete permutation operator of order j by j, showing the decomposability of P. If $p_{jj} = 1$ is found, P is also shown to be decomposable.

Otherwise, for j=n, **P*** has unit-vector columns $\mathbf{p}^*_k = \mathbf{e}_{k+1}$, for k = 1, 2, ... n-1, and $\mathbf{p}^*_n = \mathbf{e}1$ is the one remaining column already.

q.e.d.

The characteristic column vectors of a permutation operator, and their associated roots, need to obey the requirement:

$$\mathbf{P\,x} \;=\; \mathbf{x}\,\lambda \tag{8.6.7}$$

Therefore, for an <u>indecomposable</u> **P**, ordered in lower semidiagonal form:

$$x_n = x_1\,\lambda$$
$$x_{i-1} = x_i\,\lambda \;(i = 2, 3, \ldots\ldots n) \tag{8.6.8}$$

Recursive application of the second equation in (8.6.8) leads to:

$$x_{i-k} \;=\; x_{i-k+1}\,\lambda \;=\; x_n\,\lambda^{n-k} \tag{8.6.9}$$

while the first equation in (8.6.8) then requires:

$$x_n = x_1\,\lambda \;=\; x_n\,\lambda^n \tag{8.6.10}$$

The characteristic equation of an indecomposable permutation operator therefore is:

$$\lambda^n \;=\; 1 \tag{8.6.11}$$

We now come to discuss the vectors of a permutation operator.

Since all rows contain just one unity element, the sumcount of all rows is 1 = unity, i.e.:

$$\mathbf{P\,s} \;=\; \mathbf{s} \tag{8.6.12}$$

the summation-vector is a characteristic vector of a permutation operator (whether decomposable or indecomposable), and the associated root is $\lambda_1 = 1$.

The root $\lambda_1 = 1$ obviously is a root of the characteristic equation (8.6.11), but some discussion of the other roots is useful.

Theorem:

If **P** is indecomposable, then all its roots are distinct.

Proof:

On differentiation of:

$$\emptyset(\lambda) = \lambda^n - 1 \qquad\qquad (8.6.13)$$

relative to λ, we obtain:

$$d\emptyset/d\lambda = n\lambda^{n-1} \qquad\qquad (8.6.14)$$

This expression will vanish, only for $\lambda = 0$, and that is not a root of the characteristic equation (8.6.11).

Therefore $d\emptyset/d\lambda$ does not vanish for any root, and no root can be a repeated root.

q.e.d.

It follows that, apart from $\lambda_1 = 1$, and (if n is even) $\lambda_2 = -1$, <u>all roots are complex</u>.

We are now in a position, to state and prove a slightly stronger theorem on the characteristic equation and the determinant.

Theorem:

If the n by n permutation-operator **P** is indecomposable then:

(1) $\quad |-P| = -1$

and

(2) $\quad |\lambda I_n - P| = \lambda^n - 1.$

Proof:

Ad (1)

Assuming lower semidiagonal ordering, we develop P by its leading row, as follows:

$$|P| = 1 \times (-1)^{n-1} |I_{n-1}|$$ (8.6.15)

Therefore, if n is even, $|P| = |-P| = -1$, and if n is odd, we have $|P| = 1$, $|-P| = -1$.

q.e.d. ad (1)

Ad (2)

We develop $|\lambda I_n - P|$, by its leading row, as follows:

$$|I_n \lambda - P| =$$

$$\lambda \begin{vmatrix} \lambda & & & \\ -1 & \lambda & & \\ - & -1 & \lambda & \\ - & \dots & & \\ - & \dots & -1 & \lambda \end{vmatrix} - 1\,(-1)^n \begin{vmatrix} -1 & \lambda & & \\ - & -1 & \lambda & \\ - & \dots & & \\ - & \dots & -1 & \lambda \\ - & \dots & & -1 \end{vmatrix}$$

$$= \lambda(\lambda^{n-1}) + |-P| = \lambda^n - 1.$$ (8.6.16)

q.e.d. ad (2)

q.e.d.

Theorem:

An indecomposable permutation operator has no non-vanishing principal minors.

Proof:

Partition P as:

$$\mathbf{P} = \left[\begin{array}{c|c} \mathbf{P}_{11} & \mathbf{P}_{12} \\ \hline \mathbf{P}_{12} & \mathbf{P}_{22} \end{array}\right]$$

where $\mathbf{P}_{11}$ is of order n_1 by n_1.

If $\mathbf{P}_{11}$ contains its full complement of n_1 non-zero (unity) elements, $\mathbf{P}$ then is seen to be decomposable.

If $\mathbf{P}_{11}$ contains less then n_1 non-zero (unity) elements, it contains a zero row and a zero column, and no further proof is needed.

q.e.d.

Note that this theorem also confirms the characteristic equation in minors form.

The following may now be observed concerning the vectors:

First of all, we already identified one vector $\mathbf{x} = \mathbf{s}$, from the fact that all rows contain just one unity element, and therefore all add to one. In general, the elements of a vector may be calculated recursively, starting with $x_n = 1$, in the same way as was done for $\lambda = 1$.

Thus, for $n = 4$, we have:

$$\mathbf{P} = \left[\begin{array}{cccc} - & - & - & 1 \\ 1 & - & - & - \\ - & 1 & - & - \\ - & - & 1 & - \end{array}\right] \qquad \begin{array}{l} \text{with roots:} \\ \lambda_1 = 1, \quad \lambda_2 = -1, \\ \lambda_3 = i, \quad \lambda_4 = -i. \end{array}$$

Backwards recursive calculation of x_3, x_2, and x_1, initiated with $x_4 = 1$, then results in:

$$\mathbf{X} = \left[\begin{array}{cccc} 1 & -1 & -i & i \\ 1 & 1 & -1 & -1 \\ 1 & -1 & i & -i \\ 1 & 1 & 1 & 1 \end{array}\right]$$

A systematic way of calculating the row-vectors is to start with a lower semidiagonal ordering, not of P, but of P', and then impose an arbitrary requirement $w_1 = 1$. This results in an upper semidiagonal P, with

$$p_{n1} = 1, \text{ and } p_{i,i+1} = 1 \quad (i = 1, 2, \ldots\ldots n-1)$$

We shall refer to the column-vectors of this matrix as u. (= the transpose of w') Starting with $u_1=1$, we can calculate a set of column-vectors u from:

$$[I\lambda - P]u = 0$$

or in single equation-form:

$$\lambda u_i - u_{i+1} = 0 \qquad (i = 1, 2, \ldots\ldots n-1),$$

and therefore recursively:

$$u_{i+1} = \lambda u_i = \lambda^i \qquad (i = 1, 2, \ldots\ldots n-1)$$

In our example of P_4, we calculate the full matrix U as:

$$U = \begin{bmatrix} 1 & 1 & 1 & 1 \\ 1 & -1 & i & -i \\ 1 & 1 & -1 & -1 \\ 1 & -1 & -i & i \end{bmatrix}$$

and a matrix of row-vectors of a lower semi-diagonal P is found by transposition:

$$W = U' = \begin{bmatrix} 1 & 1 & 1 & 1 \\ 1 & -1 & 1 & -1 \\ 1 & i & -1 & -i \\ 1 & -i & -1 & i \end{bmatrix}$$

Note, however, that this is <u>a</u> matrix of row-vectors, and not the matrix U^{-1}.

The generalization of the results of this section to decomposable permutation-operators will be obvious: The roots of a block-diagonal matrix are those of the separate diagonal blocks.

There will therefore be repeated unity roots. But there is, for a decomposable permutation-operator with repeated roots, no problem of identifying a full set of n independent vectors. The vectors of the separate diagonal blocks only need to be completed to order n, by filling in the remaining elements, associated with other diagonal blocks, with zeros.

Thus, for $\mathbf{P} = \begin{bmatrix} 1 & - & - & - \\ - & - & - & 1 \\ - & 1 & - & - \\ - & - & 1 & - \end{bmatrix}$

we identify two independent vectors associated with $\lambda = 1$ as root, and the corresponding block of the matrix of characteristic vectors is:

$$\mathbf{X}_1 = \begin{bmatrix} 1 & - \\ - & 1 \\ - & 1 \\ - & 1 \end{bmatrix}$$

8.7 The Cayley–Hamilton theorem

This theorem states, loosely speaking, that a square matrix obeys its own characteristic equation. Before we state it more precisely, and provide a proof, it is useful to develop some additional notation.

We can develop a polynomial function, for a square matrix, in the same way as for a number or a variable.

Thus, if we have a polynomial function:

$$\phi(x) = x^3 - 7x^2 + 5x + 2,$$

then $\phi(\mathbf{A})$ means:

$$\phi(\mathbf{A}) = \mathbf{A}^3 - 7\mathbf{A}^2 + 5\mathbf{A} + 2\mathbf{A}^0,$$

where $\mathbf{A}^0$ is an identity matrix of the same order as $\mathbf{A}$.

A recursive product (power) of the same matrix can be developed either as $\mathbf{A}^p = \mathbf{A} \times \mathbf{A}^{p-1}$, or as $\mathbf{A}^{p-1} \times \mathbf{A}$.

The fact that $\mathbf{AB} = \mathbf{BA}$ is not normally true for matrix multiplication is responsible for some impossibilities to generalize from ordinary numbers or function to matrices, but in this case we can generalize the multiplication of ordinary functions to matrix polynomia.

Example: $[\mathbf{I} + 2\mathbf{A}]\,[\mathbf{I} - \mathbf{A} + 5\mathbf{A}^2 + 2\mathbf{A}^3]$

$$\mathbf{I}\,[\mathbf{I} - \mathbf{A} + 5\mathbf{A}^2 + 2\mathbf{A}^3] = \mathbf{I} - \mathbf{A} + 5\mathbf{A}^2 + 2\mathbf{A}^3$$

$$+ 2\mathbf{A}\,[\mathbf{I} - \mathbf{A} + 5\mathbf{A}^2 + 2\mathbf{A}^3] = \qquad 2\mathbf{A} - 2\mathbf{A}^2 + 10\mathbf{A}^3 + 4\mathbf{A}^4$$

$$[\mathbf{I} + 2\mathbf{A}]\,[\mathbf{I} - \mathbf{A} + 5\mathbf{A}^2 + 2\mathbf{A}^3] = \mathbf{I} + 2\mathbf{A} + 3\mathbf{A}^2 + 12\mathbf{A}^3 + 4\mathbf{A}^4$$

Note, however, that the two separate factors in this example are matrices, and that $\mathbf{A} \times \mathbf{A}^{p-1} = \mathbf{A}^{p-1} \times \mathbf{A}$ is invoked, but that the distinction between pre-multiplication and post-multiplication still applies for the separate factors:

We pre-multiplied $[\mathbf{I} + 2\mathbf{A} + 3\mathbf{A}^2 + 12\mathbf{A}^3 + 4\mathbf{A}^4]$ by $[\mathbf{I} + 2\mathbf{A}]$.

If a polynomial function $\phi(x)$ factorizes as:

$$\phi(x) = \delta(x) \times \gamma(x),$$

then $\phi(\mathbf{A})$ will be a null matrix, whenever $\delta(\mathbf{A})$ and/or $\gamma(\mathbf{A})$ is a null matrix.

Theorem (Cayley-Hamilton theorem):

If:

$$\phi(\lambda) = \sum_{p=0}^{n} c_p \, \lambda^p = 0$$

is the characteristic equation of the n by n matrix **A**,

then:

$$\phi(\mathbf{A}) = \sum_{p=0}^{n} c_p \, \lambda^p = [0]$$

is true.

Proof:

We invoke the triangularization of **A**:

$$\mathbf{A} = \mathbf{X} \, \mathbf{U} \, \mathbf{X}^{-1} \tag{8.7.1}$$

where (see section 8.3) **U** is upper triangular.

We then express any power of **A** as:

$$\mathbf{A}^p = \mathbf{X} \, \mathbf{U}^p \, \mathbf{X}^{-1} \tag{8.7.2}$$

and $\phi(\mathbf{A})$ as:

$$\phi(\mathbf{A}) = \mathbf{X} \, \phi(\mathbf{U}) \, \mathbf{X}^{-1} \tag{8.7.3}$$

We now first consider the special case, that **A** has an n-fold repeated root $\lambda = \lambda^*$.

Then $\phi(\lambda)$ has the form $\phi(\lambda) = (\lambda^* - \lambda)^n$, and $\phi(\mathbf{A})$ and $\phi(\mathbf{U})$ are:

$$\phi(\mathbf{A}) = [\mathbf{A} - \mathbf{I}\lambda^*]^n \tag{8.7.4}$$

and

$$\phi(\mathbf{U}) = [\mathbf{U} - \mathbf{I}\lambda^*]^n \tag{8.7.5}$$

In (8.7.5), $[\mathbf{U} - \mathbf{I}\lambda^*]$ is a triangular matrix, of which all the diagonal elements have been reduced to zero, by subtracting λ^* from them. Recursive development of the powers of such a matrix shows that its nth power is indeed a null matrix.

From $\emptyset(\mathbf{U}) = [0]$, and (8.7.3), we infer $\emptyset(\mathbf{A}) = [0]$.

q.e.d. for the case of an n-fold repeated root.

If there is no n-fold repeated root, we introduce the provisional assumption, that the theorem is true, for matrices of orders 1 by 1, 2 by 2, ... n-1 by n-1.

We also require $\mathbf{U}$ to have the blockdiagonal form as stated by the triangularization theorem. By assumption, the Cayley-Hamilton theorem is true for the separate $\mathbf{U}_{jj}$.

Since the characteristic equation function $\emptyset(\lambda)$ factorizes into the characteristic equation functions of the separate $\mathbf{U}_{jj}$, the $\emptyset(\mathbf{U}_{jj})$ are all null blocks, and the general proof follows by recursive induction.

q.e.d. for the Cayley-Hamilton theorem.

8.8 The inverse of a matrix of complex vectors

Complex matrices, if invertible, can be inverted by applying the arithmetic of complex numbers in the context of an otherwise conventional inversion algorithm, e.g. elementary row operations (or any other inversion method).

There are nevertheless certain advantages in being able to calculate a complex inverse while using only real arithmetic.

Graham [16], p. 255, following El Hawary [10], and source-references given by El Hawary, give the following formula for the inverse of a complex matrix:

$$[A + Bi]^{-1} = [A^2 + B^2]^{-1} [A - Bi] \qquad (8.8.1)$$

However, this formula is valid, only if the real matrices **A** and **B** obey the condition:

$$A\ B = B\ A$$

as may be verified by post-multiplication of (8.8.1) by $[A + Bi]$, to obtain:

$$[A + Bi]^{-1} [A + Bi] = I =$$

$$[A^2 + B^2]^{-1} [A - Bi][A + Bi] =$$

$$[A^2 + B^2]^{-1} [A^2 + B^2] + [AB - BA]i \qquad (8.8.2)$$

This condition (**AB** = **BA**), is <u>not</u> generally satisfied by an arbitrary complex matrix, and not by an arbitrary matrix of characteristic vectors either.

However, matrices of characteristic vectors <u>have</u> special properties, which allow us to develop their complex inverse, which will be done in the remainder of this section. We now <u>assume</u> that there is an invertible matrix of characteristic vectors, i.e. that **A** is diagonalizable. We proceed as follows:

First of all, having used **A** for an original matrix, and **X** for a matrix of characteristic vectors, we will indicate a complex matrix of characteristic vectors as:

$$X = Y + Zi \qquad (8.8.3)$$

(and not **A** + **B**i as Graham and El Hawary did)

Both **Y** and **Z** have a partitioned structure, related to the pair-wise ordering of the roots and vectors.

$$Y = [K|V|V] \qquad (8.8.4)$$

and

$$Z = [\ |C|-C] \qquad (8.8.5)$$

where **K** is the block-column of real vectors, **V** the real part of the complex vectors, and **C** the irrational part of the complex vectors.

Theorem:

If **X** is invertible,

then

$$\mathbf{M} = [\mathbf{K}|\mathbf{V}|\mathbf{C}] \qquad\qquad (8.8.6)$$

is also invertible.

Proof:

Assume by contra-assumption, the existence of an **u'**, satisfying:

$$\mathbf{u'M} = \mathbf{u'}[\mathbf{K}|\mathbf{V}|\mathbf{C}] = [\mathbf{u'K}, \mathbf{u'V}, \mathbf{u'C}] = [0]' \qquad (8.8.7)$$

Then, by implication:

$$\mathbf{u'Y} = [\mathbf{u'K}, \mathbf{u'V}, \mathbf{u'V}] = [0]' \qquad\qquad (8.8.8)$$

and

$$\mathbf{u'Z} = [\ \ , \mathbf{u'C}, -\mathbf{u'C}] = [0]' \qquad\qquad (8.8.9)$$

and therefore

$$\mathbf{u'X} = \mathbf{u'Y} + \mathbf{u'Z}i = [0]' \qquad\qquad (8.8.10)$$

contradicting the assumed non-singularity of **X**.

q.e.d.

We partition $\mathbf{M}^{-1}$ in block-rows, as follows:

$$\mathbf{M}^{-1} = \begin{bmatrix} \mathbf{R'} \\ \hline \mathbf{U'} \\ \hline \mathbf{G'} \end{bmatrix} \qquad (8.8.11)$$

where $\mathbf{R'}$ is the r by n leading block-row, r being the number of real roots of $\mathbf{A}$. The two other block-rows, $\mathbf{U'}$ and $\mathbf{G'}$ are each of order 0.5(n-r) by n. Given that complex roots occur in pairs, 0.5(n-r) is always an integer number.

Then $\mathbf{X}^{-1}$ is:

$$\mathbf{W} = \mathbf{X}^{-1} = \begin{bmatrix} \mathbf{R'} \\ \hline 0.5\mathbf{U'} \\ \hline 0.5\mathbf{U'} \end{bmatrix} + \begin{bmatrix} \ \\ \hline -0.5\mathbf{G'} \\ \hline +0.5\mathbf{G'} \end{bmatrix} i \qquad (8.8.12)$$

as may be verified per separate block as follows:

The product of the leading block-row of $\mathbf{W}$ postmultiplied by the leading block-column of $\mathbf{X}$ also is the product of the leading block-row of $\mathbf{M}^{-1}$, postmultiplied by the leading block-column of $\mathbf{M}$, therefore:

$$\mathbf{R'} \mathbf{K} = \mathbf{I}_r \qquad (8.8.13)$$

We have similar products for off-diagonal blocks of $\mathbf{M}^{-1}\mathbf{M}$:

$$\mathbf{R'} \ \mathbf{V} = \mathbf{R'} \ \mathbf{C} = [0]$$

Therefore, the product of the leading block-row of $\mathbf{W}$, postmultiplied by the second block-column of $\mathbf{X}$ is:

$$\mathbf{R'}[\mathbf{V} + \mathbf{C}i] = \mathbf{R'V} + \mathbf{R'C}i = [0] + [0]i \qquad (8.8.14)$$

For the third block-column of $\mathbf{X}$ we have then immediately:

$$\mathbf{R'}[\mathbf{V} - \mathbf{C}i] = \mathbf{R'V} - \mathbf{R'C}i = [0] - [0]i \qquad (8.8.15)$$

For the product of the second block-row of $\mathbf{W}$, postmultiplied by the leading block-column of $\mathbf{X}$, we have:

$$[0.5 \; \mathbf{U}' \; -0.5 \; \mathbf{G}'i] \; \mathbf{K} = 0.5 \; \mathbf{U}'\mathbf{K} \; -0.5 \; \mathbf{U}'\mathbf{G}i$$

$$= \; 0.5 \; [0] \; -0.5 \; [0]i \qquad (8.8.16)$$

For the product of the second block-row of $\mathbf{W}$, post-multiplied by the second block-column of $\mathbf{X}$, we have:

$$[0.5 \; \mathbf{U}' \; -0.5 \; \mathbf{G}'i] \; [\mathbf{V} + \mathbf{C}i]$$

$$= 0.5 \; \mathbf{U}'\mathbf{V} + 0.5 \; \mathbf{G}'\mathbf{C} + [0.5 \; \mathbf{U}'\mathbf{C} - 0.5 \; \mathbf{G}'\mathbf{V}]i =$$

$$0.5 \; \mathbf{I}_{n-r} + 0.5 \; \mathbf{I}_{n-r} + [0.5 \; [0] - 0.5 \; [0]]i$$

$$= \; \mathbf{I}_{n-r} \qquad (8.8.17)$$

For the product of the second block-row of $\mathbf{W}$, postmultiplied by the third block-column of $\mathbf{X}$, we have:

$$[0.5 \; \mathbf{U}' \; -0.5 \; \mathbf{G}'i] \; [\mathbf{V} - \mathbf{C}i]$$

$$= 0.5 \; \mathbf{U}'\mathbf{V} - 0.5 \; \mathbf{G}'\mathbf{C} + [-0.5 \; \mathbf{U}'\mathbf{C} - 0.5 \; \mathbf{G}'\mathbf{V}]i =$$

$$0.5 \; \mathbf{I}_{n-r} - 0.5 \; \mathbf{I}_{n-r} + [-0.5 \; [0] - 0.5 \; [0]]i$$

$$= \; [0] \qquad (8.8.18)$$

The product of the third block-row of $\mathbf{W}$, postmultiplied by the leading block-column of $\mathbf{X}$, follows immediately by generalization of (8.8.16).

The product of the third block-row of $\mathbf{W}$, postmultiplied by the third block-column of $\mathbf{X}$ is:

$$[0.5 \; \mathbf{U}' + 0.5 \; \mathbf{G}'i] \; [\mathbf{V} - \mathbf{C}i]$$

$$= 0.5 \; \mathbf{U}'\mathbf{V} + 0.5 \; \mathbf{G}'\mathbf{C} + [-0.5 \; \mathbf{U}'\mathbf{C} + 0.5 \; \mathbf{G}'\mathbf{V}]i =$$

$$0.5 \; \mathbf{I}_{n-r} + 0.5 \; \mathbf{I}_{n-r} + [-0.5 \; [0] + 0.5 \; [0]]i$$

$$= \; \mathbf{I}_{n-r} \qquad (8.8.19)$$

Which completes our verification of $\mathbf{W} = \mathbf{X}^{-1}$.

Example:

$$\mathbf{A} = \begin{bmatrix} 1 & 1 & -1 \\ - & - & 1 \\ 1 & -1 & 1 \end{bmatrix}$$

Characteristic equation: $\lambda^3 - 2\lambda^2 + 3\lambda - 2 = 0$

Roots: $\lambda_1 = 1$, $\lambda_2 = (1 + \sqrt{7})/2$, $\lambda_2 = (1 - \sqrt{7})/2$.

$$\mathbf{X} = \begin{bmatrix} 1 & 2 & 2 \\ 1 & -2 & -2 \\ 1 & -1 & -1 \end{bmatrix} + \begin{bmatrix} - & - & - \\ - & - & - \\ - & -\sqrt{7} & \sqrt{7} \end{bmatrix} i$$

$$\mathbf{M} = \begin{bmatrix} 1 & 2 & - \\ 1 & -2 & - \\ 1 & -1 & -\sqrt{7} \end{bmatrix} ; \qquad \mathbf{M}^{-1} = \begin{bmatrix} .5 & .5 & - \\ .25 & -.25 & - \\ 1/28\sqrt{7} & 3/28\sqrt{7} & -1/7\sqrt{7} \end{bmatrix}$$

$$\mathbf{X}^{-1} = \begin{bmatrix} .5 & .5 & - \\ .125 & -.125 & - \\ .125 & -.125 & - \end{bmatrix} +$$

$$\begin{bmatrix} - & - & - \\ -1/56\sqrt{7} & -3/56\sqrt{7} & 1/14\sqrt{7} \\ 1/56\sqrt{7} & 3/56\sqrt{7} & -1/14\sqrt{7} \end{bmatrix} i$$

To obtain a more manipulatable matrix of row-vectors, we multiply
the real vector by 2, and the complex vectors by 56, to obtain:

$$\mathbf{W} = \begin{bmatrix} 1 & 1 & - \\ 7 & -7 & - \\ 7 & -7 & - \end{bmatrix} + \begin{bmatrix} - & - & - \\ -\sqrt{7} & -3\sqrt{7} & 4\sqrt{7} \\ \sqrt{7} & 3\sqrt{7} & -4\sqrt{7} \end{bmatrix} i$$

8.9 Blockdiagonalization of a matrix with complex roots

We now assume that a matrix can be diagonalized, even where this involves (as, for example, in the case of most permutation-operators), complex roots and therefore a complex transformation matrix and a complex diagonal matrix.

We then have the following

Theorem:

If the n by n matrix **A** is diagonalizable conform:

$$A = [K \mid V + Ci \mid V - Ci] \begin{bmatrix} \hat{L} & & \\ \hline & \hat{P} + \hat{Q}i & \\ \hline & & \hat{P} - \hat{Q}i \end{bmatrix} \begin{bmatrix} R' \\ \hline .5\ U' - .5\ G'i \\ \hline .5\ U' + .5\ G'i \end{bmatrix} \quad (8.9.1)$$

then **A** also permits expression as:

$$A = [K \mid V \mid C] \begin{bmatrix} \hat{L} & & \\ \hline & \hat{P} & \hat{Q} \\ \hline & -\hat{Q} & \hat{P} \end{bmatrix} \begin{bmatrix} R' \\ \hline U' \\ \hline G' \end{bmatrix} \quad (8.9.2)$$

where **R'**, **U'**, and **G'** are related to **K**, **V**, and **C**, conform (8.8.6) to (8.8.13).

Proof:

Relations (8.9.1) and (8.9.2) both work out as:

$$A = K\hat{L}R' + V\hat{P}U' + V\hat{Q}G' - C\hat{Q}U' + C\hat{P}G' \quad (8.9.3)$$

(in the case of (8.9.1), after cancellation of the irrational terms relating to $V\hat{Q}U$, $V\hat{P}G$, $C\hat{P}U$, and $V\hat{Q}G$ (times $\pm\ 0.5\ i$), and aggregation of $-.5\ C\hat{P}G'i^2 = +.5\ C\hat{P}G'$, with another term $0.5\ C\hat{P}G'$).

q.e.d.

Example:

$$\mathbf{A} = \begin{bmatrix} 3 & 4 & -4 \\ -4 & 7 & - \\ 4 & - & -1 \end{bmatrix}$$

$\lambda_1 = 3$:

$$\begin{bmatrix} 3 & 4 & -4 \\ -4 & 7 & - \\ 4 & - & -1 \end{bmatrix} \begin{bmatrix} 1 \\ 1 \\ 1 \end{bmatrix} = \begin{bmatrix} 3 \\ 3 \\ 3 \end{bmatrix}$$

$\lambda_2 = 3 - 4i$:

$$\begin{bmatrix} 3 & 4 & -4 \\ -4 & 7 & - \\ 4 & - & -1 \end{bmatrix} \begin{bmatrix} 1 + i \\ 1 \\ i \end{bmatrix} = \begin{bmatrix} 7 - i \\ 3 - 4i \\ 4 + 3i \end{bmatrix} = \begin{bmatrix} 1-i \\ 1 \\ i \end{bmatrix} (3-4i)$$

$\lambda_3 = 3 + 4i$:

$$\begin{bmatrix} 3 & 4 & -4 \\ -4 & 7 & - \\ 4 & - & -1 \end{bmatrix} \begin{bmatrix} 1 - i \\ 1 \\ -i \end{bmatrix} = \begin{bmatrix} 7 + i \\ 3 + 4i \\ 4 - 3i \end{bmatrix} = \begin{bmatrix} 1-i \\ 1 \\ -i \end{bmatrix} (3+4i)$$

Accordingly, we have:

$$\mathbf{M} = [\, \mathbf{K} \mid \mathbf{V} \mid \mathbf{C} \,] = \begin{bmatrix} 1 & 1 & 1 \\ 1 & 1 & - \\ 1 & - & 1 \end{bmatrix}$$

and therefore:

$$\mathbf{M}^{-1} = \begin{bmatrix} \mathbf{R'} \\ \hline \mathbf{U'} \\ \hline \mathbf{G'} \end{bmatrix} = \begin{bmatrix} -1 & 1 & 1 \\ 1 & - & -1 \\ 1 & -1 & - \end{bmatrix}$$

and $\mathbf{A}$ is expressed as:

$$\mathbf{A} = \begin{bmatrix} 1 & 1 & 1 \\ 1 & 1 & - \\ 1 & - & 1 \end{bmatrix} \begin{bmatrix} 3 & - & - \\ - & 3 & -4 \\ - & 4 & 3 \end{bmatrix} \begin{bmatrix} -1 & 1 & 1 \\ 1 & - & -1 \\ 1 & -1 & - \end{bmatrix}$$

Exercise:

For $\mathbf{P} =$

$$\begin{bmatrix} - & - & - & 1 \\ 1 & - & - & - \\ - & 1 & - & - \\ - & - & 1 & - \end{bmatrix}$$

which is diagonalizable (see section 8.6),

find the roots and a valid set of column- and row-vectors.

Hence, grouping the column-vectors into a matrix $\mathbf{X}$, express $\mathbf{P}$ as:

$$\mathbf{P} = \mathbf{X}\,\hat{\mathbf{L}}\,\mathbf{X}^{-1}$$

as well as conform (8.9.2).

(There is an answersheet at the end of the chapter).

8.10 **Partial differentiation of a root**

In this section we consider the changes in the roots, which arise
from small marginal variations of the corresponding matrix.

We first of all differentiate the relationship between the root
and the corresponding vector, assuming that λ is a non-repeated
root of $\mathbf{A}$. From:

$$[\mathbf{A} - \mathbf{I}\lambda]\,\mathbf{x} = [0] \tag{8.10.1}$$

which needs to hold before and after any change in $\mathbf{A}$ and a
corresponding change in $\mathbf{x}$ and λ, we obtain the differential
requirement:

$$[d\mathbf{A} - \mathbf{I}d\lambda]\,\mathbf{x} + [\mathbf{A} - \mathbf{I}\lambda]\,d\mathbf{x} = [0] \tag{8.10.2}$$

Therefore, if <u>only one</u> element a_{ij} of $\mathbf{A}$ actually changes,

$$d\mathbf{A} = \mathbf{E}_{ij}\,da_{ij}$$

(8.10.2) reduces to:

$$[E_{ij} da_{ij} - I d\lambda] \, \mathbf{x} + [\mathbf{A} - I\lambda] \, d\mathbf{x} = [0] \qquad (8.10.3)$$

On pre-multiplication of (8.10.3), by the row-vector associated with λ as root we obtain:

$$w_i \, da_{ij} \, x_j - \mathbf{w'x} \, d\lambda = 0 \qquad (8.10.4)$$

Therefore, if we normalize the vectors to the requirement:

$$\mathbf{w'x} = 1 \qquad (8.10.5)$$

the partial differential quotient of any root in relation to any element of $\mathbf{A}$ is:

$$\partial\lambda/\partial a_{ij} = w_i \, x_j \qquad (8.10.6)$$

where it should be understood that $\mathbf{x}$ and $\mathbf{w'}$ are the column- and row-vectors, as corresponding to λ as root.

We now relax the restrictive assumption that only one element of A changes, and obtain the more general result:

$$\partial\lambda_i/\partial\mathbf{A} = [w_i \, x_i'] \qquad (8.10.7)$$

Note, that (8.10.7) simply is (8.10.6), stated n^2 times, for the n^2 elements of the n by n matrix $\mathbf{A}$. If $\mathbf{A}$ is itself a function of some other variables, it is often more useful to revert back to (8.10.6). For example, if $\mathbf{C}$ is the matrix of partial derivatives of the elements of $\mathbf{A}$ relative to the parameter δ:

$$\partial a_{ij}/\partial\delta = c_{ij} \qquad (8.10.8)$$

$$(j = 1, 2, \ldots n, \; i = 1, 2, \ldots n)$$

then we may develop $\partial\lambda/\partial\delta$ by the chain rule, and the curiosity of having a row-vector transposed as a column and a column-vector transposed as a row disappears:

$$\partial\lambda_i/\partial\delta = \sum_{h=1}^{n} \sum_{j=1}^{n} w_{ih} \, c_{ij} \, x_{ji} = w_i' \, C \, x_i \qquad (8.10.9)$$

where λ_i is the ith root, $\mathbf{w'}_i$ is the corresponding row-vector, and $\mathbf{x}_i$ is the corresponding column-vector, and the inner product of these two vectors obeys a generalization of (8.10.5):

$$\mathbf{w}'_i \, \mathbf{x}_i \; = \; 1 \qquad\qquad (8.10.10)$$

We now consider the situation which arises if λ is a repeated root of $\mathbf{A}$. We first of all note, that we lack a proof of the existence of any vectors satisfying (8.10.10) in that case. If no such vectors exist, then (8.10.4) becomes a contradictory requirement, and the concept of a partial derivative of the repeated root in question lacks meaning in that case.

Example:

$$\mathbf{A} \; = \; \begin{bmatrix} - & a_{12} \\ 1 & - \end{bmatrix} \quad a_{12} = 0, \qquad\qquad \mathbf{x} \; = \; \begin{bmatrix} - \\ 1 \end{bmatrix}, \quad \mathbf{w}' = [1 \quad -]$$

Then the root is $\lambda = \pm \sqrt{a_{12}} = 0$, and changes from real to irrational at the point $a_{12} = 0$.

If, on the other hand there are several pairs of vectors satisfying (8.10.10), it means that changes in $\mathbf{A}$ cause the repeated root to split into separate roots.

Example:

$$\mathbf{A} = \begin{bmatrix} 1 & - \\ - & 1 \end{bmatrix}$$

$$\partial\lambda_1/\partial\mathbf{A} = \mathbf{w}_1 \mathbf{x}'_1 = \begin{bmatrix} 1 & - \\ - & - \end{bmatrix} \; ; \qquad \partial\lambda_2/\partial\mathbf{A} = \mathbf{w}_2 \mathbf{x}'_2 = \begin{bmatrix} - & - \\ - & 1 \end{bmatrix}$$

Relation (8.10.7) gives the matrix of partial derivatives of a root, relative to the elements of $\mathbf{A}$ as such. In fact, applications often concern matrix expressions, and we will deal explicitly with a particular matrix expression.

Suppose one is concerned with the statistical estimation of a system of dynamic difference-equations:

$$\mathbf{A}\,\mathbf{y}_t \; = \; \mathbf{B}\,\mathbf{y}_{t-1} + \mathbf{C}\,\mathbf{x}_t + \mathbf{u}_t$$

where $\mathbf{x}_t$ is an a vector (are the vectors) of exogenously determined variables, u_t are random disturbances, y are the dynamic variables which are determined by the system, and we wish to estimate (some or all of) the coefficients in the matrices A, B, and C. Here A and B are both of order n by n, and the dynamic stability of the system revolves around the roots of $A^{-1}B$.

Methods for estimating these coefficients often are iterative, and require a 'starting guess', which assumes the system to be dynamically stable, which is the case if the roots of A^{-1} are all inside the unit circle (less than one in absolute value). As a completely arbitrary starting guess may well fail to meet that condition, it would be helpful to know in which direction the initial guesses should be adjusted, in order to make them meet that condition. Being able to differentiate the roots relative to the elements of A and B is by no means the full solution of that problem, but it obviously helps, and we address ourselves to that problem.

Reference to section 4.8 allows us to express $d[A^{-1}B]$ as:

$$d[A^{-1}B] = dA^{-1} B + A\, dB = -A^{-1}\, dA\, A^{-1} B + A^{-1}\, dB$$

$$(8.10.11)$$

Therefore, by (8.10.6) and (8.10.11), we evaluate any differential change in λ_k as:

$$d\lambda_k = w'_k [-A^{-1}\, dA\, A^{-1} B + A^{-1}dB]\, x_k \qquad (8.10.12)$$

and therefore for a change in an individual element of A:

$$\partial\lambda/\partial a_{ij} = w'_k [-A^{-1} E_{ij} A^{-1} B]\, x_k \qquad (8.10.13)$$

and

$$\partial\lambda_k/db_{ij} = w_k' [A^{-1} E_{ij}]\, x_k \qquad (8.10.14)$$

Exercise:

For $\mathbf{A} = \begin{bmatrix} 1 & 2 \\ 2 & 1 \end{bmatrix}$ $\mathbf{B} = \begin{bmatrix} - & 1 \\ & 1 \end{bmatrix}$

develop the partial derivatives of the dominant root of $\mathbf{A}^{-1}\mathbf{B}$ relative to a_{11}, a_{12}, and b_{12}. (There is an answersheet at the end of the chapter)

8.11 The calculation of roots by repeated multiplication

The calculation of all the roots of an arbitrary n by n matrix is not straightforward, and goes beyond the aims of this book. We can extract the characteristic equation, but that does not solve the roots: the solution of a polynomial equation of order n is difficult in itself.

The problem becomes easier if we know that all the roots are real and distinct.

It is then first of all helpful to note that, if $\mathbf{A}$ is has n independent vectors and one of its roots dominates all the others, then the higher powers of $\mathbf{A}$ will, after suitable scaling, converge to the outer product of the corresponding vector. Or, to formulate the same statement somewhat more precise, we state the following

Theorem:

If $\mathbf{x}_1$ and $\mathbf{w'}_1$ are the column and row-vectors associated with the distinct and dominant root λ_1 of $\mathbf{A}$, and the column vectors of $\mathbf{A}$ form an invertible matrix $\mathbf{V}$, ($\mathbf{A}$ is diagonalizable),

then

$$[\mathbf{A}\lambda^{-1}]^p = \mathbf{V} \, [\hat{\mathbf{L}} \, \lambda^{-1}]^p \, \mathbf{V}^{-1}$$

will, when p approaches infinity, converge to $\mathbf{v}_1 \, \mathbf{w}_1'$

Proof:

$[\hat{L} \lambda^{-1}]^p$ converges to E_{11}

(a diagonal matrix with $e_{11} = 1$ as the only non-zero element).

q.e.d.

A more or less arbitrary vector, when expressed as a linear combination of all the vectors of **A**, will usually contain a non-zero contribution from v_1, the vector associated with the dominant root. (The exception is orthogonality to w'_1.) The computational procedure incorporated in the tutorial going with this book, substitutes repeated multiplication of a vector (by default a summation vector), for the development of powers of the matrix itself, i.e. it develops $[A \ \lambda^{-1}]^p$ **s**, on the assumption that **s** is not orthogonal to v_1.

If the dominant root is complex, this procedure will obviously fail to converge.

Once a dominant real root and the associated column and row vectors have been calculated, we can make use of the following

Theorem

If **x** and **w'** obey the conditions:

$A \ x = x \ \lambda \neq [0]$ $w'A = \lambda \ w \neq [0]'$ and $w'x = 1$

then

$[A - x\lambda w'] \ y = y \ \delta, \quad \delta \neq \lambda$ implies $A \ y = y \ \delta$

Proof: **y** is orthogonal to **w'**   q.e.d.

Since $[A - x\lambda w']$ also has **x** as a vector with a zero root:

$[A - x\lambda w'] \ x = A \ x - x\lambda(w'x) = x\lambda - x \ \lambda = [0],$

the subtraction of $x\lambda w'$ from **A** amounts to setting one root at zero, after which the same procedure can be applied once more, to try to calculate the next root and its vectors.

As a number of example-matrices in this book have the summation-vector as a characteristic column-vector, this default starting-vector is not always used: If, for a second or following root, we find **A s** = [0] (as would be the case, if a root for which the summation-vector itself is the vector, had been set at zero), then one of the columns of **A** is used instead. For the initialization of the row-vector, this substitution will also arise, if the initial column-vector (which then is a column of **A**), is orthogonal to the row-summation vector.

It should be stressed that this is not a generally valid algorithm which will work for all matrices. However, all the examples in this book which involve diagonalizable matrices with real roots only, including some cases of repeated roots, were successfully after-checked by its implementation in the root-option of ILLUSTRATE, and the same applies for dominant real roots.

If the dominant root is complex, non-convergence arises, and is bound to arise, as it did.

8.12 Orthogonality in complex vectors

Theorem

Given any complex vector **x** + **u**i

there exists a complex multiplier $(\alpha + \beta i)$,

for which the real and irrational part of

[**x** + **u**i]$(\alpha + \beta i)$ = [**x**α - **u**β] + [**x**β + **u**α]i

are orthogonal to each other.

Proof

We express the inner product of the real and irrational parts of the vector, with suppression of the factor i as:

$$[\mathbf{x}\alpha - \mathbf{u}\beta]'[\mathbf{x}b + \mathbf{u}\alpha] = (\mathbf{x}'\mathbf{x} - \mathbf{u}'\mathbf{u})\alpha\beta + \mathbf{x}'\mathbf{u}(\alpha^2 - \beta^2)$$

For $\mathbf{x}'\mathbf{u} = 0$, $\alpha = 1$, $\beta = 0$ provides the proof.

For $\mathbf{x}'\mathbf{x} = \mathbf{u}'\mathbf{u}$, any $\alpha = \beta$ provides the proof.

Otherwise, we set β at the arbitrary value $\beta = 1$, and we are able to solve α from:

$$\mathbf{x}'\mathbf{u}\alpha^2 + (\mathbf{x}'\mathbf{x} - \mathbf{u}'\mathbf{u})\alpha - \mathbf{x}'\mathbf{u} = 0$$

a quadratic equation which always has two distinct real roots.

$$\alpha = \left((\mathbf{u}'\mathbf{u} - \mathbf{x}'\mathbf{x}) \pm \sqrt{(\mathbf{x}'\mathbf{x} - \mathbf{u}'\mathbf{u})^2 + 4(\mathbf{x}'\mathbf{u})^2} \right) / 2\mathbf{x}'\mathbf{u}$$

q.e.d.

The above theorem may, or may not be of any direct computational use. In addition, there are applied contexts, where it is important to know whether or not the real parts of all roots of a particular matrix are non-negative (myself [24]), and the above theorem allows us to prove the following further

Theorem

If $\mathbf{B} = [\mathbf{A} + \mathbf{A}']$ is positive (semi-)definite, then the real parts of all the roots of $\mathbf{A}$ are positive (non-negative).

Proof

Concerning the real roots:

If by contra-assumption

$$\mathbf{A}\,\mathbf{x} = \mathbf{x}\lambda, \quad \mathbf{x} \neq [0], \quad \lambda < 0 \text{ (resp. } \leq 0) \text{ were true,}$$

$$2\mathbf{x}'A\ \mathbf{x} = \mathbf{x}'B\ \mathbf{x} = 2\mathbf{x}'\mathbf{x}\lambda < 0 \ (\text{resp} \leq 0)$$

would follow, contradicting the assumed type of definiteness of B.

q.e.d. for the real roots.

Concerning the complex roots, with $\mathbf{x} \neq [0]$:

Suppose that, also contrary to what the theorem states,

$$A\ [\mathbf{x} + \mathbf{u}i] = [\mathbf{x} + \mathbf{u}i](\alpha + \beta i) = [\mathbf{x}\alpha - \mathbf{u}\beta] + [\mathbf{x}\beta + \mathbf{u}\alpha]i$$

and therefore by the real part only

$$A\ \mathbf{x} = \mathbf{x}\alpha - \mathbf{u}\beta$$

were true for $\mathbf{x} \neq [0]$, $\alpha < 0$. (resp ≤ 0)

By the previous theorem, we can require $\mathbf{x}'\mathbf{u} = 0$, and therefore

$$\mathbf{x}'A\ \mathbf{x} = \mathbf{x}'\mathbf{x}\alpha < 0 \ (\text{resp} \leq 0)$$

and the proof of this theorem, as supplied above for a real root of λ also applies for a complex root with real part α.

q.e.d. for the complex roots with $\mathbf{x} \neq [0]$.

Concerning any complex roots with $\mathbf{x} = [0]$:

For $\mathbf{x} = [0]$, and hence $\mathbf{u} \neq [0]$, the irrational part of

$$A\ [\mathbf{x} + \mathbf{u}i] = [\mathbf{x} + \mathbf{u}i](\alpha + \beta i) = [\mathbf{x}\alpha - \mathbf{u}\beta] + [\mathbf{x}\beta + \mathbf{u}\alpha]i$$

is

$$A\ \mathbf{u}\ i = \mathbf{u}\alpha\ i$$

and therefore

$$\mathbf{u}'A\ \mathbf{u} = \mathbf{u}'\mathbf{u}\alpha$$

and the proof supplied above again generalizes.

q.e.d. for any complex roots with $\mathbf{x} = [0]$.

q.e.d.

Note, however, that the criterion of the definiteness of $[\mathbf{A} + \mathbf{A'}]$ is over-strict. It is sufficient, but there are matrices of which the roots are positive, but $[\mathbf{A} + \mathbf{A'}]$ is not positive definite.

Example:

$$\mathbf{A} = \begin{bmatrix} 5 & 16 \\ 1 & 5 \end{bmatrix} \qquad\qquad \mathbf{A} + \mathbf{A'} = \begin{bmatrix} 10 & 17 \\ 17 & 10 \end{bmatrix}$$

$$\lambda = 5 \pm 4 \qquad\qquad\qquad \text{indefinite}$$

Answersheet on vectors and diagonalization

Concerning (1):
The characteristic equation is:

$$\begin{vmatrix} 2-\lambda & 8 & - \\ 2 & 2-\lambda & - \\ 1 & 1 & 1-\lambda \end{vmatrix} = 0.$$

Development of this determinant, by the third column, gives rise to the equation:

$$(1-\lambda)\left((2-\lambda)^2 - 16\right) = 0.$$

Therefore: $\lambda_1 = 1$.

The two other roots are most easily established by first solving the zero-value of the factor:

$$\left((2-\lambda)^2 - 16\right) = 0 \quad \text{as} \quad (\lambda-2)^2 = 16, \quad \text{therefore:}$$

$$\lambda-2 = \pm 4, \quad \lambda_2 = 6, \quad \lambda_3 = -2.$$

Concerning (2).

The vector associated with the root $\lambda_1 = 1$ is identified immediately as:

$$\mathbf{x} = \begin{bmatrix} - \\ - \\ 1 \end{bmatrix}, \quad \begin{bmatrix} 2 & 8 & - \\ 2 & 2 & - \\ 1 & 1 & 1 \end{bmatrix} \begin{bmatrix} - \\ - \\ 1 \end{bmatrix} = \begin{bmatrix} - \\ - \\ 1 \end{bmatrix} = \begin{bmatrix} - \\ - \\ 1 \end{bmatrix} \times 1$$

For $\lambda_2 = 6$, we consider first of all the block-triangular structure of $\mathbf{A}$, and of $[\mathbf{A} - \mathbf{I}\lambda]$.

$[\mathbf{A} - \mathbf{I}\lambda]\mathbf{x}$ partitions as:

$$\left.\begin{array}{l} [\mathbf{A}_{11} - \mathbf{I}\lambda]\mathbf{x}_1 = [0] \\[2mm] \mathbf{A}_{21}\mathbf{x}_1 + [\mathbf{A}_{22} - \mathbf{I}\lambda]\mathbf{x}_2 = [0] \end{array}\right\}$$

and the first block-equation needs to hold on its own.

Therefore,   for $\mathbf{A}_{11} = \begin{bmatrix} 2 & 8 \\ 2 & 2 \end{bmatrix}$

$$(2-6)x_1 + 8x_2 = -4x_1 + 8x_2 = 0 \quad \rightarrow \quad x_1 = 2x_2$$

and

$$2x_1 + (2-6)x_2 = 2x_1 - 4x_2 = 0 \quad \rightarrow \quad x_1 = 2x_2$$

The corresponding value for x_3 is now found with the help of the third equation:

$$x_1 + x_2 + (1-6)x_3 = 0 \rightarrow 5x_3 = x_1 + x_2 .$$

In the interest of avoiding fractional numbers, the arbitrary requirement is therefore set as:

$$x_2 = 5, \text{ and therefore: } x_1 = 2x_2 = 10,$$

The third equation, already transformed to: $5x_3 = x_1 + x_2$, now yields:

$$5x_3 = x_1 + x_2 = 15 \rightarrow x_3 = 3 .$$

Similarly, for $\lambda_3 = -2$:

$$(2+2)x_1 + 8x_2 = 0 \rightarrow 4x_1 + 8x_2 = 0 \rightarrow x_1 = -2x_2$$

as well as by the second equation:

$$2x_2 + (2+2)x_2 = 0 \rightarrow 2x_1 + 4x_2 = 0 \rightarrow x_1 = -2x_2$$

A requirement on x_3 is now found from the third equation:

$$x_1 + x_2 + (1+2)x_3 = 0 \rightarrow 3x_3 = -x_1 - x_2$$

In the interest of avoiding fractions, we now put the arbitrary requirement as $x_2 = 3$.

Therefore: $x_1 = -2x_2 = -6$, and x_3 is solved from:

$$3x_3 = -x_1 - x_2 = -(-6) -3 = 6 - 3 = 3 \rightarrow x_3 = 1.$$

Summary of the results so far:

$$\begin{bmatrix} 2 & 8 & - \\ 2 & 2 & - \\ 1 & 1 & 1 \end{bmatrix} \begin{bmatrix} 10 & -6 & - \\ 5 & 3 & - \\ 3 & 1 & 1 \end{bmatrix} = \begin{bmatrix} 60 & 12 & - \\ 30 & -6 & - \\ 18 & -2 & 1 \end{bmatrix} =$$

$$\begin{bmatrix} 10 & -6 & - \\ 5 & 3 & - \\ 3 & 1 & 1 \end{bmatrix} \begin{bmatrix} 6 & - & - \\ - & -2 & - \\ - & - & 1 \end{bmatrix}$$

Concerning (3), we first of all need to calculate the inverse $\mathbf{X}^{-1}$

v_1	v_2	v_3	=	u_1	u_2	u_3	Σ
$\lvert\underline{10}\rvert$	-6	-		1	-	-	5
5	3	-		-	1	-	9
3	1	1		-	-	1	6

v_1	v_2	v_3	=	u_1	u_2	u_3	Σ
1	-0.6	-		0.1	-	-	0.5
-	$\lvert\underline{6}\rvert$	-		-0.5	1	-	6.5
-	2.8	1		-0.3	-	1	4.5

$$
\begin{array}{ccc}
v_1 & v_2 & v_3
\end{array}
=
\begin{array}{ccc}
u_1 & u_2 & u_3
\end{array}
\quad \Sigma
$$

$$
\begin{bmatrix}
1 & - & - \\
- & 1 & - \\
- & - & 1
\end{bmatrix}
\begin{bmatrix}
0.05 & 0.1 & - \\
-0.0833 & 0.1667 & - \\
-0.0667 & -0.4667 & 1
\end{bmatrix}
\begin{array}{c}
1.15 \\
1.0833 \\
1.4667
\end{array}
$$

We are now able to express $\mathbf{A}$ as:

$$
\begin{bmatrix}
10 & -6 & - \\
5 & 3 & - \\
3 & 1 & 1
\end{bmatrix}
\begin{bmatrix}
6 & - & - \\
- & -2 & - \\
- & - & 1
\end{bmatrix}
\begin{bmatrix}
0.05 & 0.1 & - \\
-0.0833 & 0.1667 & - \\
-0.0667 & -0.4667 & 1
\end{bmatrix}
=
$$

$$
\begin{bmatrix}
60 & 12 & - \\
30 & -6 & - \\
18 & -2 & 1
\end{bmatrix}
\begin{bmatrix}
0.05 & 0.1 & - \\
-0.0833 & 0.1667 & - \\
-0.0667 & -0.4667 & 1
\end{bmatrix}
=
\begin{bmatrix}
2 & 8 & - \\
2 & 2 & - \\
1 & 1 & 1
\end{bmatrix}
$$

Answersheet on the powers of a Jordan block

We use the recursion formula:

$$
[\mathbf{U}^p]_{ij} = ((\lambda p)/(p-j+i)) \, [\mathbf{U}^{p-1}]_{ij} \qquad (p > j-i) \qquad (8.4.13)
$$

for

$$
\mathbf{U} =
\begin{bmatrix}
2 & 1 & - & - & - \\
- & 2 & 1 & - & - \\
- & - & 2 & 1 & - \\
- & - & - & 2 & 1 \\
- & - & - & - & 2
\end{bmatrix}
$$

We have $\mathbf{U}$ itself already, therefore we start with $\mathbf{U}^2$:

$$
\lambda p = 2 \times 2 = 4
$$

For the diagonal cells, $j = i$ we have:

$$
(\lambda p)/(p-j+i) = (\lambda p)/p = \lambda
$$

i.e. we multiply by $\lambda = 2$, and the diagonal cells of $\mathbf{U}^2$ are all equal to:

$$[\mathbf{U}^2]_{ii} = \lambda \times [\mathbf{U}]_{ii} = 2 \times 2 = 4.$$

For j = i+1 we divide by p-1 = 1:

$$(\lambda p)/(p-i+j) = (2 \times 2) / (2 - 1) = 4$$

These cells in **U** are so far all unity elements, therefore in $\mathbf{U}^2$, they all become equal to:

$$[\mathbf{U}^2]_{i,i+1} = (\lambda \times 2) \times \mathbf{U}_{i,i+1} = 4.$$

Adding a new row of unity elements, we obtain:

$$\mathbf{U}^2 = \begin{bmatrix} 4 & 4 & 1 & - & - \\ - & 4 & 4 & 1 & - \\ - & - & 4 & 4 & 1 \\ - & - & - & 4 & 4 \\ - & - & - & - & 4 \end{bmatrix}$$

We now again multiply the diagonal elements by $\lambda = 2$, and they will become:

$$[\mathbf{U}^3]_{ii} = \lambda \times [\mathbf{U}^2]_{ii} = 2 \times 4 = 8.$$

For the line one above the diagonal we have:

$$(\lambda p)/(p-1) = (2 \times 3) / (3 - 1) = 6 / 2 = 3,$$

and these elements become $3 \times 4 = 12$.

For j - i = 2, there is no division by 2:

$$(\lambda p)/(p-i+j) = (2 \times 3) / (3 - 2) = 6 / 1 = 6,$$

and these elements become:

$$[\mathbf{U}^3]_{i,i+2} = (\lambda \times 3) \times [\mathbf{U}^2]_{i,i+2} = 6.$$

Therefore, once more adding a line of unity elements:

$$U^4 = \begin{bmatrix} 8 & 12 & 6 & 1 & - \\ - & 8 & 12 & 6 & 1 \\ - & - & 8 & 12 & 6 \\ - & - & - & 8 & 12 \\ - & - & - & - & 8 \end{bmatrix}$$

Answersheet on idem-(nil)-potent matrices and periodics

$$\begin{bmatrix} 0.5 & -0.5 \\ -0.5 & 0.5 \end{bmatrix} \begin{bmatrix} 0.5 & -0.5 \\ -0.5 & 0.5 \end{bmatrix} = \begin{bmatrix} 0.5 & -0.5 \\ -0.5 & 0.5 \end{bmatrix}$$

immediately proves that this is an idempotent matrix.

For $A \neq [0]$, the same matrix cannot then also be nilpotent: all its powers are the matrix itself, which is not a null matrix. The definition of a periodic is included in that of being idempotent, therefore this _is_ a periodic.

Exactly the same arguments and conclusions apply for:

$$\begin{bmatrix} 0.5 & 0.5 \\ 0.5 & 0.5 \end{bmatrix} \begin{bmatrix} 0.5 & 0.5 \\ 0.5 & 0.5 \end{bmatrix} = \begin{bmatrix} 0.5 & 0.5 \\ 0.5 & 0.5 \end{bmatrix}$$

and, even without symmetry for:

$$\begin{bmatrix} 0.5 & 1 \\ 0.25 & 0.5 \end{bmatrix} \begin{bmatrix} 0.5 & 1 \\ 0.25 & 0.5 \end{bmatrix} = \begin{bmatrix} 0.5 & 1 \\ 0.25 & 0.5 \end{bmatrix}$$

These three matrices all have roots 1 and 0.

For $A = \begin{bmatrix} 0.25 & 0.25 \\ 0.25 & 0.25 \end{bmatrix}$

of which the roots obviously are half those of $\begin{bmatrix} 0.5 & 0.5 \\ 0.5 & 0.5 \end{bmatrix}$

i.e $\lambda_1 = 0.5$, $\begin{bmatrix} 0.25 & 0.25 \\ 0.25 & 0.25 \end{bmatrix} \begin{bmatrix} 1 \\ 1 \end{bmatrix} = \begin{bmatrix} 0.5 \\ 0.5 \end{bmatrix}$

and $\lambda_2 = 0$, $\begin{bmatrix} 0.25 & 0.25 \\ 0.25 & 0.25 \end{bmatrix} \begin{bmatrix} 1 \\ -1 \end{bmatrix} = \begin{bmatrix} - \\ - \end{bmatrix}$

no power can be **A** itself, i.e. this matrix is neither idempotent nor a periodic, nor can it be nilpotent, as that requires all roots to be zero.

$\mathbf{A} = \begin{bmatrix} 1 & - \\ - & - \end{bmatrix}$ is again idempotent, and by implication a periodic.

$\mathbf{A} = \begin{bmatrix} - & - \\ 1 & - \end{bmatrix}$ is nilpotent:

its second power is a null matrix. As its third and following powers then also are null matrices, this matrix cannot be a periodic, let alone idempotent,

For $\mathbf{A} = \begin{bmatrix} - & 1 \\ 1 & \end{bmatrix}$

we have

$\begin{bmatrix} - & 1 \\ 1 & - \end{bmatrix} \begin{bmatrix} - & 1 \\ 1 & - \end{bmatrix} = \begin{bmatrix} 1 & - \\ - & 1 \end{bmatrix}$

therefore this matrix is not idempotent, but its third power is **A** itself, i.e. it is a periodic.

In fact all odd powers are **A** itself, and all even powers are I_2. This matrix is therefore not nilpotent (The roots are 1 and -1, and not 0.)

Finally, since we did not actually exclude the null matrix from the definition of either an idempotent matrix or a periodic, the null matrix is idempotent: $\mathbf{A}^2 = \mathbf{A} = [0]$ as well as a periodic, and, of course nilpotent.

Answersheet on the (complex) roots and vectors of P.

For $\mathbf{P} = \begin{bmatrix} - & - & - & 1 \\ 1 & - & - & - \\ - & 1 & - & - \\ - & - & 1 & - \end{bmatrix}$

the characteristic equation is (see section 8.6) $\lambda^4 = 1$, and the roots are:

$\lambda_1 = 1, \quad \lambda_2 = -1, \quad \lambda_3 = i, \quad \lambda_4 = -i.$

The corresponding matrix of column vectors (also from section 8.6) is:

$$\mathbf{X} = \begin{bmatrix} 1 & -1 & -i & i \\ 1 & 1 & -1 & -1 \\ 1 & -1 & i & -i \\ 1 & 1 & 1 & 1 \end{bmatrix}$$

However, the matrix of row-vectors, offered in section 8.6, is not $\mathbf{X}^{-1}$.

We proceed to invert, using a generalization of the all-integer elimination method, as follows:

u_1	u_2	u_3	u_4	=	w_1	w_2	w_3	w_4	Σ
$\lvert 1 \rvert$	-1	$-i$	i		1	$-$	$-$	$-$	1
1	1	-1	-1		$-$	1	$-$	$-$	1
1	-1	i	$-i$		$-$	$-$	1	$-$	1
1	1	1	1		$-$	$-$	$-$	1	5

Our first inversion-step consists simply of subtracting the leading row from the three other ones.

u_1	u_2	u_3	u_4	=	w_1	w_2	w_3	w_4	Σ
1	-1	-i	i		1	-	-	-	1
-	2	-1+i	-1-i		-1	1	-	-	-
-	-	2i	-2i		-1	-	1	-	-
-	2	1+i	1-i		-1	-	-	1	4

In the interest of avoiding fractions, and, if at all possible, complex division, we multiply the leading row by 2, rather than dividing the pivotal row by 2. (Strictly, the all-integer method also calls for multiplication of rows 3 and 4 by 2).

u_1	u_2	u_3	u_4	=	w_1	w_2	w_3	w_4	Σ
2	-2	-2i	2i		2	-	-	-	2
-	|2|	-1+i	-1-i		-1	1	-	-	-
-	-	2i	-2i		-1	-	1	-	-
-	2	1+i	1-i		-1	-	-	1	4

We now add the second row to the leading row, and subtract the second row from the last row, to obtain:

u_1	u_2	u_3	u_4	=	w_1	w_2	w_3	w_4	Σ
2	-	-1-i	-1+i		1	1	-	-	2
-	2	-1+i	-1-i		-1	1	-	-	-
-	-	2i	-2i		-1	-	1	-	-
-	-	2	2		-	-1	-	1	4

We now multiply rows 1 and 2 by 2i, and row 4 by i, to obtain:

u_1	u_2	u_3	u_4	=	w_1	w_2	w_3	w_4	Σ
4i	-	2-2i	-2-2i		2i	2i	-	-	4i
-	4i	-2+2i	2-2i		-2i	2i	-	-	-
-	-	|2i|	-2i		-1	-	1	-	-
-	-	2i	2i		-	-i	-	i	4i

The third elimination-step is now made, by adding the third row, to rows 1 and 2 with multiples of 1+i, and 1-i, respectively, and subtracting the third row from the fourth row (adding with a multiple of -1):

The following recapitulation of the complex arithmetic is useful in that context:

For the addition of $(1+i)$ times the second row to the first row:

$(1+i)2i = 2i-2 = -2+2i$, causing, as usual in any method of row-operations, the vanishment of the figure of $2-2i$ in the leading row cell of the u_3 column.

$(1+i)(-2i) = -2i+2 = 2-2i$. Added to the figure of $-2-2i$ in the leading row of the u_4 column we get $-4i$ as result.

$(1+i)(-1) = -1-i$. Added to the figure of $2i$ in the leading row of the w_1 column, we get $-1+i$ as result.

$(1+i) \times 1 = 1+i$. Added to the zero in the leading cell of the w_3 column, we simply make the entry of $1+i$.

For the addition of $(1-i)$ times the third row to the second row:

$(1-i)2i = -2i+2 = 2-2i$, again causing vanishment of the figure in the pivotal column.

$(1-i)(-2i) = -2i-2 = -2-2i$. Added to the $2-2i$ in the second row of the u_4 column, we get $-4i$

$(1-i)(-1) = -1+i$. Added to the $-2i$ in the second row entry of the w_1 column, we get $-1-i$.

Finally, adding the 1 in the w_3 column, multiplied by $(1-i)$ to zero, amounts to making an entry of $1-i$.

The straightforward subtraction of the third row from the fourth row should be obvious, and the successor tableau is:

u_1	u_2	u_3	u_4	=	w_1	w_2	w_3	w_4	Σ
4i	–	–	-4i		-1+i	2i	1+i	–	4i
–	4i	–	-4i		-1-i	2i	1-i	–	–
–	–	2i	-2i		-1	–	1	–	–
–	–	–	4i		1	-i	-1	i	4i

we now multiply the third row by 2, to obtain the intermediate tableau:

u_1	u_2	u_3	u_4	=	w_1	w_2	w_3	w_4	Σ
4i	-	-	-4i		-1+i	2i	1+i	-	4i
-	4i	-	-4i		-1-i	2i	1-i	-	-
-	-	4i	-4i		-2	-	2	-	-
-	-	-	|4i|		1	-i	-1	i	4i

Adding the fourth row to the three other rows, we obtain:

u_1	u_2	u_3	u_4	=	w_1	w_2	w_3	w_4	Σ
4i	-	-	-		i	i	i	i	8i
-	4i	-	-		-i	i	-i	i	4i
-	-	4i	-		-1	-i	1	i	4i
-	-	-	4i		1	-i	-1	i	4i

To obtain the inverse, we need to divide everything by 4i, or, equivalently, to multiply by -0.25 i:

u_1	u_2	u_3	u_4	=	w_1	w_2	w_3	w_4	Σ
1	-	-	-		.25	.25	.25	.25	2
-	1	-	-		-.25	.25	-.25	.25	1
-	-	1	-		.25i	-.25	-.25i	.25	1
-	-	-	1		-.25i	-.25	.25i	.25	1

Alternatively, we can calculate $\mathbf{X}^{-1}$, by using the method of section 8.8:

We first form the matrix

$$\mathbf{M} = \begin{bmatrix} 1 & -1 & - & -1 \\ 1 & 1 & -1 & - \\ 1 & -1 & - & 1 \\ 1 & 1 & 1 & - \end{bmatrix}$$

and proceed to invert it, now using the standard method of row-operations:

u_1	u_2	u_3	u_4	=	v_1	v_2	v_3	v_4	Σ
|1|	-1	-	-1		1	-	-	-	-
1	1	-1	-		-	1	-	-	2
1	-1	-	1		-	-	1	-	2
1	1	1	-		-	-	-	1	4

u_1	u_2	u_3	u_4	=	v_1	v_2	v_3	v_4	Σ
1	-1	-	-1		1	-	-	-	-
-	|2|	-1	1		-1	1	-	-	2
-	-	-	2		-1	-	1	-	2
-	2	1	1		-1	-	-	1	4

u_1	u_2	u_3	u_4	=	v_1	v_2	v_3	v_4	Σ
1	-	-0.5	-0.5		0.5	0.5	-	-	1
-	1	-0.5	0.5		-0.5	0.5	-	-	1
-	-	-	2		-1	-	1	-	2
-	-	2	-		-	-1	-	1	2

Before we can make the third step, we need to swap rows 3 and 4:

u_1	u_2	u_3	u_4	=	v_1	v_2	v_3	v_4	Σ
1	-	-.5	-.5		.5	.5	-	-	1
-	1	-.5	.5		-.5	.5	-	-	1
-	-	|2|	-		-	-1	-	1	2
-	-	-	2		-1	-	1	-	2

u_1	u_2	u_3	u_4	=	v_1	v_2	v_3	v_4	Σ
1	-	-	-.5		.5	.25	-	.25	1.5
-	1	-	.5		-.5	.25	-	.25	1.5
-	-	1	-		-	-.5	-	.5	1
-	-	-	|2|		-1	-	1	-	2

$$
\begin{array}{cccc}
u_1 & u_2 & u_3 & u_4
\end{array} =
\begin{array}{cccc}
v_1 & v_2 & v_3 & v_4
\end{array} \quad \Sigma
$$

$$
\begin{bmatrix}
1 & - & - & - \\
- & 1 & - & - \\
- & - & 1 & - \\
- & - & - & 1
\end{bmatrix}
\begin{bmatrix}
.25 & .25 & .25 & .25 \\
-.25 & .25 & -.25 & .25 \\
- & -.50 & - & .50 \\
-.50 & - & .50 & -
\end{bmatrix}
\begin{array}{c}
2 \\ 1 \\ 1 \\ 1
\end{array}
$$

Thus $\mathbf{M}^{-1}$ is:

$$
\mathbf{M}^{-1} =
\begin{bmatrix}
.25 & .25 & .25 & .25 \\
-.25 & .25 & -.25 & .25 \\
- & -.50 & - & .50 \\
-.50 & - & .50 & -
\end{bmatrix}
$$

The two leading rows are immediately copied as the leading rows of $\mathbf{W} = \mathbf{X}^{-1}$.

For the complex vectors, we have a combination of rows 3 and 4, as follows:

$$
\mathbf{w}'_3 = .50 \, [\; - \;\; -.50 \;\; - \;\; .50 \;] - .50 \, [\; -.50 \;\; - \;\; .50 \;\; - \;] \, i
$$

$$
= \quad [\; .50 \, i \quad -.50 \quad -.50 \, i \quad .50 \;]
$$

(Half the third row of $\mathbf{M}^{-1}$ <u>minus</u> half the fourth row, multiplied by i)

and with sign-inversion of the irrational component:

$$
\mathbf{w}'_4 = .50 \, [\; - \;\; -.50 \;\; - \;\; .50 \;] + .50 \, [\; -.50 \;\; - \;\; .50 \;\; -] \, i
$$

$$
= \quad [-0.50 \, i \quad -0.50 \quad 0.50 \, i \quad 0.50 \;]
$$

Which is the same result as obtained above, by directly applying row-operations on the complex matrix.

The diagonal expression for $\mathbf{P}_4$ therefore is:

$$\mathbf{P}_4 =$$

$$\begin{bmatrix} 1 & -1 & -i & i \\ 1 & 1 & -1 & -1 \\ 1 & -1 & i & -i \\ 1 & 1 & 1 & 1 \end{bmatrix} \begin{bmatrix} 1 & - & - & - \\ - & -1 & - & - \\ - & - & i & - \\ - & - & - & -i \end{bmatrix} \begin{bmatrix} .25 & .25 & .25 & .25 \\ -.25 & .25 & -.25 & .25 \\ .25i & -.25 & -.25i & .25 \\ -.25i & -.25 & .25i & .25 \end{bmatrix}$$

$$\begin{bmatrix} 1 & 1 & 1 & 1 \\ 1 & -1 & -i & i \\ 1 & 1 & -1 & -1 \\ 1 & -1 & i & -i \end{bmatrix} \begin{bmatrix} .25 & .25 & .25 & .25 \\ -.25 & .25 & -.25 & .25 \\ .25i & -.25 & -.25i & .25 \\ -.25i & -.25 & .25i & .25 \end{bmatrix} = \begin{bmatrix} - & - & - & 1 \\ 1 & - & - & - \\ - & 1 & - & - \\ - & - & 1 & - \end{bmatrix}$$

The blockdiagonalization conform section 8.9 is:

$$\begin{bmatrix} 1 & -1 & - & -1 \\ 1 & 1 & -1 & - \\ 1 & -1 & - & 1 \\ 1 & 1 & 1 & - \end{bmatrix} \begin{bmatrix} 1 & - & - & - \\ - & -1 & - & - \\ - & - & - & 1 \\ - & - & -1 & - \end{bmatrix} \begin{bmatrix} .25 & .25 & .25 & .25 \\ -.25 & .25 & -.25 & .25 \\ - & -.50 & - & .50 \\ -.50 & - & .50 & - \end{bmatrix} =$$

$$\begin{bmatrix} 1 & 1 & 1 & - \\ 1 & -1 & - & -1 \\ 1 & 1 & -1 & - \\ 1 & -1 & - & 1 \end{bmatrix} \begin{bmatrix} .25 & .25 & .25 & .25 \\ -.25 & .25 & -.25 & .25 \\ - & -.50 & - & .50 \\ -.50 & - & .50 & - \end{bmatrix} = \begin{bmatrix} - & - & - & 1 \\ 1 & - & - & - \\ - & 1 & - & - \\ - & - & 1 & - \end{bmatrix}$$

Answersheet on the differentiation of roots

We are asked to differentiate the dominant root of $\mathbf{A}^{-1}\mathbf{B}$, relative to some elements of

$$\mathbf{A} = \begin{bmatrix} 1 & 2 \\ 2 & 1 \end{bmatrix} \qquad \text{and} \qquad \mathbf{B} = \begin{bmatrix} - & 1 \\ 1 & - \end{bmatrix}$$

We first of all need to calculate:

$$\mathbf{A}^{-1} = \begin{bmatrix} -.33 & .67 \\ .67 & -.33 \end{bmatrix} \quad \text{and} \quad \mathbf{A}^{-1}\mathbf{B} = \begin{bmatrix} .67 & -.33 \\ -.33 & .67 \end{bmatrix}$$

The characteristic equation of $\mathbf{A}^{-1}\mathbf{B}$ therefore is:

$$\begin{vmatrix} .67-\lambda & -.33 \\ -.33 & .67-\lambda \end{vmatrix} = (.67-\lambda)^2 - (-.33)^2 = 0$$

and this characteristic equation resolves as $\lambda = 0.67 \pm 0.33$.

The dominant root therefore is: $\lambda = 1$, as may be confirmed by:

$$\begin{bmatrix} .67 & -.33 \\ -.33 & .67 \end{bmatrix} \begin{bmatrix} 1 \\ -1 \end{bmatrix} = \begin{bmatrix} 1 \\ -1 \end{bmatrix}$$

As this is a symmetric matrix, the transposed column vector is also a row-vector, but to conform with the requirement $\mathbf{w'x} = 1$, we divide the elements of the row-vector by 2, and state:

$$\mathbf{w}\,\mathbf{x'} = \begin{bmatrix} .5 \\ -.5 \end{bmatrix} \begin{bmatrix} 1 & -1 \end{bmatrix} = \begin{bmatrix} .5 & -.5 \\ -.5 & .5 \end{bmatrix}$$

Therefore, applying (8.10.13), for i=1, j=1:

$$\partial\lambda/\partial a_{ij} = \mathbf{w'}_k \left[-\mathbf{A}^{-1}\,\mathbf{E}_{ij}\,\mathbf{A}^{-1}\,\mathbf{B}\right]\mathbf{x}_k =$$

$$\partial\lambda/\partial a_{11} = \mathbf{w'}_k \left[-\mathbf{A}^{-1}\,\mathbf{E}_{11}\,\mathbf{A}^{-1}\,\mathbf{B}\right]\mathbf{x}_k =$$

$$\begin{bmatrix} .5 & -.5 \end{bmatrix} \begin{bmatrix} -.33 & .67 \\ .67 & -.33 \end{bmatrix} \begin{bmatrix} 1 & - \\ - & - \end{bmatrix} \begin{bmatrix} -.33 & .67 \\ .67 & -.33 \end{bmatrix} \begin{bmatrix} - & 1 \\ 1 & - \end{bmatrix} \begin{bmatrix} 1 \\ -1 \end{bmatrix} =$$

$$\begin{bmatrix} .5 & -.5 \end{bmatrix} \begin{bmatrix} -.33 & - \\ .67 & - \end{bmatrix} \begin{bmatrix} -.33 & .67 \\ .67 & -.33 \end{bmatrix} \begin{bmatrix} -1 \\ 1 \end{bmatrix} =$$

$$\begin{bmatrix} -.5 & - \end{bmatrix} \begin{bmatrix} 1 \\ -1 \end{bmatrix} = -.5.$$

For i=1, j=2, the similar calculation for a_{12} is:

$$\begin{bmatrix} .5 & -.5 \end{bmatrix} \begin{bmatrix} -.33 & .67 \\ .67 & -.33 \end{bmatrix} \begin{bmatrix} - & 1 \\ - & - \end{bmatrix} \begin{bmatrix} -.33 & .67 \\ .67 & -.33 \end{bmatrix} \begin{bmatrix} - & 1 \\ 1 & - \end{bmatrix} \begin{bmatrix} 1 \\ -1 \end{bmatrix}$$

$$= \begin{bmatrix} .5 & -.5 \end{bmatrix} \begin{bmatrix} - & -.33 \\ - & .67 \end{bmatrix} \begin{bmatrix} -.33 & .67 \\ .67 & -.33 \end{bmatrix} \begin{bmatrix} -1 \\ 1 \end{bmatrix} =$$

$$[\ -\quad -0.5\ ]\begin{bmatrix} 1 \\ -1 \end{bmatrix} = \ 0.5\ .$$

Now differentiating with respect to b_{12}, the second, not the first term in the expression within [ ] on the righthand side of (8.10.12) is non-trivial, and we need to apply (8.10.14):

$$\partial\lambda_k/\partial b_{ij} = \mathbf{w'}_k\ [\mathbf{A}^{-1}\ \mathbf{E}_{ij}]\ \mathbf{x}k = \mathbf{w'}_k\ [\mathbf{A}^{-1}\ \mathbf{E}_{12}]\ \mathbf{x}_k =$$

$$[.5 \quad -.5\ ]\begin{bmatrix} -.33 & .67 \\ .67 & -.33 \end{bmatrix}\begin{bmatrix} - & 1 \\ - & - \end{bmatrix}\begin{bmatrix} 1 \\ -1 \end{bmatrix} =$$

$$[-.5 \quad .5\ ]\begin{bmatrix} -1 \\ - \end{bmatrix} = \ 0.5\ .$$

CHAPTER IX

SYMMETRIC EIGENVALUE PROBLEMS

9.1 A recapitulation of the eigenvalue problem

The general eigenvalue problem has been dealt with in Chapter VII, and we recapitulate some of the main points here, as introduction to our discussion of the properties of symmetric matrices, where, as we shall see below, a number of simplifying features arise.

A <u>characteristic vector</u> (of a square matrix), is a vector, by which a matrix may be multiplied, to yield as product a vector, which is proportional to the characteristic vector itself.

$$\mathbf{A} \, \mathbf{x} = \mathbf{x} \, \lambda \qquad\qquad (9.1.1)$$

for example:

$$\begin{bmatrix} 2 & 1 \\ 1 & 2 \end{bmatrix} \begin{bmatrix} 1 \\ 1 \end{bmatrix} = \begin{bmatrix} 3 \\ 3 \end{bmatrix} = \begin{bmatrix} 1 \\ 1 \end{bmatrix} \times 3$$

$$\mathbf{x} = \begin{bmatrix} 1 \\ 1 \end{bmatrix} \qquad \lambda = 3,$$

but also:

$$\begin{bmatrix} 2 & 1 \\ 1 & 2 \end{bmatrix} \begin{bmatrix} 1 \\ -1 \end{bmatrix} = \begin{bmatrix} 1 \\ -1 \end{bmatrix} = \begin{bmatrix} 1 \\ -1 \end{bmatrix} \times 1$$

$$\mathbf{x} = \begin{bmatrix} 1 \\ -1 \end{bmatrix} \qquad \lambda = 1.$$

For $\mathbf{x} = [0]$, (9.1.1) does not imply a requirement on λ. We therefore exclude vectors $\mathbf{x} = [0]$.

Relation (9.1.1) gives rise to the requirement:

$$\mathbf{Ax} - \mathbf{x}\lambda = [\mathbf{A} - \mathbf{I}\lambda]\mathbf{x} \tag{9.1.2}$$

therefore, for $\mathbf{x} \neq [0]$:

$$| \mathbf{A} - \mathbf{I}\lambda| = 0 \tag{9.1.3}$$

or in this particular example:

$$\begin{vmatrix} 2-\lambda & 1 \\ 1 & 2-\lambda \end{vmatrix} = (2-\lambda)(2-\lambda) - 1 = 0$$

Therefore:

$$(2-\lambda)^2 = 1 \quad \rightarrow \quad 2 - \lambda = \pm 1 \quad \rightarrow \quad \lambda = 2 \pm 1$$

9.2 Symmetric matrices have real roots

In the general case, the roots of the characteristic equation may be complex. If so, the roots always occur as pairs of complex numbers, and the corresponding vectors can also be presented as vectors of pairs of complex numbers. (A proof of this statement was supplied is section 7.5)

However, in the case of a symmetric matrix, this complication does not arise.

Theorem:

The latent roots and characteristic vectors of a symmetric matrix are real only. (Not complex as may be the case with a non-symmetric matrix)

Proof:

Suppose by contra-assumption:

$$A [x + ui] = [x + ui] (\alpha+\beta i)$$

$$= x\alpha - u\beta + x\beta i + u\alpha i \qquad (9.2.1)$$

For a symmetric matrix, which is its own transpose, the transposed column vectors are also row-vectors.

Therefore:

$$[x + ui]'A = (\alpha+\beta i) [x + ui]'$$

$$= \alpha x' - \beta u' + i\beta x' + i\alpha u' \qquad (9.2.2)$$

But since latent roots, if complex, occur as pairs of complex numbers, and the corresponding vectors may also be required to be pairs of complex numbers, (9.2.2) implies:

$$[x - ui]' A = (\alpha - \beta i) [x - u]' \qquad (9.2.3)$$

Therefore, on pre-multiplication of (9.2.1) by $[x - ui]'$:

$$[x-ui]'A [x+ui] = \Big([x-ui]'A\Big)[x+ui] = (\alpha-\beta i)[x-ui]'[x+ui]$$

$$= [x-ui]'\Big(A[x+ui]\Big) = [x-ui]'[x+ui](\alpha+\beta i) \qquad (9.2.4)$$

The inner product $[x-ui]'[x+ui]$, as it occurs in two of the members of (9.2.4), is evaluated as:

$$[x - ui]'[x + ui] = x'x + x'ui - u'xi - u'u \, i^2$$

$$= x'x \qquad\qquad + u'u \qquad (9.2.5)$$

For $x = [0]$, $u = [0]$, we would not have specified a characteristic vector at all. Therefore:

$$x'x + u'u > 0 \qquad (9.2.6)$$

Therefore, by the equality of the third and the fifth member of (9.2.4):

$$\alpha + \beta i = \alpha - \beta i \qquad (9.2.7)$$

From which:

$$\beta = 0 \tag{9.2.8}$$

q.e.d.

Thus, if, on trying to solve a characteristic equation of a symmetric matrix, one appears to obtain a pair of complex roots, one can be sure that a mistake has been made somewhere.

9.3 A symmetric matrix has n independent vectors

In the general, i.e. non-symmetric case, it is possible that a repeated root of a the characteristic equation is associated with only one vector. We dealt with that complication in Chapter VIII, but a reminder is provided here:

Example:

$$\mathbf{A} = \begin{bmatrix} 1 & - \\ 1 & 1 \end{bmatrix}$$

This matrix has two equal roots $\lambda_1 = \lambda_2 = 1$, but any vector which is not proportional to $\mathbf{x} = \begin{bmatrix} - \\ 1 \end{bmatrix}$, is not a characteristic vector.

This complication does not arise with a symmetric matrix. We first of all state and will proceed to prove the following

Theorem:

If the n by n matrix $\mathbf{A}$ is symmetric then:

If λ_j is a k-fold repeated root of $|\mathbf{A} - \mathbf{I}\lambda_j| = 0$, then $[\mathbf{A} - \mathbf{I}\lambda_j]$ is of rank n-k.

Proof:

For n=1 the theorem is neither applicable (there being no repeated roots), nor contradictable, as it merely states $r \geq 0$ in that case.

q.e.d. for n=1.

In the general, n > 1 case, we invoke our results from section 7.7, where it was shown that for a symmetric matrix, a k-fold repeated root is associated with the separate vanishment of all principal minors $[\mathbf{I}\lambda - \mathbf{A}]$, of which the orders of the minor-matrices are in excess of n-k by n-k.

It was shown in section 5.11, that an n-k by n block-row which is of (full) rank n-k, implies (if the matrix is symmetric), an n-k by n-k invertible diagonal block.

In the absence of such invertible diagonal blocks of $[\mathbf{I}\lambda - \mathbf{A}]$, a symmetric matrix cannot then have such full rank block-rows either.

q.e.d.

Since there cannot be <u>more</u> than n roots, there cannot be more than n independent vectors either, and we have the general result:

The characteristic column vectors of a symmetric matrix can be required to form an invertible matrix.

It is generally true (for symmetric and non-symmetric matrices alike), that inversion of a matrix of column vectors, -assuming that such an inverse exists-, yields a matrix of row-vectors as the resulting inverse:

$$\mathbf{A}\,\mathbf{X} = \mathbf{X}\,\hat{\mathbf{L}} \quad \rightarrow \quad \mathbf{A} = \mathbf{X}\,\hat{\mathbf{L}}\,\mathbf{X}^{-1} \quad \rightarrow \quad \mathbf{X}^{-1}\,\mathbf{A} = \hat{\mathbf{L}}\,\mathbf{X}^{-1}$$

But to ensure that that inverse is, in the symmetric case, also the transpose, normalization conform:

$$\mathbf{x}'_j\,\mathbf{x}_j = 1 \quad (j = 1, 2, \ldots n) \tag{9.3.7}$$

is required.

Example:

For $\mathbf{A} = \begin{bmatrix} 2 & 1 \\ 1 & 2 \end{bmatrix}$ we already established $\mathbf{X} = \begin{bmatrix} 1 & 1 \\ 1 & -1 \end{bmatrix}$.

Simply turning the column vectors into row vectors by transposition, does indeed produce a matrix of row vectors, that immediately follows from the symmetry of $\mathbf{A}$.

But the inverse is: $\mathbf{X}^{-1} = \begin{bmatrix} 0.5 & 0.5 \\ 0.5 & -0.5 \end{bmatrix}$

To obtain a matrix of which the transpose is also the inverse, we first normalize $\mathbf{X}$ to become:

$$\mathbf{X} = \begin{bmatrix} 0.5\sqrt{2} & 0.5\sqrt{2} \\ 0.5\sqrt{2} & -0.5\sqrt{2} \end{bmatrix}$$

9.4 Definiteness and the roots of a matrix

The fact that a symmetric matrix always has a full set of n independent vectors, and that all the roots are real, permits us to investigate its definiteness by way of its roots, and to do so in a more straightforward way than in the general case.

We group the vectors, normalized to length 1, in a matrix:

$$\mathbf{A}\,\mathbf{X} = \mathbf{X}\,\hat{\mathbf{L}} \tag{9.4.1}$$

and $\mathbf{X}$ has its own transpose, which is a matrix of row-vectors, as inverse:

$$\mathbf{X}'\,\mathbf{A} = \mathbf{W}\,\mathbf{A} = \mathbf{X}^{-1}\,\mathbf{A} = \hat{\mathbf{L}}\,\mathbf{X}' \tag{9.4.2}$$

Using the invertability of $\mathbf{X}$, we can express any $\mathbf{v}$ as a linear combination of the n independent vectors, by resolving:

$$\mathbf{X}'\,\mathbf{v} = \mathbf{u} \tag{9.4.3}$$

as

$$u = X'^{-1} v = X' v = \qquad\qquad (9.4.4)$$

Therefore, substituting $X u$ for v, and $u'X'$ for v:

$$v'A v = u'X' A X u = u' X' X \hat{L} X' X u = u' \hat{L} u$$

$$= \sum_{j=1}^{n} u_j \, \lambda_j \, u_j = \Sigma \, u_j{}^2 \, \lambda_j \qquad\qquad (9.4.5)$$

Conversely, given any X, and any combination of roots, we can decide which terms of the last member of (9.4.5) will actually contribute to the quadratic form $v'Av$, and by how much, by choosing a u, and then setting v as:

$$v = X u \qquad\qquad (9.4.6)$$

Since $u_j{}^2 > 0$ will apply for $u_j \neq 0$, these relationships allow us to infer the type of definiteness of A from the roots:

If the roots are all positive non-zero, $v'Av > 0$ applies for any $u \neq 0$, therefore for any $v \neq 0$, and A is positive definite, if some are negative and some positive, then A is indefinite, if they are all negative non-zero, then A is negative definite. Semidefiniteness will arise if the roots all have the same sign, but one or more roots are zeros.

Note that this theorem is specific for a symmetric matrix. If we generalize the definition of definiteness to non-symmetric matrices, then there are matrices which display various characteristics of (symmetric) definite matrices, but for which $x'A x$ does not necessarily match its type of 'definiteness' according to other criteria for establishing the type of definiteness of a symmetric matrix.

For example, for

$$A = \begin{bmatrix} 4 & 2 \\ 8 & 5 \end{bmatrix}$$

the characteristic equation

$$(4-\lambda)(5-\lambda) - 16 = \lambda^2 - 9\lambda + 4 = 0$$

has 2 positive roots $\lambda_1 = 8.531$, and $\lambda = 0.469$.

The similar symmetric matrix which arises when the second column is multiplied by 2 and the second row divided by 2,

$$\mathbf{A} = \begin{bmatrix} 4 & 4 \\ 4 & 5 \end{bmatrix}$$

is indeed positive definite, but for the non-symmetric matrix as given we have $\mathbf{x'A\,x} = -1$ for $\mathbf{x'} = [1 \quad -1]$ (as was already illustrated in section 6.5).

Nor is inversion by positive pivots any guarantee for having positive roots:

$$\mathbf{A} = \begin{bmatrix} 0.1 & - & - & 1 \\ 1 & 0.1 & - & - \\ - & 1 & 0.1 & - \\ - & - & 1 & 0.1 \end{bmatrix}$$

can be inverted along the main diagonal, with three pivots of 0.1 and a fourth one of 0.09, but its roots are: 1.1, -0.9 and a pair of complex roots, 0.1 $\pm$i.

The clear association between positive pivots, positive roots and positive definiteness, is indeed a property of symmetric matrices.

Exercise:

$$\text{For } \mathbf{A} = \begin{bmatrix} 3 & 3 & 1 \\ 3 & 1 & 3 \\ 1 & 3 & 3 \end{bmatrix}$$

find the characteristic equation, and (noting the one root $\lambda=7$ associated with the summation-vector as vector), find the other roots by factorization, and establish the type of definiteness of

A. (There is an answersheet at the end of the chapter).

9.5 **The roots and vectors of product expressions**

Let **A** be an arbitrary real matrix, of order m by n and rank r.

We consider the matrix expressions **A A'** i.e. **A** postmultiplied by its own transpose. We shall refer to such an expression as a <u>symmetric product expression</u>.

Theorem

If

$$\mathbf{A} \mathbf{A}' \, \mathbf{v}_j \; = \; \mathbf{v}_j \, \lambda_j \; \neq \; [0] \tag{9.5.1}$$

holds for $\mathbf{v} \neq [0]$,

then

λ_j is a root of **A'A**

Proof:

We <u>denote</u>:

$$\mathbf{A}' \, \mathbf{v}_j \; = \; \mathbf{u}_j \tag{9.5.2}$$

Given the assumed $\mathbf{A} \mathbf{A}' \, \mathbf{v}_j \neq [0]$, we must also assume $\mathbf{A}' \mathbf{v}_j \neq [0]$.

Pre-multiplication of (9.5.2) by **A'A** leads to:

$$\mathbf{A}'\mathbf{A} \mathbf{A}' \, \mathbf{v}_j \; = \; \mathbf{A}'\mathbf{A} \, \mathbf{u}_j \tag{9.5.3}$$

From which, by (9.5.1) and (9.5.2), and $\mathbf{A}' \mathbf{v}_j \neq [0]$:

$$\mathbf{A}' \, \mathbf{v}_j \, \lambda_j \; = \; \mathbf{u}_j \, \lambda_j \; = \; \mathbf{A}'\mathbf{A} \, \mathbf{u}_j \neq [0] \tag{9.5.4}$$

q.e.d.

The above theorem immediately generalizes to:

$$\mathbf{A}\,\mathbf{A}'\,\mathbf{V}_1 \;=\; \mathbf{V}_1\,\hat{\mathbf{L}}_1 \qquad\qquad (9.5.5)$$

implies:

$$\mathbf{A}'\mathbf{A}\,\mathbf{U}_1 \;=\; \mathbf{U}_1\,\hat{\mathbf{L}}_1 \qquad\qquad (9.5.6)$$

where $\mathbf{V}_1$ is an m by r matrix (block) of characteristic column vectors of $\mathbf{AA}'$, $\hat{\mathbf{L}}_1$ the corresponding diagonal matrix of the non-zero roots of $\mathbf{AA}'$, the second block-column of $\mathbf{V}$ relating to the zero roots. (By the correlation matrix theorem, section 5.12, both $\mathbf{AA}'$ and $\mathbf{A}'\mathbf{A}$ are of rank r, and we saw in section 9.3, that this implies the existence of r independent vectors associated with the non-zero roots, regardless of repeated roots.)

The vectors $\mathbf{v}_j$ associated with the zero roots of $\mathbf{AA}'$ obviously are the m-r independent vectors satisfying:

$$\mathbf{A}'\mathbf{v}_j = [0], \quad \mathbf{v}_j \neq [0].$$

For r < m-1, there is, as usual with multiple roots of symmetric matrices, an element of choice in determining a particular vector.

The above theorem has an only slightly weaker equivalent in relation to the product of two matrices, whose product does not form a symmetric product expression:

Theorem:

If the m by n matrix $\mathbf{A}$ and the n by m matrix $\mathbf{B}$ obey the condition:

$$\mathbf{A}\,\mathbf{B}\,\mathbf{x} = \mathbf{x}\,\lambda \neq [0]$$

then we have, for

$$\mathbf{y} = \mathbf{B}\,\mathbf{x} \neq [0]$$

$$\mathbf{B}\,\mathbf{A}\,\mathbf{y} = \mathbf{y}\,\lambda \neq [0]$$

Proof:

$$\mathbf{B\,A\,B\,x} = \mathbf{B\,[A\,B]\,x} = \mathbf{B\,x}\,\lambda = \mathbf{y}\,\lambda = \mathbf{[B\,A]\,B\,x} = \mathbf{B\,A\,y}$$

q.e.d.

Now returning to symmetric product expressions, we consider the diagonalization of $\mathbf{AA'}$, assuming normalization of each characteristic vector $\mathbf{v}_j$:

$$\mathbf{A\,A'} = \mathbf{V\,\hat{L}\,V'} = \mathbf{V_1\,\hat{L}_1\,V'_1} \qquad (9.5.7)$$

Given the equivalence between the matrix expression $\mathbf{V\hat{L}V'}$ and the equivalent summation expression:

$$\mathbf{V\,\hat{L}\,V'} = \sum_{i=1}^{n} \mathbf{v}_i\,\lambda_i\,\mathbf{v}'_i = \sum_{i=1}^{r} \mathbf{v}_i\,\lambda_i\,\mathbf{v}'_i$$

(see section 3.4, multiplication by outer product, involving a diagonal matrix), we can legitimately suppress reference to the zero roots and of $\mathbf{AA'}$ and their associated vectors, in the most righthand member of (9.5.7).

Note however, that the generalization of (9.5.7) to $\mathbf{A'A}$:

$$\mathbf{A'A} = \mathbf{U_1\hat{L}_1 U_1'} \qquad (9.5.8)$$

is valid only if the columns of $\mathbf{U_1}$ are normalized, and may be untrue, if $\mathbf{U_1}$ is obtained from a matrix form of (9.5.2):

$$\mathbf{U_1} = \mathbf{A'V_1} \qquad (9.5.9)$$

The assumed normalization of all the vectors, implies for the vectors associated with the non-zero roots:

$$\mathbf{V_1'\,V_1} = \mathbf{I_r}$$

as usual for normalized vectors.

However, $\mathbf{U_1}$, as defined by (9.5.9) obeys the condition:

$$\mathbf{U_1'\,U_1} = \mathbf{V_1'A\,A'V_1}$$

and therefore:

$$U_1' \, U_1 \;=\; V_1' A \, A' V_1 \;=\; V_1' \, V_1 \hat{L}_1 \;=\; \hat{L}_1 \qquad (9.5.10)$$

To obtain a block of normalized vectors, by which we may express $A'A$ in diagonalized form, we therefore need:

$$P_1 \;=\; U_1 \, \hat{L}_1^{-\frac{1}{2}} \qquad\qquad (9.5.11)$$

and we can express $[A'A]$ on the usual way as:

$$[A'A] \;=\; P_1 \hat{L}_1 P_1' \qquad\qquad (9.5.12)$$

Example:

$$A \;=\; \begin{bmatrix} 1 & 2 & 3 \\ 2 & 4 & 6 \end{bmatrix} \qquad\qquad AA' \;=\; \begin{bmatrix} 14 & 28 \\ 28 & 56 \end{bmatrix}$$

$$\begin{bmatrix} 14 & 28 \\ 28 & 56 \end{bmatrix} \begin{bmatrix} 1 \\ 2 \end{bmatrix} \;=\; \begin{bmatrix} 70 \\ 140 \end{bmatrix} \qquad \lambda_1 = 70$$

Therefore, normalizing the vector v_1 to $v_1 = \begin{bmatrix} 1/\sqrt{5} \\ 2/\sqrt{5} \end{bmatrix}$

$$AA' \;=\; \begin{bmatrix} 1/\sqrt{5} \\ 2/\sqrt{5} \end{bmatrix} \times 70 \times [1/\sqrt{5} \quad 2/\sqrt{5}] \;=\; \begin{bmatrix} 1/5 & 2/5 \\ 2/5 & 4/5 \end{bmatrix} \times 70$$

To obtain $A'A$, we first calculate:

$$u_1 \;=\; A' v_1 \;=\; \begin{bmatrix} 1 & 2 \\ 2 & 4 \\ 3 & 6 \end{bmatrix} \begin{bmatrix} 1/\sqrt{5} \\ 2/\sqrt{5} \end{bmatrix} \;=\; \begin{bmatrix} 5/\sqrt{5} \\ 10/\sqrt{5} \\ 15/\sqrt{5} \end{bmatrix}$$

The inner product $u_1' u_1$ is:

$$(5/\sqrt{5})^2 + (10/\sqrt{5})^2 + (15/\sqrt{5})^2 = (25 + 100 + 225) / 5 =$$

$$350 / 5 \;=\; 70 \;=\; \lambda_1$$

After normalization to $p_1' p_1 = 1$, this becomes: $\quad p_1 = \begin{bmatrix} 1/\sqrt{14} \\ 2/\sqrt{14} \\ 3/\sqrt{14} \end{bmatrix}$

Therefore $\mathbf{A'A} = \begin{bmatrix} 1/\sqrt{14} \\ 2/\sqrt{14} \\ 3/\sqrt{14} \end{bmatrix} \times 70 \times [1/\sqrt{14} \quad 2/\sqrt{14} \quad 3/\sqrt{14}]$

$$= \begin{bmatrix} 1/14 & 2/14 & 3/14 \\ 2/14 & 4/14 & 6/14 \\ 3/14 & 6/14 & 9/14 \end{bmatrix} \times 70 = \begin{bmatrix} 5 & 10 & 15 \\ 10 & 20 & 30 \\ 15 & 30 & 45 \end{bmatrix}$$

9.6 The components of a matrix

The term <u>principal component</u> arises on account of the use in econometrics, of the vectors $\mathbf{p_j}$ as discussed in the previous section, and the designator $\mathbf{p}$ was chosen for that reason.

Econometric data-matrices are conventionally presented with each row containing all the observations for one variable, i.e. the row-indices refer to the observed economic variables, and the column-indices indicate the separate observations. The <u>principal component</u> is an attempt to represent a matrix of data-variables as just multiples of one variable, which represents both the trend and any systematic fluctuations, which are common to all the variables. To the extent that that actually is so, the matrix of stylized hypothetical observations is:

$$\mathbf{H} = \mathbf{m}\,\mathbf{p'} \tag{9.6.1}$$

where $\mathbf{m}$ is the vector of multiplicative factors, by which the stylized hypothetical observations for the separate variables differ from each other, while $\mathbf{p}$ is required to meet the normalization condition:

$$\mathbf{p'p} = 1 \tag{9.6.2}$$

(See Theil [46], pp. 48-56, but note that the convention that data-matrices have the observation-index as column-index was not at that stage firmly established, and Theil does not follow it.) The remaining unrepresented part of the real observations is:

$$\mathbf{R} = \mathbf{A} - \mathbf{H} = \mathbf{A} - \mathbf{m}\,\mathbf{p'} \tag{9.6.3}$$

The criterion for the most suitable vector **p** is the minimization of the total sum of squares remaining as the unrepresented part of the actual observations. Thus, we seek the minimization of:

$$\mathcal{M} = \sum_{i=1}^{m} \mathbf{r'}_i [\mathbf{r'}_i]' \qquad (9.6.4)$$

i.e. the sum of the lengths of the rows of **R**, each of which is $\mathbf{r'}_i$, postmultiplied by its transpose $[\mathbf{r'}_i]'$.

That is therefore the sum of the diagonal elements of **R R'**. The sum of the diagonal elements of a square matrix is (see section 7.2) known as the <u>trace</u>.

Recall (also from section 7.2), that the roots of a matrix add to the trace.

It was shown in the previous section, that **A A'** and **A'A** have the same non-zero roots.

Therefore:

$$\text{tr} (\mathbf{A A'}) = \text{tr} (\mathbf{A'A}) \qquad (9.6.5)$$

We now seek the minimization of:

$$\mathcal{M} = \text{tr} (\mathbf{R R'}) = \text{tr} ([\mathbf{A} - \mathbf{m} \mathbf{p'}] [\mathbf{A} - \mathbf{m} \mathbf{p'}]')$$

$$= \text{tr} (\mathbf{A A'} - \mathbf{A} \mathbf{p} \mathbf{m'} - \mathbf{m} \mathbf{p'} \mathbf{A'} + \mathbf{m} \mathbf{p'} \mathbf{p} \mathbf{m'})$$

$$(9.6.6)$$

In the last term on the righthand side of (9.6.6), the inner product **p'p** is a scalar number, and its value is given by (9.6.2) as **p'p** = 1.

The last three terms in (9.6.6) therefore are together:

tr (**m m'** -[**A p**] **m'** -**m** [**p'A'**]) =

tr ([**m** - **A p**] [**m'** - **p'A'**] - [**A p**] [**p'A'**]) =

tr ([**m'** - **p'A'**][**m** - **A p**] - **p'A'A p**) = **m'm** -2**p'A m**   (9.6.7)

where the trace of the outer product [**m'** - **p'A**][**m** - **A p**] has been replaced by the trace of the corresponding inner product, which is the inner product itself.

Therefore:

$$\mu = \text{tr} (\mathbf{A} \mathbf{A}') - 2 \mathbf{p}'\mathbf{A}' \mathbf{m} + \mathbf{m}'\mathbf{m} \qquad (9.6.8)$$

From which we obtain a necessary first-order condition for the minimization of μ:

$$\partial\mu/\partial\mathbf{m} = 2 \mathbf{m} - 2 \mathbf{A} \mathbf{p} = [0] \qquad (9.6.9)$$

Substitution of **A p** for **m**, according to (9.6.9), into (9.6.8), leads to:

$$\mu = \text{tr} (\mathbf{A} \mathbf{A}') - \mathbf{p}'\mathbf{A}'\mathbf{A} \mathbf{p} \qquad (9.6.10)$$

We cannot determine **p** by simply differentiating (9.6.10).

This is clear both from the unboundedness of (9.6.10) as such, and from the absence of a reference to the side condition (9.6.2). This apparent unboundedness arises, because we used (9.6.2) to evaluate μ, but now have an expression in which conformity with (9.6.2) is not assured.

As usual in the calculus of constrained variations, we form a Lagrangean expression, adding a term which is zero-valued, whenever the side-condition is met.

In our case, that Lagrangean expression is:

$$L (\mu, \lambda) = \mu + \lambda (\mathbf{p}'\mathbf{p} - 1) =$$

$$\text{tr} (\mathbf{A} \mathbf{A}') + \mathbf{p}'[\mathbf{I}\lambda - \mathbf{A}'\mathbf{A}] \mathbf{p} - \lambda \qquad (9.6.11)$$

and a necessary first-order condition is:

$$\partial L/\partial p = 2 \lambda p - 2 A'A p = [0] \tag{9.6.12}$$

In (9.6.11) and (9.6.12) λ is in first instance a Lagrangean multiplier, but (9.6.12) makes clear that that multiplier is a root of $A'A$.

The econometric problem of minimizing μ while using only <u>one</u> vector p, requires λ to be the <u>dominant</u> root of $A'A$. This <u>is</u> clear from (9.6.11): in the constrained minimum of μ, where the term $\lambda(p'p - 1)$ has the value zero, the constrained minimum is:

$$\text{Min } \mu = \text{tr} (A A') - \lambda \tag{9.6.13}$$

Apparently, there is a local minimum for each root $\lambda_i \neq 0$, and the global minimum is attained by taking the dominant root. Hence the name <u>principal</u> component.

In the wider context of matrix algebra, we are interested in <u>all</u> the vectors and the roots of both $A'A$, and of $[A A']$.

On diagonalizing $A'A$, we note that each column of A' is a linear combination of the p_j:

$$A'A = P_1 \hat{L}_1 P_1' \tag{9.6.14}$$

The rows of A are therefore linear combinations of the transposes of the p_j (insofar as relating to non-zero roots of $A'A$), and we shall refer to the p_j as the <u>row-components</u>. To distinguish them from the vectors associated with the zero roots of $A'A$, the (proper) row-components of A.

The normalized vectors of $[A A']$ on the other hand, occur in various combination as the columns of $[A A']$, and by implication the same applies to the columns of A. We therefore refer to those vectors (again, insofar as relating to the <u>non-zero</u> roots of $[A A']$), as the <u>column-components</u> of A.

Comparison of (9.6.9), (9.5.2) and (9.5.11) indicates that the multiplier-vectors m_j as used in this section, are related to the v_j and the u_j, and to the p_j, by:

$$m_j = A p_j = A u_j \lambda_j^{1/2} = A A' v_j \lambda_j^{-1/2} = v_j \lambda_j^{1/2} \tag{9.6.15}$$

Note also, that if the outer product $\mathbf{v}_j' \lambda_j \mathbf{v}_j$ is subtracted from the symmetric product expression $\mathbf{A}\,\mathbf{A}'$, the other roots are undisturbed:

$$[\mathbf{A}\,\mathbf{A}' - \mathbf{v}'_j \lambda_j \mathbf{v}_j]\,\mathbf{v}_i = \mathbf{A}\,\mathbf{A}'\,\mathbf{v}_i = \mathbf{v}_i \lambda_i \qquad (9.6.16)$$

Recursive subtraction of the outer product $\mathbf{m}_j\,\mathbf{p}'_j$ from $\mathbf{A}$ therefore expresses $\mathbf{A}$ as:

$$\mathbf{A} = \sum_{j=1}^{r} \mathbf{m}_j\,\mathbf{p}'_j = \sum_{j=1}^{r} \mathbf{v}_j\,\mathbf{p}'_j\,\lambda_j^{1/2} = \sum_{j=1}^{r} \mathbf{v}_j\,\mathbf{u}'_j \qquad (9.6.17)$$

9.7 The generalized inverse

A number of people, notably Moore [34], Penrose [36] and Rao [38], [39], have suggested that it would be useful to have a definition of a matrix, which is in some sense the inverse of another matrix, even if that matrix is <u>not</u> square and nonsingular.

The most widely used definition has become known as the Moore-Penrose inverse, and a systematic survey of the relevant theory at textbook level is provided by Rao [40], as well as by Ben Israel [5]. Its applications, in particular in statistical estimation problems, have also been summarized at textbook level, by Albert [4].

The definition of a generalized inverse does not specifically refer to a symmetric matrix. Rather, the reason for dealing with the generalized inverse in this chapter, is that the development of the relevant theory depends heavily on the properties of symmetric matrices.

The n by m matrix $\mathbf{B}$ is the generalized inverse of the m by n matrix $\mathbf{A}$, if

$$\mathbf{A}\,\mathbf{x} = \mathbf{y} \qquad (9.7.1)$$

and

$$\mathbf{B}\,\mathbf{y} = \mathbf{x} \tag{9.7.2}$$

are both true for every $\mathbf{x}$ for which (9.7.2) permits an expression of $\mathbf{y}$ in $\mathbf{x}$, and for every $\mathbf{y}$, for which (9.7.1) permits an expression of $\mathbf{x}$ in $\mathbf{y}$.

Compared to section 4.4, we drop the requirement that (9.7.1) and/or (9.7.2) are generally resolvable in the first place.

Recall our definition of the inverse from section 4.4: if we can find, for any n by column $\mathbf{y}$, a corresponding $\mathbf{x}$, for which $\mathbf{x}$ and $\mathbf{y}$ satisfy both (4.4.1) and (4.4.3) -. Once (9.7.1) is resolvable for some $\mathbf{y}$, each corresponding vector $\mathbf{x}$, (there is no requirement that the resolution of either relation determines the lefthand vector uniquely) must fit the other relation as well. Similarly, once (9.7.2) is resolvable for some $\mathbf{x}$, all the corresponding vectors $\mathbf{y}$ must also fit (9.7.1).

Note the subtle difference: In Chapter IV we argued that, since it was a requirement that we could find a corresponding $\mathbf{x}$ for any $\mathbf{y}$, we were at liberty to choose $\mathbf{y}$ as the unit vectors $\mathbf{y} = \mathbf{e}_1$, $\mathbf{e}_2$, etc. That cannot be required now, because there is no requirement that $\mathbf{A}\,\mathbf{x} = \mathbf{y}$ is generally resolvable.

For example, for $\mathbf{A} = \begin{bmatrix} - & - & - \\ - & 2 & 1 \\ - & 1 & 2 \end{bmatrix}$

$\mathbf{A}\,\mathbf{x} = \mathbf{y}$ is resolvable for vectors $\mathbf{y}$, which are combinations of the two non-zero columns, for $\mathbf{y} = [0]$, but not for any vector of which the leading element is non-zero.

An implicit restriction on $\mathbf{B}$ and the vectors $\mathbf{x}$ for which $\mathbf{B}\,\mathbf{y} = \mathbf{x}$ is resolvable is obvious in this example. Relation (9.7.2) must not be resolvable for an $\mathbf{x} \neq [0]$, for which $\mathbf{A}\,\mathbf{x} = [0]$ applies.

For example, if $\mathbf{B}\,\mathbf{y} = \mathbf{e}_1$ were resolvable (and hence consistent with $\mathbf{y} \neq [0]$), then $\mathbf{A}\,\mathbf{x} = \mathbf{A}\,\mathbf{e}_1 = \mathbf{y} \neq [0]$, is also required, and plainly this is impossible.

There is no need to require $\mathbf{B} \neq [0]$, $\mathbf{x} \neq [0]$, or $\mathbf{y} \neq [0]$ by definition. Indeed, there are advantages in keeping the definition general. Some proofs are provided more readily, if $\mathbf{A}\,[0] = [0]$, and $\mathbf{B}\,[0] = [0]$ are included in the definition as resolvable systems.

For $\mathbf{A} = [0]$, $\mathbf{B} = [0]$ is the generalized inverse, with $\mathbf{x} = [0]$ and $\mathbf{y} = [0]$, as the only pair of vectors which fit both relations.

For $\mathbf{A} \neq [0]$, $\mathbf{B} \neq [0]$ is implied.

Following Rao [40], we shall use the notation $\mathbf{A}^-$ for the generalized inverse of $\mathbf{A}$.

This is not a firm convention: Albert [4] uses $\mathbf{A}^+$, as well as a different definition. Albert's definition does not include the ordinary inverse (as our definition does, with the set of vectors for which (9.7.1) is contradictory being empty) but instead pre-assumes a definition of $\mathbf{A}^{-1}$, and we shall not pursue this point, except to observe that the results are obviously equivalent.

Before we deal with the calculation of the generalized inverse as such, it is useful to develop some necessary conditions, which any matrix $\mathbf{B}$ needs to meet, if it is indeed $\mathbf{A}^-$ as defined.

Substitution of $\mathbf{By}$ for $\mathbf{x}$, conform (9.7.2) into (9.7.1) leads to:

$$\mathbf{A}\,\mathbf{B}\,\mathbf{y} \;=\; \mathbf{y} \tag{9.7.3}$$

whereas substitution of $\mathbf{Ax}$ for $\mathbf{y}$, conform (9.7.1), into (9.7.2) results in:

$$\mathbf{B}\,\mathbf{A}\,\mathbf{x} \;=\; \mathbf{x} \tag{9.7.4}$$

These substitutions are a straight application of the definition of a generalized inverse, and their result is therefore both necessary and sufficient, to ensure that $\mathbf{B}$ is a generalized inverse of $\mathbf{A}$. But the question of the uniqueness of $\mathbf{B}$, and indeed, of the existence of such a matrix remains as yet open.

The order-requirements of (9.7.1) and (9.7.2) imply that a matrix $\mathbf{B}$, which is a generalized inverse of the m by n matrix $\mathbf{A}$, must be of order n by m, but we will postpone any further discussion of its existence, content, and uniqueness, and develop some more generally relevant theory first.

9.8 The components of a matrix and its G-inverse

To develop suitable formulae for the generalized inverse of a matrix, it is useful to develop the notion of the components of a matrix, as discussed in section 9.6, one stage further.

Recall first of all, (from section 9.5):

$$P_1 = U_1 \, \hat{L}_1^{-1/2} = A'V \, \hat{L}_1^{-1/2}$$

From which, for an individual component:

$$p_j = A'v_j \, \lambda_j^{-1/2} \quad (j = 1, 2, \ldots r) \tag{9.8.1}$$

and by analogy, for the column components:

$$v_j = A \, p_j \, \lambda_j^{-1/2} \quad (j = 1, 2, \ldots r) \tag{9.8.2}$$

Secondly, in the interest of being able to express any vector as a linear combination of components, we shall introduce the term dummy-components:

$$p_j \quad (j = r + 1, r + 2, \ldots n)$$

the dummy row-components are the normalized vectors associated with the zero roots of $[A'A]$, thereby completing the n by r block-column P_1 to an n by n invertible matrix P, and we similarly complete the m by r block-column V_1 into an m by m invertible matrix V, by specifying m-r dummy column-components v_j.

We now return to the definition of the generalized inverse.

If y, as referred to in:

$$A \, x = y \tag{9.7.1}$$

is a combination of the proper column components of A:

$$y = V_1 \, z \tag{9.8.3}$$

then y can be expressed as:

$$\mathbf{A} \; \mathbf{U}_1 \hat{\mathbf{L}}_1{}^{-1} \mathbf{z} = \mathbf{A} \; \mathbf{A}' \mathbf{V}_1 \hat{\mathbf{L}}_1{}^{-1} \mathbf{z}$$

$$= \mathbf{V}_1 \; \hat{\mathbf{L}}_1 \; \hat{\mathbf{L}}_1{}^{-1} \; \mathbf{z} = \mathbf{V}_1 \; \mathbf{z} = \mathbf{y} \qquad (9.8.4)$$

and (9.7.1) is seen to be resolved as

$$\mathbf{x} = \mathbf{U}_1 \; \hat{\mathbf{L}}_1{}^{-1} \; \mathbf{z} = \mathbf{P}_1 \; \hat{\mathbf{L}}_1 \; \mathbf{z} \qquad (9.8.5)$$

If $\mathbf{y}$ cannot be expressed as a combination of the (proper) column components of $\mathbf{A}$, (and hence not directly in the columns of $\mathbf{A}$ either), i.e. $\mathbf{V}^{-1} \mathbf{y}$ contains non-zero elements in position r+1 or beyond, then (9.7.1) is contradictory.

We are now in a position to say some fairly specific things about the generalized inverse:

If the n by m matrix $\mathbf{B}$ is the generalized inverse of the m by n matrix $\mathbf{A}$, then:

- 1) $\mathbf{B} \; \mathbf{y} = \mathbf{x}$ must be resolvable for any $\mathbf{x}$ calculated according to (9.8.5), for any $\mathbf{z}$, and therefore for $\mathbf{x} = \mathbf{p}_i$, i = 1 to r and therefore for any combination of the $\mathbf{p}_i$.

($\mathbf{A} \; \mathbf{x} = \mathbf{y}$ is resolvable for $\mathbf{y} = \mathbf{V}_1 \mathbf{z}$, any $\mathbf{z}$)

- 2) Each column of $\mathbf{B}$ must be a linear combination of <u>only</u> the (proper) row-components of $\mathbf{A}$.

If $\mathbf{b}_j = \mathbf{P}_1 \mathbf{w}_1 + \mathbf{P}_2 \mathbf{w}_2$ were true for $\mathbf{w}_2 \neq [0]$, we would come to an impossible conclusion:

$\mathbf{B} \; \mathbf{y} = \mathbf{x} = \mathbf{b}_j$ is resolvable as $\mathbf{x} = \mathbf{e}_j$, and $\mathbf{B} \; \mathbf{y} = \mathbf{x} = \mathbf{P}_1 \; \mathbf{w}_1$ must be resolvable by 1) above.

If that were indeed so, then $\mathbf{B} \; \mathbf{y} = \mathbf{x} = \mathbf{b}_j - \mathbf{P}_1 \; \mathbf{w}_1 = \mathbf{P}_2 \mathbf{w}_2$ must also be solvable, and the solution must conform to $\mathbf{A} \; \mathbf{x} = \mathbf{y}$.

But since $\mathbf{A} \; \mathbf{x} = \mathbf{A} \; \mathbf{P}_2 \mathbf{w}_2 = [0] \; \mathbf{w}_2 = [0]$ holds, $\mathbf{B} \; \mathbf{y} = \mathbf{P}_2 \mathbf{w}_2 \neq [0]$ cannot be true.

(note that $\mathbf{w}$ is here used for 'weights', even while this letter usually, but not this time, indicates a row-vector.)

Since by 2), the rank of $\mathbf{B}$ cannot exceed the rank r of $\mathbf{A}$, and 1) states the existence of r independent vectors $\mathbf{x}$, satisfying

$$\mathbf{B}\,\mathbf{A}\,\mathbf{x}\ =\ \mathbf{x} \qquad\qquad (9.7.4)$$

we also infer:

- 3) $\mathbf{B}\,\mathbf{A}$ is of rank r, and idempotent, and therefore:

- 4) $\mathbf{B}$ is of rank r.

Since any vector $\mathbf{v}$, which is a linear combination of the columns of $\mathbf{V}_2$ (of the dummy column components of $\mathbf{A}$), obeys the condition:

$$\mathbf{v}'\mathbf{A} = [0]'$$

the transposes of the (proper) column components $\mathbf{v}_1$ to $\mathbf{v}_r$ are the row-vectors associated with the unity-roots of the idempotent matrix $\mathbf{A}\,\mathbf{B}$.

Therefore:

- 5) $\mathbf{A}\,\mathbf{B}$ is symmetric.

$$\mathbf{A}\,\mathbf{B}\ =\ [\mathbf{V}_1]'\mathbf{V}_1 \qquad\qquad (9.8.6)$$

By the same argument, applied to $\mathbf{x} = \mathbf{p}_1$ to $\mathbf{p}_r$ as vectors of $\mathbf{B}\,\mathbf{A}$, we also infer:

- 6) $\mathbf{B}\,\mathbf{A}$ is symmetric.

$$\mathbf{B}\,\mathbf{A}\ =\ \mathbf{P}_1\,\mathbf{P}'_1 \qquad\qquad (9.8.7)$$

Post-multiplication of $\mathbf{B}$, (wile noting that $\mathbf{A}\,\mathbf{u}_j = \mathbf{A}\mathbf{A}'\mathbf{v}_j = \mathbf{v}_j\lambda_j$, and therefore $\mathbf{B}\,\mathbf{v}_j = \mathbf{u}_j\,\lambda_j^{-1}$ needs to apply for j = 1, 2, .. r, and the orthogonality of the rows of $\mathbf{B}$, which is of rank r, to the other $\mathbf{v}_j$), by $\mathbf{V}$, leads to:

$$\mathbf{B}\,\mathbf{V}\ =\ [\mathbf{U}_1\hat{\mathbf{L}}_1{}^{-1}\mid\quad] \qquad\qquad (9.8.8)$$

At the same time, we have:

$$\mathbf{A}'\mathbf{V}_1\hat{\mathbf{L}}_1{}^{-1}\mathbf{V}'_1\,\mathbf{V} = \mathbf{A}'\mathbf{V}_1\hat{\mathbf{L}}_1{}^{-1}\,[\mathbf{I}_r\mid\quad] = [\mathbf{U}_1\hat{\mathbf{L}}_1{}^{-1}\mid\quad] \qquad\qquad (9.8.9)$$

Therefore, on post-multiplication of both (9.8.8) and (9.8.9) by $V^{-1} = V'$:

$$B = A^- = A'V_1\hat{L}_1^{-1}V_1' \qquad (9.8.10)$$

(where it should be noted that V_1', the transpose of the leading block-column of V, is also the leading block-row of the matrix of row-vectors of $[A\ A']$, also expressed in normalized form, and as such also the leading block-row of V^{-1}.)

It therefore follows, that if a matrix B which satisfies the condition of the generalized inverse, actually exists, this must be $B = A^-$, as stated by (9.8.10).

We now verify the remaining conditions, notably:

$$A\ B\ v_j = A\ A'V_1\hat{L}_1^{-1}\ e_j = V_1\ \hat{L}_1\ \hat{L}_1^{-1}\ e_j = v_j$$

confirming (9.7.3), for r independent vectors $y = v_j$.

Similarly, for $B\ A$:

$$B\ A\ u_j = A'V_1\hat{L}_1^{-1}V'_1A\ u_j = U_1\hat{L}_1^{-1}U'_1u_j = U_1\hat{L}_1^{-1}\ e_j\lambda_j = u_j$$

confirming (9.7.4), for r independent vectors $x = u_j$.

These last conditions are also sufficient, and B, as calculated according to (9.8.10) is seen to be the one and only generalized inverse of A.

Although (9.8.10) gives the formula for the one and only generalized inverse of A, there is an equivalent formula, which is:

$$A^- = P_1\hat{L}_1^{-1}P'_1A' \qquad (9.8.11)$$

Post-multiplication of (9.8.10) and of (9.8.11) by the full n by n invertible matrix V turns the righthand-sides of both relations into:

$$A^-V = [V_1\hat{L}_1^{-1/2}|\quad] \qquad (9.8.12)$$

and therefore proves the equality of both sides of (9.8.11).

9.9 Some G-inverses with particular characteristics

If A is a symmetric matrix, A itself diagonalizes as:

$$A = V \hat{L} V' = V_1 \hat{D}_1 V'_1 \qquad (9.9.1)$$

where, to avoid confusion with the roots of the symmetric product expression, we have used the symbol $\hat{D}$ rather than $\hat{L}$, as we did in the four previous sections.

Therefore:

$$A A' = A'A = V \hat{D}^2 V' \qquad (9.9.2)$$

and

$$A^- = A' V_1 \hat{D}_1^{-2} V'_1 = V_1 \hat{D}_1 V'_1 V_1 \hat{D}_1^{-2} V'_1$$

$$= V_1 \hat{D}_1^{-1} V'_1 \qquad (9.9.3)$$

Now referring back to the formulae for the general, i.e. non-symmetric case, (9.8.10) and (9.8.11), as developed section 9.8, we see that the expressions $V_1 \hat{L}_1^{-1} V_1'$ and $P_1 \hat{L}_1^{-1} P'_1$, as they occur there, are the generalized inverses of the symmetric product expressions AA' and $A'A$.

Therefore (9.8.10) and (9.8.11) are equivalently written as:

$$A^- = A'[A A']^- = [A'A]^- A' \qquad (9.9.4)$$

If A is of order n by n and invertible, then the generalized inverse is the ordinary inverse.

This fact often provides a practical approach towards calculating the generalized inverse of a full rank matrix.

If A is of full rank m, $[A A']$ is invertible, and we have:

$$A^- = A' [A A']^- = A' [A A']^{-1} \qquad (9.9.5)$$

If A is of full rank n, $[A'A]$ is invertible, and we have:

$$A^- = [A'A]^- A' = [A'A]^{-1} A' \qquad (9.9.6)$$

9.10 **The calculation of a g-inverse by rank completion**

Theorem:

If **A** is of order m by n and of rank r, and the n by n-r matrix **C** is of full rank n-r and satisfies:

$$\mathbf{A}\,\mathbf{C} = [0] \tag{9.10.1}$$

then

$$\mathbf{F} = \mathbf{A'A} + \mathbf{C}\,\mathbf{C'} \tag{9.10.2}$$

is invertible.

Proof:

By assumption we have:

$$\mathbf{A}\,c_j = [0] \quad (j = 1, 2, \ldots n-r)$$

and therefore

$$\mathbf{A'A}\,c_j = [0] \quad (j = 1, 2, \ldots n-r) \tag{9.10.3}$$

and **C** is seen to complete **V**, the m by r matrix of the characteristic column vectors associated with the non-zero roots of **A'A**, to a full n by n matrix of column vectors of **A'A**.

From which we infer that (see also section 9.4), for $x \neq [0]$, there are multiplicative factors μ_j and δ_j for which

$$x'[\mathbf{A'A} + \mathbf{C}\,\mathbf{C'}]x = \sum_{j=1}^{r} \lambda_j\,\mu_j^{\,2} + \sum_{h=1}^{n-r} c'_h c_h\,\delta_j^{\,2} > 0$$

will apply, i.e. **F** is positive definite, and therefore invertible.

q.e.d.

Given [**A** **C**] = [0], we also note the property:

$$[\mathbf{A'} + \mathbf{C}]\,[\mathbf{A} + \mathbf{C'}] \;=\; [\mathbf{A'A} + \mathbf{C\,C'}] \;=\; \mathbf{F} \qquad\qquad (9.10.4)$$

Theorem:

If $\mathbf{A}$ and $\mathbf{C'}$ both contain n columns and $\mathbf{C}$ obeys, besides (9.10.1), also the condition:

$$\mathbf{C'C} \;=\; \mathbf{I}_{n-r} \qquad\qquad (9.10.5)$$

then

$$[\mathbf{A'A}]^{-} \;=\; \mathbf{F}^{-1} - \mathbf{C\,C'} \;=\; [\mathbf{A'A} + \mathbf{C\,C'}]^{-1} - \mathbf{C\,C'} \qquad (9.10.6)$$

holds.

Proof:

$\mathbf{F}$ diagonalizes as:

$$\mathbf{F} = \mathbf{P_1\hat{L}_1P'_1} + \mathbf{C\,C'} = [\mathbf{P_1}\ \ \mathbf{C}]\left[\begin{array}{c|c}\mathbf{\hat{L}_1} & \\ \hline & \mathbf{I}_{n-r}\end{array}\right]\left[\begin{array}{c}\mathbf{P_1'}\\ \hline \mathbf{C'}\end{array}\right] \qquad (9.10.7)$$

where $\mathbf{P_1}$ is the matrix of normalized vectors associated with the non-zero roots of $\mathbf{A'A}$, and $\mathbf{C} = \mathbf{P_2}$ completes $\mathbf{P}$ to a full matrix of normalized vectors of $\mathbf{A'A}$.

Therefore (9.10.6) follows immediately, on subtraction of $\mathbf{CC'}$ from both sides of:

$$\mathbf{F}^{-1} = [\mathbf{P_1}\ \ \mathbf{C}]\left[\begin{array}{c|c}\mathbf{\hat{L}_1}^{-1} & \\ \hline & \mathbf{I}_{n-r}\end{array}\right]\left[\begin{array}{c}\mathbf{P_1'}\\ \hline \mathbf{C'}\end{array}\right] \qquad (9.10.8)$$

q.e.d.

For matrices $\mathbf{C}$ obeying (9.10.1), but not (9.10.5), we need a transformation first:

$\mathbf{C'C}$ is positive definite and can be expressed as:

$$\mathbf{C'C} \;=\; \mathbf{J'J} \qquad\qquad (9.10.9)$$

where

$$J = H \hat{G}^{1/2} H'$$

and where H is the matrix of the normalized column vectors of $C'C$, which conform to $H H' = H'H = I_{n-r}$, i.e. $J'J$ is the diagonalization of the n-r by n-r matrix $C'C$, and as the roots are all positive we can take their half powers and write them as a diagonal matrix $\hat{G}$.

Since C is of full rank n-r, both $C'C$, and D, which are of order n-r by n-r are invertible.

In that case

$$P_2 = C J'^{-1} \tag{9.10.10}$$

obeys the condition

$$P_2'P_2 = [J']^{-1} C'C J^{-1} = [J']^{-1} J' J J^{-1}$$

$$= I_{n-r} \tag{9.10.11}$$

It is obviously possible to develop a suitable adaptation of (9.10.7) and (9.10.8), but if we indeed want to calculate $[A'A]^-$ and not A^-, imposing (9.10.5) on the transformed matrix C, making each column of C into a column of the full matrix P of the vectors of $A'A$, which inverts by transposition, is a more practical approach.

However, that emphasis on ensuring (9.10.5) needs to be qualified, if we are after A^- rather than $[A'A]^-$.

<u>Any</u> set of n-r independent columns c_j satisfying (9.10.3), any matrix C, which is of order n by n-r and of full rank n-r and satisfies $A C = [0]$ provides an F which will do as a 'substitute' for $[A'A]^-$ in applying

$$A^- = [A'A]^- A' = F^- A' = [A'A + C C']^{-1} A' \tag{9.10.12}$$

To confirm that (9.10.12) is indeed a correct calculation of the generalized inverse, it is useful to explore the relationship between F^{-1} and $[A'A]^-$ more fully.

Comparing (9.10.2), (9.10.1), and the diagonalized form of the symmetric product expression:

$$\mathbf{A'A} = \mathbf{P_1} \, \hat{\mathbf{L}}_1 \, \mathbf{P_1'} \qquad\qquad (9.6.13)$$

we express $\mathbf{F}$ as:

$$\mathbf{F} = \mathbf{P_1} \, \hat{\mathbf{L}}_1 \, \mathbf{P'}_1 + \mathbf{C} \, \mathbf{C'} =$$

$$\mathbf{P_1} \, \hat{\mathbf{L}}_1 \, \mathbf{P'}_1 + \mathbf{P_2} \, \mathbf{J'J} \, \mathbf{P'}_2 \qquad\qquad (9.10.14)$$

and the corresponding inverse is:

$$\mathbf{F^{-1}} = \mathbf{P_1} \, \hat{\mathbf{L}}_1^{-1} \, \mathbf{P'}_1 + \mathbf{P_2} \, \mathbf{J^{-1}} \, \mathbf{J'^{-1}} \, \mathbf{P'}_2 =$$

$$[\mathbf{A'A}]^{-} \qquad\quad + \mathbf{P_2} \, \mathbf{J^{-1}} \, \mathbf{J'^{-1}} \, \mathbf{P'}_2 \qquad\qquad (9.10.15)$$

as may be verified by pre-multiplication of (9.10.14) by $\mathbf{F}$, and consideration of the orthogonality of all columns of both $\mathbf{C}$ and $\mathbf{P_2}$ to the rows of $\mathbf{P'}_1$.

Relation (9.10.15) does not directly tell us what $\mathbf{A^-}$ is, because we don't know the two terms on the righthand side separately, except in the special case where (9.10.5) has been imposed beforehand, and $\mathbf{C} = \mathbf{P_2}$ and $\mathbf{J} = \mathbf{I_{n-r}}$ apply.

However, substitution of the righthand side of (9.10.15) for $\mathbf{F^{-1}}$ into the middle (third) member of (9.10.12), does indeed confirm the identity of the middle member of (9.10.12) with the others:

$$\mathbf{F^{-1}} \, \mathbf{A'} = [\mathbf{A'A}]^{-} \, \mathbf{A'} + \mathbf{P_2} \, \mathbf{J^{-1}} \, \mathbf{J'^{-1}} \, \mathbf{P'}_2 \, \mathbf{A'}$$

$$= [\mathbf{A'A}]^{-} \, \mathbf{A} \qquad\qquad (9.10.16)$$

Given $\mathbf{A} \, \mathbf{P_2} = [0]$ and hence $\mathbf{P'}_2 \, \mathbf{A'} = [0]$, the precise configuration of $\mathbf{C} = \mathbf{P_2} \, \mathbf{J'^{-1}}$ is not of any real importance. Application of the rank algorithm from section 5.10 is an obvious way to find a matrix $\mathbf{C}$.

Even if we do indeed want $[\mathbf{A'A}]^{-}$ for its own sake, it is not actually necessary to calculate $\mathbf{P_2}$ -i.e. $\mathbf{C}$ with (9.10.5) imposed on it- explicitly. The symmetric product $\mathbf{P_2} \, \mathbf{P'}_2$ is rather easier to obtain as:

$$P_2 \, P'_2 \;=\; C \, [C'C]^{-1} \, C' \qquad\qquad (9.10.17)$$

and that is all we need.

Before we proceed to numerical illustration, it is worth to mention that the formulae as developed in this section so far, using the condition $A \, C = [0]$, relate to the symmetric product expression $A'A$. We are dealing with the vectors associated with the zero roots of $A'A$, hence the use of the symbol P_2 (the dummy row-components), for their normalized form. This form of the various formulae slots in directly with the use of $[A'A]^-A'$ as calculation-formula for A^-, to which we shall refer as rank-completion (of the symmetric product expression) on the left. If we want $A^- = A'[A \, A']^-$ we need generalizations of (9.10.1), (9.10.2), (9.10.5), (9.10.14) and (9.10.15):

Our generalization of (9.10.1) is:

$$Z \, A = [0] \qquad\qquad (9.10.18)$$

where Z needs to be of order m-r by m, and of full rank m-r, but (as in the case of C) no further requirement on its numerical content applies.

The generalization of (9.10.2), the rank-completion of the symmetric product expression on the right in general is:

$$R = A \, A' + Z'Z \qquad\qquad (9.10.19)$$

If the requirement:

$$Z \, Z' = I_{m-r} \qquad\qquad (9.10.20)$$

is imposed, then Z' becomes the m by m-r matrix of dummy column components V_2, and an R which meets that condition is related to $[A \, A']^-$ by:

$$[A \, A']^- \;=\; [A \, A' + V_2 \, V'_2]^{-1} - V_2 \, V'_2 \qquad\qquad (9.10.21)$$

and A^- is calculated by rank-completion on the right as:

$$\mathbf{A}^- = \mathbf{A}'\,[\mathbf{A}\,\mathbf{A}']^- = \mathbf{A}'\,\mathbf{R}^{-1} \qquad\qquad (9.10.22)$$

and it makes no difference to the calculation of $\mathbf{A}^-$, whether (9.10.20) has been imposed or not.

Example:

$$\mathbf{A} = \begin{bmatrix} 1 & 2 \\ 2 & 4 \\ 3 & 6 \end{bmatrix}$$

In this example of a 3 by 2 matrix $\mathbf{A}$, rank-completion on the right, $\mathbf{A}^- = \mathbf{A}'\,[\mathbf{A}\,\mathbf{A}']^-$ is not the most efficient method.

It requires a 3 by 3 symmetric product expression $[\mathbf{A}\,\mathbf{A}']$, whereas for rank-completion on the left, $\mathbf{A}^- = [\mathbf{A}'\mathbf{A}]\,\mathbf{A}'$, the 2 by 2 symmetric product expression $[\mathbf{A}'\mathbf{A}]$ would suffice.

Rank completion on the right has been chosen here, partly because most of the formulae in the text above have been cast, in that form, and partly because as an illustration it has the advantage of the calculations not being apparently trivial.

We first calculate the symmetric product expression :

$$\mathbf{A}\,\mathbf{A}' = \begin{bmatrix} 1 & 2 \\ 2 & 4 \\ 3 & 6 \end{bmatrix} \begin{bmatrix} 1 & 2 & 3 \\ 2 & 4 & 6 \end{bmatrix} = \begin{bmatrix} 5 & 10 & 15 \\ 10 & 20 & 30 \\ 15 & 30 & 45 \end{bmatrix}$$

The rank of $\mathbf{A}$ is also the rank of the symmetric product expression, and in this example the rank is 1, and non-trivial rank completion is required.

Since we are using $\mathbf{A}\,\mathbf{A}'$, the rank needs to be completed with $\mathbf{Z}'\mathbf{Z}$ i.e. we need the vectors $z_j \neq [0]$, which satisfy $z'_j \mathbf{A} = [0]$, and hence $\mathbf{A}'z_j = [0]$, and $\mathbf{A}\,\mathbf{A}'z_j = [0]$. To find the appropriate vectors which satisfy $\mathbf{A}\,\mathbf{A}'\,z_j = [0]$, we employ the rank algorithm from section 5.10. The one step on the $\mathbf{A}\,\mathbf{A}'z_j = [0]$ system is summarized below:

$$
\begin{array}{c}
\begin{array}{ccc} \mathbf{z_1} & \mathbf{z_2} & \mathbf{z_3} \end{array} \\
\begin{array}{l}
\text{eq 1} \\ \text{eq 2} \\ \text{eq 3}
\end{array}
\left[\begin{array}{c|cc}
5 & 10 & 15 \\ \hline
10 & 20 & 30 \\
15 & 30 & 45
\end{array} \right]
\end{array}
$$

$$
\begin{array}{c}
\begin{array}{ccc} \mathbf{z_1} & \mathbf{z_2} & \mathbf{z_3} \end{array} \\
\begin{array}{l}
\text{eq 1} \\ \text{eq 2} \\ \text{eq 3}
\end{array}
\left[\begin{array}{ccc}
1 & 2 & 3 \\
- & - & - \\
- & - & -
\end{array} \right]
\end{array}
$$

and accordingly, the two independent columns $\mathbf{c_j}$, representing respectively :

$$z_1 = 3, \ z_3 = -1 \ (z_2 = 0), \text{ and } z_1 = 2, \ z_2 = -1 \ (z_3 = 0)$$

are grouped in the matrix:

$$
\begin{array}{c}
\begin{array}{cc} \mathbf{z_3} & \mathbf{z_2} \end{array} \\
\begin{array}{l} \mathbf{z_1} \\ \mathbf{z_2} \\ \mathbf{z_3} \end{array}
\left[\begin{array}{cc}
3 & 2 \\
- & -1 \\
-1 & -
\end{array} \right]
\end{array}
\quad \text{i.e.} \quad
\mathbf{Z} =
\left[\begin{array}{cc}
3 & 2 \\
- & -1 \\
-1 & -
\end{array} \right]
$$

(The interchanging of the z_2 and z_3 columns arose because the rank-option of ILLUSTRATE, which was employed to calculate these vectors has interchanged columns, in an attempt to find a second pivot, and does not concern is any further.)

We now first of all deal with the explicit calculation of $\mathbf{V_2} \, \mathbf{V'_2}$.

We begin with calculating the symmetric product expression $\mathbf{Z} \, \mathbf{Z'}$.

$$
\left[\begin{array}{ccc}
3 & - & -1 \\
2 & -1 & -
\end{array} \right]
\left[\begin{array}{cc}
3 & 2 \\
- & -1 \\
-1 & -
\end{array} \right]
=
\left[\begin{array}{cc}
10 & 6 \\
6 & 5
\end{array} \right]
$$

The inverse of this symmetric product expression is:

$$
\left[\begin{array}{cc}
10 & 6 \\
6 & 5
\end{array} \right]^{-1}
=
\left[\begin{array}{cc}
0.357 & -0.429 \\
-0.429 & 0.714
\end{array} \right]
$$

Therefore:

$$\mathbf{V}_2 \ \mathbf{V'}_2 \ = \ \mathbf{Z'} [\mathbf{Z} \ \mathbf{Z'}]^{-1} \ \mathbf{Z} \ =$$

$$\begin{bmatrix} 3 & 2 \\ - & -1 \\ -1 & - \end{bmatrix} \begin{bmatrix} 0.357 & -0.429 \\ -0.429 & 0.714 \end{bmatrix} \begin{bmatrix} 3 & - & -1 \\ 2 & -1 & - \end{bmatrix} \ =$$

$$\begin{bmatrix} 0.213 & 0.141 \\ 0.429 & -0.714 \\ -0.357 & 0.429 \end{bmatrix} \begin{bmatrix} 3 & - & -1 \\ 2 & -1 & - \end{bmatrix} = \begin{bmatrix} 0.929 & -0.143 & -0.214 \\ -0.143 & 0.714 & -0.429 \\ -0.214 & -0.429 & 0.357 \end{bmatrix}$$

Therefore, if $\mathbf{R}$ is calculated on the basis of $\mathbf{Z'} = \mathbf{V}_2$, this matrix becomes:

$$\mathbf{R} = \mathbf{A} \ \mathbf{A'} + \mathbf{V}_2 \ \mathbf{V'}_2 \ =$$

$$\begin{bmatrix} 5 & 10 & 15 \\ 10 & 20 & 30 \\ 15 & 30 & 45 \end{bmatrix} + \begin{bmatrix} 0.929 & -0.143 & -0.214 \\ -0.143 & 0.714 & -0.429 \\ -0.214 & -0.429 & 0.357 \end{bmatrix} \ =$$

$$\begin{bmatrix} 5.929 & 9.857 & 14.786 \\ 9.857 & 20.714 & 29.571 \\ 14.786 & 29.571 & 45.357 \end{bmatrix}$$

Therefore $\mathbf{R}^{-1}$ is:

$$\begin{bmatrix} 5.929 & 9.857 & 14.786 \\ 9.857 & 20.714 & 29.571 \\ 14.786 & 29.571 & 45.357 \end{bmatrix}^{-1} = \begin{bmatrix} 0.930 & -0.141 & -0.211 \\ -0.141 & 0.718 & -0.422 \\ -0.211 & -0.422 & 0.366 \end{bmatrix}$$

As this matrix $\mathbf{R}$ was calculated, with a $\mathbf{Z} = \mathbf{V'}_2$, which respects (9.10.20), we can also use it directly to evaluate $[\mathbf{A} \ \mathbf{A'}]^-$:

$$[\mathbf{A}\ \mathbf{A'}]^- = \mathbf{R}^{-1} - \mathbf{V}_2\ \mathbf{V'}_2 =$$

$$\begin{bmatrix} 0.930 & -0.141 & -0.211 \\ -0.141 & 0.718 & -0.422 \\ -0.211 & -0.422 & 0.366 \end{bmatrix} - \begin{bmatrix} 0.929 & -0.143 & -0.214 \\ -0.143 & 0.714 & -0.429 \\ -0.214 & -0.429 & 0.357 \end{bmatrix}$$

$$= \begin{bmatrix} 0.001 & 0.002 & 0.003 \\ 0.002 & 0.004 & 0.006 \\ 0.003 & 0.006 & 0.009 \end{bmatrix}$$

Therefore $\mathbf{A}^- = \mathbf{A'}\ [\mathbf{A}\ \mathbf{A'}]^-$ is:

$$\begin{bmatrix} 1 & 2 & 3 \\ 2 & 4 & 6 \end{bmatrix} \begin{bmatrix} 0.001 & 0.002 & 0.003 \\ 0.002 & 0.004 & 0.006 \\ 0.003 & 0.006 & 0.009 \end{bmatrix}$$

$$= \begin{bmatrix} 0.014 & 0.029 & 0.043 \\ 0.029 & 0.057 & 0.086 \end{bmatrix}$$

But $\mathbf{A}^-$ can also be calculated as $\mathbf{A}^- = \mathbf{A'R}^{-1}$, either with, or without imposing (9.10.20)

$$\begin{bmatrix} 1 & 2 & 3 \\ 2 & 4 & 6 \end{bmatrix} \begin{bmatrix} 0.929 & -0.143 & -0.214 \\ -0.143 & 0.714 & -0.429 \\ -0.214 & -0.429 & 0.357 \end{bmatrix}$$

$$= \begin{bmatrix} 0.014 & 0.029 & 0.043 \\ 0.029 & 0.057 & 0.086 \end{bmatrix}$$

with a matrix $\mathbf{R}$ which is also suitable for calculating $[\mathbf{A}\ \mathbf{A'}]^-$.

If (9.10.20) is <u>not</u> imposed, we calculate first of all $\mathbf{R}$ as:

312   *Matrices and Their Roots*

$$R = A\,A' + Z'Z =$$

$$
\begin{bmatrix} 5 & 10 & 15 \\ 10 & 20 & 30 \\ 15 & 30 & 45 \end{bmatrix}
+
\begin{bmatrix} 3 & 2 \\ - & -1 \\ -1 & - \end{bmatrix}
\begin{bmatrix} 3 & - & -1 \\ 2 & -1 & - \end{bmatrix}
=
$$

$$
\begin{bmatrix} 5 & 10 & 15 \\ 10 & 20 & 30 \\ 15 & 30 & 45 \end{bmatrix}
+
\begin{bmatrix} 13 & -2 & -3 \\ -2 & 1 & - \\ -3 & - & 1 \end{bmatrix}
=
\begin{bmatrix} 18 & 8 & 12 \\ 8 & 21 & 30 \\ 12 & 30 & 46 \end{bmatrix}
$$

Therefore, if (9.10.20) is not imposed R^{-1} is (among others):

$$
\begin{bmatrix} 18 & 8 & 12 \\ 8 & 21 & 30 \\ 12 & 30 & 46 \end{bmatrix}^{-1}
=
\begin{bmatrix} 0.067 & -0.008 & -0.012 \\ -0.008 & 0.698 & -0.453 \\ -0.012 & -0.453 & 0.320 \end{bmatrix}
$$

Therefore A^- is:

$$A^- = A'\,R^{-1} =$$

$$
\begin{bmatrix} 1 & 2 & 3 \\ 2 & 4 & 6 \end{bmatrix}
\begin{bmatrix} 0.067 & -0.008 & -0.012 \\ -0.008 & 0.698 & -0.453 \\ -0.012 & -0.453 & 0.320 \end{bmatrix}
$$

$$
=
\begin{bmatrix} 0.014 & 0.029 & 0.043 \\ 0.029 & 0.057 & 0.086 \end{bmatrix}
$$

Exercise:

Find the generalized inverse of $A = \begin{bmatrix} 2 & -1 & 3 \\ 1 & 1 & - \end{bmatrix}$

and also of $B = \begin{bmatrix} 2 & -1 & 3 \\ 1 & 1 & - \\ 3 & - & 3 \end{bmatrix}$

Use both lefthand and righthand rank completion (for A, rank-completion on the right is trivial, $[A\,A']$ is itself invertible), and cross-check the results of the various calculations against each other. (There is a partial answersheet at the end of the chapter anyhow.)

9.11 **The differentiation of a generalized inverse**

The differentiation of a generalized inverse of the m by n matrix
A is meaningful, only if there is no change in rank, a condition
to which we shall refer as consistent rank.

No restriction on a vector or matrix of differential variations,
can guarantee that the rank does not suddenly become less, but,
as in other applications of calculus, we normally assume that
existing non-zero variables and functions don't suddenly become
zero because of a small change. We will therefore, in practice,
interpret the requirement of consistent rank, as the rank not
becoming <u>more</u> than it was.

A sufficient condition to that purpose is:

$$[d\mathbf{A}] \; \mathbf{P}_2 \;\; = \;\; [0] \tag{9.11.1}$$

where $\mathbf{P}_2$ is the matrix containing the dummy row-components.

The restriction stated by (9.11.1) is indeed sufficient to ensure
consistent rank, even on finite changes:

$$[\mathbf{A} + \Delta\mathbf{A}] \; \mathbf{P}_2 \;\; = \;\; [0] \tag{9.11.2}$$

applies before and after change.

We shall refer to a change in **A**, which is constrained by
(9.11.1), or by its finite equivalent (9.11.2), as one which
maintains consistent rank, <u>by row-orthogonality</u>.

The alternative requirement:

$$\mathbf{V}' \; [d\mathbf{A}] \;\; = \;\; [0] \tag{9.11.3}$$

then requires consistent rank <u>by column-orthogonality</u>.

When applying any results of the two separate formulae which will arise from these side-conditions, in finite approximation, it is important to note that they cannot generally be added together, as the second-order effect may invalidate the condition of consistent rank.

Example:

$$\mathbf{A} = \begin{bmatrix} 1 & 1 \\ 1 & 1 \end{bmatrix} \qquad \mathbf{A} + \mathbf{\Delta A} = \begin{bmatrix} 1 & 1.2 \\ 1 & 1.2 \end{bmatrix}$$

The rank is consistent at 1 (by column-orthogonality),

$$\mathbf{V'}_2 = [1 \quad -1] \times \tfrac{1}{2}\sqrt{2}$$

$$\mathbf{A}^- = \begin{bmatrix} 0.25 & 0.25 \\ 0.25 & 0.25 \end{bmatrix} \qquad [\mathbf{A} + \mathbf{\Delta A}]^- = \begin{bmatrix} 0.2049 & 0.2049 \\ 0.2459 & 0.2459 \end{bmatrix}$$

$$\mathbf{A} = \begin{bmatrix} 1 & 1 \\ 1 & 1 \end{bmatrix} \qquad \mathbf{A} + \mathbf{\Delta A} = \begin{bmatrix} 1 & 1 \\ 1.2 & 1.2 \end{bmatrix}$$

The rank is consistent at 1 (by row-orthogonality),

$$\mathbf{A}^- = \begin{bmatrix} 0.25 & 0.25 \\ 0.25 & 0.25 \end{bmatrix} \qquad [\mathbf{A} + \mathbf{\Delta A}]^- = \begin{bmatrix} 0.2049 & 0.2459 \\ 0.2049 & 0.2459 \end{bmatrix}$$

but for:

$$\mathbf{A} = \begin{bmatrix} 1 & 1 \\ 1 & 1 \end{bmatrix} \qquad \mathbf{A} + \mathbf{\Delta A} = \begin{bmatrix} 1 & 1.2 \\ 1.2 & 1.4 \end{bmatrix}$$

the rank is not consistent, but changes from 1 to 2, and $\mathbf{A}$ becomes invertible, but is ill-conditioned (= <u>almost</u> singular, giving rise to very large elements of the inverse):

$$[\mathbf{A} + \mathbf{\Delta A}]^- = [\mathbf{A} + \mathbf{\Delta A}]^{-1} = \begin{bmatrix} -35 & 30 \\ 30 & -25 \end{bmatrix}$$

$$\text{For } d\mathbf{A} = \begin{bmatrix} - & 0.01 \\ 0.01 & 0.02 \end{bmatrix} \qquad \mathbf{A} + d\mathbf{A} = \begin{bmatrix} 1 & 1.01 \\ 1.01 & 1.02 \end{bmatrix}$$

the true generalized inverse (= the ordinary inverse) is:

$$[\mathbf{A} + \Delta\mathbf{A}]^- = [\mathbf{A} + \Delta\mathbf{A}]^{-1} = \begin{bmatrix} -10200 & 10100 \\ 10100 & -10000 \end{bmatrix}$$

but the generalized inverse option of ILLUSTRATE could not be employed for its calculation. Illustrate is programmed to only except numbers of an absolute value above 0.00001 as non-zero, and since the generalized inverse option works on the symmetric product-expression, it found, given:

$$\begin{vmatrix} 1 & 1.01 \\ 1.01 & 1.02 \end{vmatrix} = 0.0001$$

the determinant of the symmetric product expression (= $\mathbf{A}^2$ in this case), to be 0.00000001. (The determinant of the product of two square matrices is the product of their determinants, see section 5.7). Illustrate therefore concluded to a rank of 1, and offered

$$\begin{bmatrix} 0.2451 & 0.2457 \\ 0.2457 & 0.2500 \end{bmatrix}$$

as the generalized inverse.

Theorem:

If d$\mathbf{A}$ obeys (9.11.1),

then the symmetric product expression [$\mathbf{A'A}$] and its g-inverse obey the differential restriction:

$$d[\mathbf{A'A}]^- = -[\mathbf{A'A}]^- d[\mathbf{A'A}]^- [\mathbf{A'A}] \qquad (9.11.4)$$

Proof:

We first express [$\mathbf{A'A}$]$^-$ as:

$$[\mathbf{A\,A'}]^- = [\mathbf{A\,A'} + \mathbf{P}_2\,\mathbf{P'}_2]^{-1} - \mathbf{P}_2\,\mathbf{P'}_2$$

$$= \mathbf{F}^{-1} - \mathbf{P}_2\,\mathbf{P'}_2 \qquad (9.11.5)$$

(See section 9.10)

Given that the side-condition (9.11.1) ensures that the same dummy-components stay valid, the differentiation of $[A'A]^-$ is equivalent to the differentiation of the inverse of the positive definite matrix F.

$$d[A'A]^- = dF^{-1} = F^{-1} dF F^{-1} = F^{-1} d[A'A] F^{-1}$$

$$= [[A'A]^- + P_2 P'_2] d[A'A] [[A'A]^- + P_2 P'_2]$$

$$= [A'A]^- d[A'A] [A'A]^- \qquad (9.11.6)$$

q.e.d.

(Refer to section 4.8 for the differentiation of an ordinary inverse.)

For an arbitrary A, we now differentiate both sides of

$$A^- = [A'A]^- A'$$

to obtain:

$$dA^- = d[A'A]^- A' + [A'A]^- dA'$$

$$= -[A'A]^- d[A'A] [A'A]^- A' + [A'A]^- dA'$$

$$= dF^{-1} A' + F^{-1} dA' \qquad (9.11.7)$$

as the formula for the differential of a g-inverse, if consistent rank is ensured by row-orthogonality.

If consistent rank is ensured by column-orthogonality

$$V'_2 [dA] = [0]$$

we have, by analogy:

$$dA^- = A' dR^{-1} + dA R^{-1}$$

$$= -A' [AA']^- d[AA'] [AA']^- + dA' [AA']^- \qquad (9.11.8)$$

Note, that, as was illustrated by the example at the beginning of this section, these two formulae do <u>not</u> imply each other, but apply to two separate sets of legitimate matrices of differential variation.

Answersheet on assessing definiteness by the roots

We begin with the characteristic equation

a) Directly, by developing $|A - I\lambda|$:

$$\begin{vmatrix} 3-\lambda & 3 & 1 \\ 3 & 1-\lambda & 3 \\ 1 & 3 & 3-\lambda \end{vmatrix} =$$

$$(3-\lambda)\begin{vmatrix} 1-\lambda & 3 \\ 3 & 3-\lambda \end{vmatrix} - 3\begin{vmatrix} 3 & 3 \\ 1 & 3-\lambda \end{vmatrix} + \begin{vmatrix} 3 & 1-\lambda \\ 1 & 3 \end{vmatrix} =$$

$$(3-\lambda)\Big((1-\lambda)(3-\lambda) -9\Big) \quad -3\Big(3(3-\lambda) -3\Big) \quad + \Big(9 - (1-\lambda)\Big) =$$

$$(3-\lambda)\Big(3 -4\lambda +\lambda^2 -9\Big) \quad -3\Big(9 -3\lambda -3\Big) \quad + \Big(9 - 1 + \lambda\Big) =$$

$$(3-\lambda)\Big(-6 -4\lambda + \lambda^2\Big) \quad -3\Big(6 - 3\lambda\Big) \quad + \Big(8 + \lambda\Big) =$$

$$-18 -12\lambda + 3\lambda^2$$

$$+ 6\lambda + 4\lambda^2 - \lambda^3 \qquad -18 + 9\lambda \qquad + 8 + \lambda \qquad =$$

$$-28 + 4\lambda + 7\lambda^2 - \lambda^3 = 0.$$

The characteristic equation therefore is:

$$\lambda^3 - 7\lambda^2 - 4\lambda + 28 = 0.$$

b) Using principal minors (section 7.2):

$$\begin{vmatrix} -3 & -3 & -1 \\ -3 & -1 & -3 \\ -1 & -3 & -3 \end{vmatrix} =$$

$$- 3 \times \begin{vmatrix} -1 & -3 \\ -3 & -3 \end{vmatrix} \; -(-3) \begin{vmatrix} -3 & -3 \\ -1 & -3 \end{vmatrix} \; +(-1) \begin{vmatrix} -3 & -1 \\ -1 & -3 \end{vmatrix} =$$

$-3 \times (-6) + 3 \times 6 - 8 = 28$, confirming the constant.

$$\begin{vmatrix} -3 & -3 \\ -3 & -1 \end{vmatrix} + \begin{vmatrix} -1 & -3 \\ -3 & -3 \end{vmatrix} + \begin{vmatrix} -3 & -1 \\ -1 & -3 \end{vmatrix} = -6 \quad -6 \quad + 8 = -4,$$

confirming the coefficient of λ.

$|-3| + |-1| + |-3| = -7$, confirming the coefficient for λ^2.

Now dividing the function $\phi(\lambda) = \lambda^3 - 7\lambda^2 - 4\lambda + 28$ by $\lambda-7$ in view of the known root $\lambda=7$:

$$\lambda - 7 \; / \; \lambda^3 - 7\lambda^2 - 4\lambda + 28 \; \backslash \; \lambda^2 - 4$$

$$\underline{\lambda^2 \quad - 7\lambda^2} \qquad\qquad\qquad (= \lambda^2 \, (\lambda - 7) \;)$$

$$- 4\lambda \quad + 28$$

$$\underline{- 4\lambda \quad + 28} \qquad\qquad (= -4 \times (\lambda - 7) \;)$$

$$0$$

Therefore:

$$\lambda^3 - 7\lambda^2 - 4\lambda + 28 = (\lambda - 7) (\lambda^2 - 4)$$

$$= (\lambda - 7) (\lambda - 2) (\lambda + 2)$$

and the roots are: $\lambda_1 = 7$, $\lambda_2 = 2$, and $\lambda_3 = -2$.

Therefore **A** has both positive and negative roots, and is indefinite.

Answersheet on calculating a generalized inverse

We are asked to find the generalized inverse of:

$$\mathbf{A} = \begin{bmatrix} 2 & -1 & 3 \\ 1 & 1 & - \end{bmatrix}$$

There are several formulae for the calculation of a generalized inverse.

For

$$\mathbf{G} = \mathbf{A}' \, [\mathbf{A}\mathbf{A}']^-$$

the g-inverse of the symmetric product expression $[\mathbf{A}\mathbf{A}']$ is the ordinary inverse, and neither rank-completion, nor the evaluation of any roots is required.

We first calculate $\mathbf{A}\mathbf{A}' =$

$$\begin{bmatrix} 2 & -1 & 3 \\ 1 & 1 & - \end{bmatrix} \begin{bmatrix} 2 & 1 \\ -1 & 1 \\ 3 & - \end{bmatrix} = \begin{bmatrix} 14 & 1 \\ 1 & 2 \end{bmatrix}$$

Therefore $[\mathbf{A}\,\mathbf{A}']^- = [\mathbf{A}\,\mathbf{A}']^{-1} =$

$$\begin{bmatrix} 14 & 1 \\ 1 & 2 \end{bmatrix}^{-1} = 1/27 \times \begin{bmatrix} 2 & -1 \\ -1 & 14 \end{bmatrix} = \begin{bmatrix} 0.076 & -0.037 \\ -0.037 & 0.519 \end{bmatrix}$$

Therefore: $\mathbf{A}^- = \mathbf{A}' \, [\mathbf{A}\mathbf{A}']^-$

$$\begin{bmatrix} 2 & -1 & 3 \\ 1 & 1 & - \end{bmatrix}^- = \begin{bmatrix} 2 & 1 \\ -1 & 1 \\ 3 & - \end{bmatrix} \begin{bmatrix} 2 & -1 \\ -1 & 14 \end{bmatrix} \times (1/27)$$

$$= \begin{bmatrix} 3 & 12 \\ -3 & 15 \\ 6 & -3 \end{bmatrix} \times (1/27) = \begin{bmatrix} 0.111 & 0.444 \\ -0.111 & 0.556 \\ 0.222 & -0.111 \end{bmatrix}$$

We now deal with the in this case slightly more difficult for-mula:

$$G = [A'A]^{-} A'$$

We still have a number of methods of calculation available.

We first deal with taking reciprocals of the non-zero roots of of $[A'A]$. Simply because that is easier in terms of a 2 by 2 matrix, we extract those as the non-zero roots of $[AA']$.

We develop the characteristic equation of $[AA']$:

$$\begin{vmatrix} 14-\lambda & 1 \\ 1 & 2-\lambda \end{vmatrix} = (14-\lambda)(2-\lambda) - 1 = \lambda^2 - 16\lambda + 27 = 0$$

Therefore :
$$\lambda = \left(16 \pm \sqrt{16^2 - 4 \times 27} \right) / 2 =$$
$$\left(16 \pm \sqrt{256 - 108} \right) / 2 =$$
$$\left(16 \pm \sqrt{148} \right) / 2 =$$
$$(16 \pm 12.166) / 2 = 8 \pm 6.083$$

We initially calculate the vectors of $[AA']$ conform:

$$\begin{bmatrix} 14 & 1 \\ 1 & 2 \end{bmatrix} \begin{bmatrix} 12.083 \\ 1 \end{bmatrix} = \begin{bmatrix} 170.162 \\ 14.083 \end{bmatrix} = \begin{bmatrix} 12.083 \\ 1 \end{bmatrix} \times 14.083$$

for $\lambda_1 = 8 + 6.083 = 14.083$, and:

$$\begin{bmatrix} 14 & 1 \\ 1 & 2 \end{bmatrix} \begin{bmatrix} -1 \\ 12.083 \end{bmatrix} = \begin{bmatrix} -1.917 \\ 23.166 \end{bmatrix} = \begin{bmatrix} -1 \\ 12.083 \end{bmatrix} \times 1.917$$

for $\lambda_2 = 8 - 6.083 = 1.917$

To obtain the vectors associated with the non-zero roots of $[A'A]$, we need to pre-multiply by A', to obtain:

$$\mathbf{u}_1 = \begin{bmatrix} 2 & 1 \\ -1 & 1 \\ 3 & - \end{bmatrix} \begin{bmatrix} 12.083 \\ 1 \end{bmatrix} = \begin{bmatrix} 25.166 \\ -11.083 \\ 36.249 \end{bmatrix} = \begin{bmatrix} 0.553 \\ -0.243 \\ 0.797 \end{bmatrix} \times 45.5$$

for λ_1, where the most righthand vector has been normalized.

Similarly, for λ_2:

$$\mathbf{u}_2 = \begin{bmatrix} 2 & 1 \\ -1 & 1 \\ 3 & - \end{bmatrix} \begin{bmatrix} -1 \\ 12.083 \end{bmatrix} = \begin{bmatrix} 10.083 \\ 13.083 \\ -3 \end{bmatrix} = \begin{bmatrix} 0.601 \\ 0.779 \\ -0.179 \end{bmatrix} \times 16.788$$

Note, that, as it does not figure in $[\mathbf{A'A}]^-$, we don't strictly need the vector associated with the third (zero) root of $[\mathbf{A'A}]$.

However, in the interest of being able to check the internal consistency of $\mathbf{P}$, this third vector <u>has</u> been calculated.

To obtain the vector associated with the zero root of $[\mathbf{A'A}]$, we could use the requirement of orthogonality of the vectors already calculated, but it is here done directly from $[\mathbf{A'A}]$ itself, and we first of all calculate that matrix:

$$\begin{bmatrix} 2 & 1 \\ -1 & 1 \\ 3 & - \end{bmatrix} \begin{bmatrix} 2 & -1 & 3 \\ 1 & 1 & - \end{bmatrix} = \begin{bmatrix} 5 & -1 & 6 \\ -1 & 2 & -3 \\ 6 & -3 & 9 \end{bmatrix}$$

The corresponding vector conforms to:

$$\begin{bmatrix} 5 & -1 & 6 \\ -1 & 2 & -3 \\ 6 & -3 & 9 \end{bmatrix} \begin{bmatrix} -1 \\ 1 \\ 1 \end{bmatrix} = \begin{bmatrix} - \\ - \\ - \end{bmatrix} \qquad \text{or} \qquad \mathbf{p}_3 = \begin{bmatrix} -0.577 \\ 0.577 \\ 0.577 \end{bmatrix}$$

where the most righthand presentation represents normalization to the requirement $\mathbf{p'p} = 1$.

We are now in a position to express $[A'A]$ in diagonalized form:

$A A' =$

$$
\begin{bmatrix} 0.553 & 0.601 & -0.577 \\ -0.243 & 0.779 & 0.577 \\ 0.797 & -0.179 & 0.577 \end{bmatrix}
\begin{bmatrix} 14.083 & - & - \\ - & 1.917 & - \\ - & - & - \end{bmatrix}
\begin{bmatrix} 0.553 & -0.243 & 0.797 \\ 0.601 & 0.779 & -0.179 \\ -0.577 & 0.577 & 0.577 \end{bmatrix}
$$

$$
= \begin{bmatrix} 7.788 & 1.152 & - \\ -3.422 & 1.493 & - \\ -11.224 & -0.343 & - \end{bmatrix}
\begin{bmatrix} 0.553 & -0.243 & 0.797 \\ 0.601 & 0.779 & -0.179 \\ -0.577 & 0.577 & 0.577 \end{bmatrix}
= \begin{bmatrix} 5 & -1 & 6 \\ -1 & 2 & -3 \\ 6 & 2 & -3 \end{bmatrix}
$$

Therefore, on taking the reciprocals of the non-zero roots, the generalized inverse $[A'A]^-$ of $[A'A]$ is calculated as:

$[A A']^- =$

$$
\begin{bmatrix} 0.553 & 0.601 & -0.577 \\ -0.243 & 0.779 & 0.577 \\ 0.797 & -0.179 & 0.577 \end{bmatrix}
\begin{bmatrix} 0.071 & - & - \\ - & 0.522 & - \\ - & - & - \end{bmatrix}
\begin{bmatrix} 0.553 & -0.243 & 0.797 \\ 0.601 & 0.779 & -0.179 \\ -0.577 & 0.577 & 0.577 \end{bmatrix}
$$

$$
= \begin{bmatrix} 0.0393 & 0.3135 & - \\ -0.0173 & 0.4064 & - \\ 0.0566 & -0.0934 & - \end{bmatrix}
\begin{bmatrix} 0.553 & -0.243 & 0.797 \\ 0.601 & 0.779 & -0.179 \\ -0.577 & 0.577 & 0.577 \end{bmatrix}
$$

$$
= \begin{bmatrix} 0.210 & 0.235 & -0.025 \\ 0.235 & 0.321 & -0.087 \\ -0.025 & -0.087 & 0.062 \end{bmatrix}
$$

Therefore:

$$
A^- = [A'A]^- A' = \begin{bmatrix} 0.210 & 0.235 & -0.025 \\ 0.235 & 0.321 & -0.087 \\ -0.025 & -0.087 & 0.062 \end{bmatrix}
\begin{bmatrix} 2 & 1 \\ -1 & 1 \\ 3 & - \end{bmatrix}
$$

$$
= \begin{bmatrix} 0.111 & 0.444 \\ -0.111 & 0.556 \\ 0.222 & -0.111 \end{bmatrix} \qquad \text{the same as above}
$$

The symmetric product expression $A'A$ was already calculated above, and the same applies to the vector associated with its one

zero root, but for completeness sake we once more summarize the
calculation of the vector associated with the zero root by the
rank algorithm of section 5.10.

$$
\begin{array}{c}
\begin{array}{ccc} z1 & z2 & z3 \end{array} \\
\begin{array}{l}
eq\ 1 \\ eq\ 2 \\ eq\ 3
\end{array}
\left[\begin{array}{ccc}
\underline{5} & -1 & 6 \\
-1 & 2 & -3 \\
6 & -3 & 9
\end{array}\right]
\end{array}
$$

$$
\begin{array}{c}
\begin{array}{ccc} z1 & z2 & z3 \end{array} \\
\begin{array}{l}
eq\ 1 \\ eq\ 2 \\ eq\ 3
\end{array}
\left[\begin{array}{ccc}
1 & -0.200 & 1.200 \\
- & \underline{1.800} & -1.800 \\
- & -1.800 & 1.800
\end{array}\right]
\end{array}
$$

$$
\begin{array}{c}
\begin{array}{ccc} z1 & z2 & z3 \end{array} \\
\begin{array}{l}
eq\ 1 \\ eq\ 2 \\ eq\ 3
\end{array}
\left[\begin{array}{ccc}
1 & - & 1 \\
- & 1 & -1 \\
- & - & -
\end{array}\right]
\end{array}
$$

On adding the completing -1 element, we calculate the vector
associated with the zero root of the symmetric product expression
A'A as:

$$
\begin{array}{c}
z3 \\
\begin{array}{l} z1 \\ z2 \\ z3 \end{array}
\left[\begin{array}{c}
1 \\ -1 \\ -1
\end{array}\right]
\end{array}
$$

After normalization (= division by $\sqrt{3}$), the outer product **v v'**
then is:

$$
\left[\begin{array}{ccc}
0.333 & -0.333 & -0.333 \\
-0.333 & 0.333 & 0.333 \\
-0.333 & 0.333 & 0.333
\end{array}\right]
$$

The rank-completed symmetric product expression (prepared with a
view of also calculating $[\mathbf{A'A}]^-$, hence rank-completed with the
normalized vector associated with the zero root) is:

$$\begin{bmatrix} 5 & -1 & 6 \\ -1 & 2 & -3 \\ 6 & -3 & 9 \end{bmatrix} + \begin{bmatrix} 0.333 & -0.333 & -0.333 \\ -0.333 & 0.333 & 0.333 \\ -0.333 & 0.333 & 0.333 \end{bmatrix} = \mathbf{F}$$

$$= \begin{bmatrix} 5.333 & -1.333 & 5.667 \\ -1.333 & 2.333 & -2.667 \\ 5.667 & -2.667 & 9.333 \end{bmatrix}$$

From which on inversion:

$$\mathbf{F}^{-1} = \begin{bmatrix} 0.543 & -0.099 & -0.358 \\ -0.099 & 0.654 & 0.247 \\ -0.358 & 0.247 & 0.395 \end{bmatrix}$$

Therefore $[\mathbf{A'A}]^-$ is:

$$[\mathbf{A'}\,\mathbf{A}]^- = \mathbf{F}^{-1} - \mathbf{v}\,\mathbf{v'} =$$

$$\begin{bmatrix} 0.543 & -0.099 & -0.358 \\ -0.099 & 0.654 & 0.247 \\ -0.358 & 0.247 & 0.395 \end{bmatrix} - \begin{bmatrix} 0.333 & -0.333 & -0.333 \\ -0.333 & 0.333 & 0.333 \\ -0.333 & 0.333 & 0.333 \end{bmatrix}$$

$$= \begin{bmatrix} 0.210 & 0.235 & -0.025 \\ 0.235 & 0.321 & -0.087 \\ -0.025 & -0.087 & 0.062 \end{bmatrix}$$

If the object of the exercise is simply to calculate $\mathbf{A}^-$ efficiently, then the scaling of the vector of the zero root of $\mathbf{A'A}$ would not have been necessary.

$$\mathbf{F} = \begin{bmatrix} 5 & -1 & 6 \\ -1 & 2 & -3 \\ 6 & -3 & 9 \end{bmatrix} + \begin{bmatrix} 1 & -1 & -1 \\ -1 & 1 & 1 \\ -1 & 1 & 1 \end{bmatrix} = \begin{bmatrix} 6 & -2 & 5 \\ -2 & 3 & -2 \\ -5 & -2 & 10 \end{bmatrix}$$

will do the job.

Nor was the subtraction of the outer product from $\mathbf{F}$ necessary, as far as the calculation of $\mathbf{A}^-$ is concerned.

The verification of $\mathbf{F}\,\mathbf{A}'$ (for either calculation of $\mathbf{F}$), and rank-completion on the right, is left to the reader. Or use ILLUSTRATE, both for the remaining bit on $\mathbf{A}^- = \mathbf{F}^{-1}\,\mathbf{A}'$, and for B.

CHAPTER X

GEOMETRICAL INTERPRETATIONS

10.1 Vectors and coordinate-spaces

The two-dimensional coordinate plane can be used to visually represent pairs of points, i.e. vectors of order 2. We associate one variable with the vertical direction, and one with the horizontal direction. To be able to read pairs of number from a graph, we use a rectangular pair of lines, the <u>coordinate-axes</u>. The point $x_1 = 0$, $x_2 = 0$ is called the <u>origin</u>.

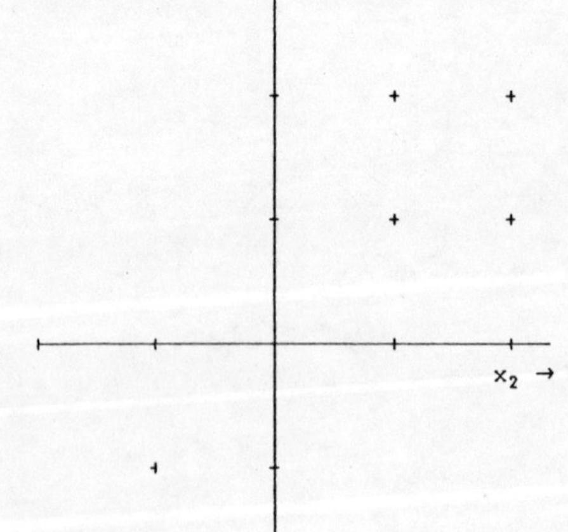

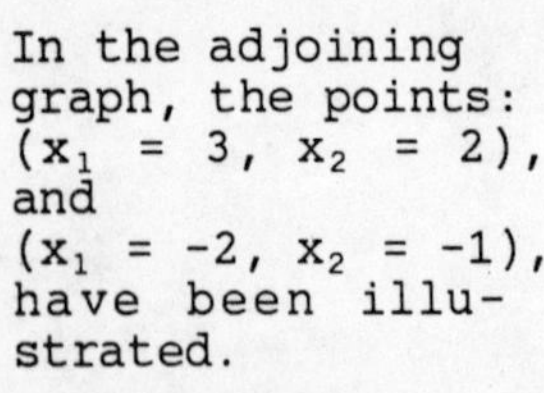

In the adjoining graph, the points:
$(x_1 = 3, x_2 = 2)$, and
$(x_1 = -2, x_2 = -1)$, have been illustrated.

The two points are both marked with a small circle.

Note the crosses as gridmarkings.

The idea of associating a variable with a direction, permits immediate generalization to the three-dimensional case, where we associate the third variable with vertical height above the paper, while we have a 'forward' and 'rightward' direction within a horizontal plane. While we lack that sort of visual analogy in the case of four or more variables/dimensions, there are nevertheless certain geometrical concepts, which can be formulated in purely algebraic terms, and which have to some extent an analogous relevance in the n-dimensional case.

10.2 Distance and straight lines

The distance between two points in two-dimensional space, i.e. the flat coordinate plane, is the sum of squares of their distance in the rightward and forward directions. In three-dimensional space, the upward direction comes to it.

In the graph besides,
the distance between:
$(x_1 = 1, x_2 = 2)$,
and:
$(x_1 = 2, x_2 = 4)$,
is 2-1 = 1 horizontally,
and 4-2 = 2 vertically.

Therefore, by Pythogoras
Theorem, the distance is
in total:

$$\sqrt{(2 - 1)^2 + (4 - 2)^2} = \sqrt{5} .$$

The same argument applies to the relation between an angular distance in the horizontal two-dimensional coordinate plane, and a vertical distance above it.

In the n-dimensional case, it is more convenient, to <u>define</u> the distance between two points (vectors) as:

$$|\mathbf{x} - \mathbf{y}| = \sqrt{\Sigma(x_j - y_j)^2} = \sqrt{[\mathbf{x}' - \mathbf{y}'][\mathbf{x} - \mathbf{y}]} \qquad (10.2.1)$$

By Euclid's postulate, the line-segment between $\mathbf{x}$ and $\mathbf{y}$ is the set of points $\mathbf{z}$, for which the the sum of the distances between $\mathbf{x}$ and $\mathbf{z}$, and $\mathbf{y}$ and $\mathbf{z}$ attains a minimum-value, the minimum being the distance between $\mathbf{x}$ and $\mathbf{y}$.

Distances, as defined above <u>do</u> permit an n-dimensional generalization of the concept of a <u>line-segment</u> between two points. To show that this is indeed the case, we first of all <u>denote</u> (the sum of the two distances):

$$\emptyset = |\mathbf{x} - \mathbf{z}| + |\mathbf{y} - \mathbf{z}|$$

$$= \left([\mathbf{x} - \mathbf{z}]'[\mathbf{x} - \mathbf{z}]\right)^{1/2} + \left([\mathbf{y} - \mathbf{z}]'[\mathbf{y} - \mathbf{z}]\right)^{1/2} \qquad (10.2.2)$$

Therefore:

$$\emptyset^2 = \left([\mathbf{x}-\mathbf{z}]'[\mathbf{x}-\mathbf{z}]\right) + \left([\mathbf{y}-\mathbf{z}]'[\mathbf{y}-\mathbf{z}]\right)$$

$$+ 2\left([\mathbf{x}-\mathbf{z}]'[\mathbf{x}-\mathbf{z}]\;[\mathbf{y}-\mathbf{z}]'[\mathbf{y}-\mathbf{z}]\right)^{1/2} \qquad (10.2.3)$$

For any $\mathbf{z}$ we may require:

$$\mathbf{z} = \mathbf{v} + \mu\,\mathbf{x} + (1-\mu)\,\mathbf{y} \qquad (10.2.4)$$

As (10.2.4) stands, there is no requirement, $\mathbf{v}$ being unspecified. An implied restriction on μ arises, once we additionally require:

$$[\mathbf{x}-\mathbf{y}]'\mathbf{v} = [0]$$

The implied value of μ is found by pre-multiplication of (10.2.4) by $[\mathbf{x}-\mathbf{y}]'$:

$$[\mathbf{x}-\mathbf{y}]'\mathbf{z} = [\mathbf{x}-\mathbf{y}]'\mathbf{v} + [\mathbf{x}-\mathbf{y}]'\mathbf{y} + \mu\,[\mathbf{x}-\mathbf{y}]'[\mathbf{x}-\mathbf{y}] \qquad (10.2.5)$$

Once $\mathbf{v}$ is chosen in such a way that the first term on the right-hand side vanishes, a specific value of μ is implied, and can be resolved from:

$$[\mathbf{x}-\mathbf{y}]'[\mathbf{z}-\mathbf{y}] = [\mathbf{x}-\mathbf{y}]'[\mathbf{x}-\mathbf{y}]$$

and is:

$$\mu = \Big([x-y]'[z-y] \Big) / \Big([x-y]'[x-y] \Big) \tag{10.2.6}$$

(we are obviously assuming $x \neq y$)

Substitution of the righthand side of (10.2.4), for z into (10.2.3), leads in first instance to:

$$\phi^2 = \Big([x - \mu x - y + \mu y - v]' [x - \mu x - y + \mu y - v] \Big)$$

$$+ \Big([y - \mu x - y + \mu y - v]' [y - \mu x - y + \mu y - v] \Big)$$

$$+ 2 \Big([x - \mu x - y + \mu y - v]' [x - \mu x - y + \mu y - v]$$

$$[y - \mu x - y + \mu y - v]' [y - \mu x - y + \mu y - v] \Big)^{1/2}$$

$$\tag{10.2.7}$$

In working out the product-expressions in (10.2.7), the terms involving $v'[x-y]$ or its transpose $[x-y]'v$ can be dropped, and we find (10.2.7) to be equivalent to:

$$\phi^2 = \Big((1-\mu)^2 + \mu^2 \Big) \Big([x-y]'[x-y] \Big) + 2\, v'v$$

$$+ 2 \Big(((1-\mu)^2\, \mu^2)\, ([x-y]'[x-y])^2$$

$$+ ((1-\mu)^2 + \mu^2)\, ([x-y]'[x-y])\, (v'v) + (v'v) \Big)^{1/2}$$

$$\tag{10.2.8}$$

As will be shown below, unconstrained minimization of ϕ^2, relative to v only, leads to a solution, which also satisfies the condition for a minimum relative to μ:

$$\partial\phi^2 / \partial\mu = 0 \tag{10.2.9}$$

As we used $v'[x-y] = 0$ to eliminate direct reference to z, we also need to verify $v'[x-y] = 0$, -and by implication (10.2.4)-.

The first-order condition relative to **v** on its own is:

$$\partial \phi^2 / \partial \mathbf{v} = 2\,\mathbf{v}$$

$$+ \Big(((1-\mu)^2\,\mu^2)\,([\mathbf{x}-\mathbf{y}]'[\mathbf{x}-\mathbf{y}])^2$$

$$+ ((1-\mu)^2 + \mu^2)\,([\mathbf{x}-\mathbf{y}]'[\mathbf{x}-\mathbf{y}])\,(\mathbf{v}'\mathbf{v}) \;+\; (\mathbf{v}'\mathbf{v}) \Big)^{-\frac{1}{2}}$$

$$\Big(2\,((1-\mu)^2 + \mu^2)\,([\mathbf{x}-\mathbf{y}]'[\mathbf{x}-\mathbf{y}])\mathbf{v} \;+\; 2\,\mathbf{v} \Big) = [0]$$

$$(10.2.10)$$

and is obviously satisfied by:

$$\mathbf{v} = [0] \qquad\qquad (10.2.11)$$

Therefore, by (10.2.8) and (10.2.11):

$$\phi^2 = \Big((1-\mu)^2 + \mu^2 + 2\,(1-\mu)\,\mu \Big)\Big([\mathbf{x}-\mathbf{y}]'[\mathbf{x}-\mathbf{y}] \Big) =$$

$$(1-2\mu+\mu^2 + \mu^2 + 2\mu - 2\mu^2)\,[\mathbf{x}-\mathbf{y}]'[\mathbf{x}-\mathbf{y}] = [\mathbf{x}-\mathbf{y}]'[\mathbf{x}-\mathbf{y}] \quad (10.2.12)$$

As reference to μ has disappeared form (10.2.12) altogether, we note that the first-order derivative of ϕ^2, relative to μ, vanishes automatically, thereby satisfying (10.2.9).

The null vector <u>is</u> orthogonal to [**x**-**y**], and (10.2.4) is now satisfied as:

$$\mathbf{z} = \mu\,\mathbf{x} + (1-\mu)\,\mathbf{y} \qquad\qquad (10.2.13)$$

Which is a usual geometrical definition for a straight line through the points **x** and **y**.

As far as the line-segment <u>between</u> **x** and **y** is concerned, we also need to require:

$$0 \leq \mu \leq 1 \qquad\qquad (10.2.14)$$

There is a special name for the distance between vector and the origin, which is the <u>length</u> of the vector.

10.3 **Redefinition of the coordinate system**

A calculation rule, which transforms any vector **x** into a uniquely determined vector **y**, and for which there exist a converse rule, which transforms **y** again into the original vector **x**, is called a mapping of **x** into the **y** space. We are here exclusively concerned with linear transformations, and the calculation rule needs to be:

$$\mathbf{y} = \mathbf{T}\,\mathbf{x} + \mathbf{c} \tag{10.3.1}$$

where **T** is an n by n invertible matrix, and **c** an n by 1 column-vector.

The converse calculation-rule is in that case:

$$\mathbf{x} = \mathbf{T}^{-1}\,\mathbf{y} + \mathbf{T}^{-1}\,\mathbf{c} \tag{10.3.2}$$

There is no inherent logical requirement, for (10.3.1) as such to imply the invertability of **T**. For example, the projection into one of the ordinate planes, e.g.:

$$\mathbf{T} = \begin{bmatrix} 1 & - & - \\ - & 1 & - \\ - & - & - \end{bmatrix}$$

i.e. the first (let us say: backwards), and the second (let us say: rightward) coordinates are copied, but the third (let us say: vertically upward), is reduced to zero. The result of this transformation is the projection of everything in the horizontal coordinate plane. However, we shall not be using such transformations. There is to some extent a tradition to define a linear transformation as invertible (for example Gantmacher [14], section 3.4). This may, or may not be desirable, but our analysis is in any case mainly concerned with invertible trans-formation-matrices.

At this stage, we concentrate on the re-definition of the coordinate-directions and units of measurement.

We consider the lefthand of the two graphs below, as the 'original' graph. As the gridmarkings indicate, the graph contains a circle, with the point $x_1=1.5$, $x_2=1.5$ as origin. The radius of this circle is in fact 1.2. In the righthand graph,

the units of measurement have been redefined. The horizontal
unit has been redefined as x_2(right)=1.5 x_2(left), and the
vertical unit has been redefined as x_1(right)=2x_1(left).

We therefore first of all need to remove half of the grid-markers
from the vertical axis, thereby causing one new unit-distance to
cover two old ones, and also to space the ones on the horizontal
axis wider. As long as there is no change in the layout of the
graph itself, this does not make any real difference: we just
re-name the distances, as was done in the middle one of the two
graphs below. However, the notion of a unit is that it is a
uniform distance. We therefore need to 'squeeze' (or 'elongate')
the coordinate directions, causing a unit of measurement to
be again a unit of measurement.

If we now map some kind of shape or function, e.g. the circle
according to these redefined units, we apply the <u>inverse</u>, e.g.
two original vertical units are squeezed into one new vertical
unit, and 1.5 original horizontal units are squeezed into one new
horizontal unit. After the transformation the original circle is
no longer is a circle, but an ellipse. This is because, what was
originally the radius when measured as the distance of the top to
the midpoint, has been reduced to 0.6 units, while the distance
of the most righthand point to the midpoint has been reduced to
1.2/1.5 = 0.8 units.

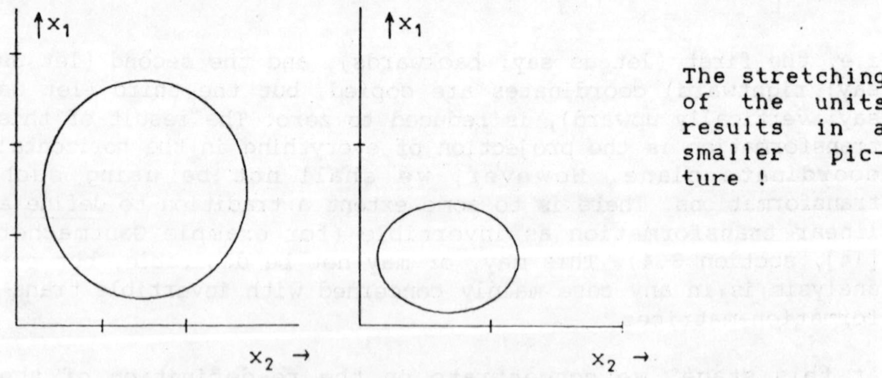

The stretching
of the units
results in a
smaller pic-
ture !

A more complicated transformation arises, if we consider a
transformation-matrix, in which there are also non-zero off-
diagonal elements. The slanted ellipse below is obtained, by
transforming the circle with origin (1.5, 1.5) and radius 1.2,
according to:

x_1(new) = 2x_1(old) and x_2(new) = -1.5 x_1(old) + 1.5 x_2(old).

The inverse calculation conform (10.3.2) is now applicable, i.e. the old points, when mapped in the new coordinate space, are transformed conform:

$$T^{-1} = \begin{bmatrix} 2 & - \\ -1.5 & 1.5 \end{bmatrix}^{-1} = \begin{bmatrix} 0.5 & - \\ 0.5 & 0.67 \end{bmatrix}$$

The corresponding graphical transformation is illustrated below:

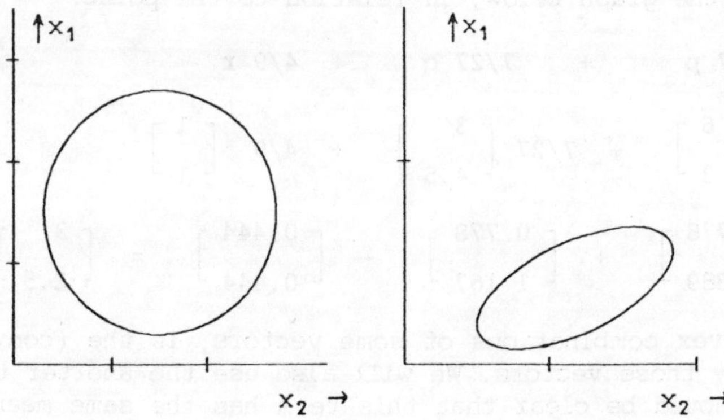

10.4 Content and area

There is a close relationship between determinants and content or area. Before we can explore that relationship, we first need to establish the relationship between vectors as the algebraic equivalents of points in space, and an area in which they may, or may not be. The n-dimensional vector **x** is said to be a <u>convex combination</u> of k other n-dimensional vectors $\mathbf{v}_1$ to $\mathbf{v}_k$, if there are k numbers $w_j \geq 0$ (j = 1, 2, ... k), for which the following requirements hold:

$$\left.\begin{array}{l} \displaystyle\sum_{j=1}^{k} w_j \ = \ 1 \\[2em] \text{and} \\[1em] \mathbf{x} \ = \ \displaystyle\sum_{j=1}^{n} w_j \, \mathbf{v}_j \end{array}\right\} \qquad (10.4.1)$$

The geometrical interpretation of the concept of area is illustrated in the graph below, in relation to the point:

$$\mathbf{c} \ = \ 8/27 \ \mathbf{p} \quad + \quad 7/27 \ \mathbf{q} \quad + \ 4/9 \ \mathbf{r}$$

$$= \ 8/27 \begin{bmatrix} 6 \\ 3 \end{bmatrix} \ + \ 7/27 \begin{bmatrix} 3 \\ 4.5 \end{bmatrix} \ + \ 4/9 \begin{bmatrix} 1 \\ 1 \end{bmatrix}$$

$$= \begin{bmatrix} 1.778 \\ 0.889 \end{bmatrix} \ + \ \begin{bmatrix} 0.778 \\ 1.167 \end{bmatrix} \ + \ \begin{bmatrix} 0.444 \\ 0.444 \end{bmatrix} \ = \ \begin{bmatrix} 3 \\ 2.5 \end{bmatrix}$$

The set of convex combinations of some vectors, is the (convex) area spanned by those vectors. We will also use the shorter term area, and it should be clear that this term has the same meaning as the more elaborate term (convex) area spanned. We do not need a theorem which states that the area inside the triangle pqr is the set of convex combinations of the points p, q, and r. We define an area spanned by some vectors algebraically as the set of their convex combinations, we geometrically interpret the area spanned as a geometrical surface or content.

For example, we may algebraically say that an area spanned is contained in some other area, i.e. each convex combination of the points defining the area are also convex combinations of the points defining the other area. (The triangle pqc is contained in the triangle pqr.)

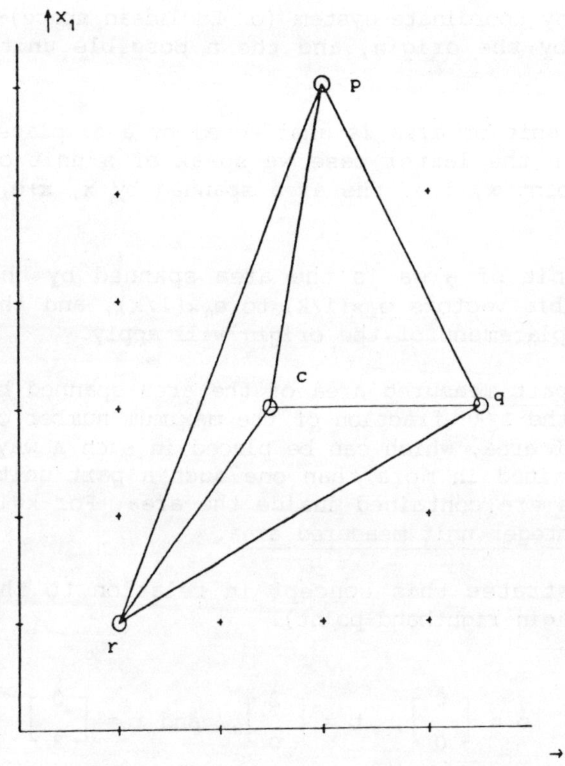

To avoid ambiguity, we generally assume that some points spanning an area have the property that none of them can itself be expressed as a convex combination of the others. For example, the definition of an area spanned as stated above, is also met by the set of convex combinations of **p**, **q**, **r**, and **c**. But the inclusion of **c** in the list of vectors spanning the area which geometrically is the triangle pqr, is not required, as **c** can itself be expressed as a convex combination of **p**, **q**, and **r**. A vector which, though listed as one of the vectors defining an area, is itself a convex combination of other vectors in the list, is said to be <u>redundant</u> to its definition. Vectors which cannot themselves be expressed as convex combinations of other (points) in the area, are said to be <u>extreme</u> vectors (points).

Before we can make use of the link between the area-concept, as geometrically interpreted, and the corresponding algebraic definition, we need to discuss the measurement of area.

The <u>unit of area</u> -in any coordinate-system (or Euclidean space)-, is the area spanned by the origin, and the n possible unit-vectors.

We postulate that the unit of area is unaffected by a displacement of the origin. In the latter case we speak of a unit of area, <u>placed at</u> the point x, i.e. the area spanned by x, $x+e_1$, $x+e_2$, etc.

The <u>k^n part</u> of the unit of area is the area spanned by the origin and the n possible vectors $e_1 \times (1/k)$ to $e_n \times (1/k)$, and the similar concept of displacement of the origin will apply.

We now define the <u>k^n part measured area</u> of the area spanned by v_1, v_2, ... v_k, as the $1/k^n$ fraction of the maximum number of k^n parts of the unit of area, which can be placed in such a way, that no point is contained in more than one such a part-unit, while all the k^n parts are contained inside the area. For k=1, we will speak of the <u>integer unit measured area</u>.

The graph below illustrates this concept in relation to the triangle tor (= **t**op **o**rigin **r**ighthand-point).

The extreme points are $\quad o = \begin{bmatrix} 0 \\ 0 \end{bmatrix}$, $\quad t = \begin{bmatrix} 2 \\ 0 \end{bmatrix}$, $\quad$ and $r = \begin{bmatrix} 0 \\ 3 \end{bmatrix}$

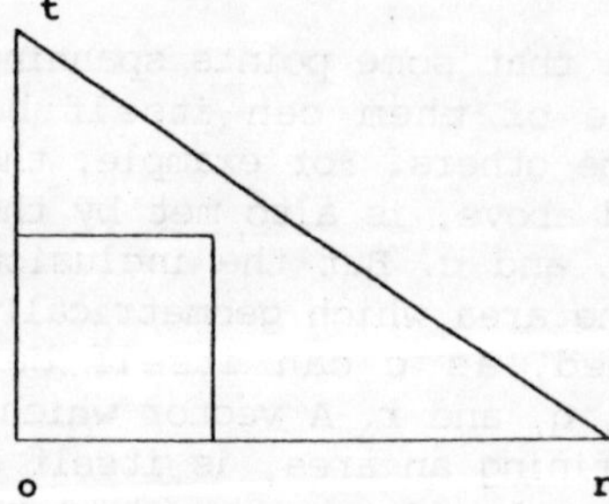

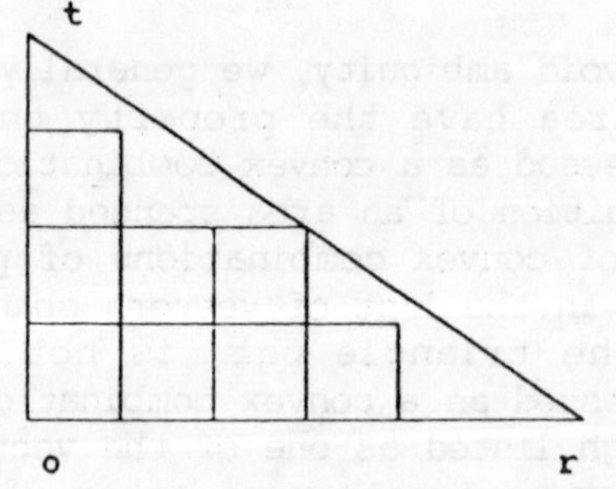

The left half of the graph illustrates the integer unit measured area of the triangle tor, which is just 1, leaving a significant part of the actual geometrical area uncovered. For k=2, we find 8 1/4 area units, and the 1/4 part measured area is 2. An obvious approach towards expressing the true area is calculus. We therefore define the measured area of any convex area (we are not discussing non-convex areas with holes or dents in them, but the generalization is obvious), as the limit which the $1/k^n$ part measured area will approximate, if k goes to infinity.

The measured area of the triangle tor therefore is is found by integration. Going downwards from the maximum height of 2 at the point t, along the slope of 2/3 , the area is:

$$\int_0^3 (2 - 2/3\ x)dx \quad = \quad 2 \times 3 - 1/3 \times 9 = \quad 3$$

We now proceed to analyse the effect of a particular type of coordinate transformation on a measured area. The transformation to which we refer is:

$$x^*{}_k = x_k + \sum_{j=1}^{k} a_j x_j \qquad\qquad (10.4.2)$$

and therefore:

$$x_k = x^*{}_k - \sum_{j=1}^{k} a_j x_j \qquad\qquad (10.4.3)$$

The graph below illustrates the effect of renaming the vertical direction by which the triangle tor above is measured, from x_1 to $x^*{}_1 = x_1 + x_2$. The new horizontal $x^*{}_1 = 0$ axis expressed in the old coordinate system slopes $45°$ downwards, the old horizontal axis expressed in the new coordinate system slopes $45°$ upwards.

Recursive application of transformations of this type, in relation to different axes, increases the class of areas that may be obtained by means of transformation of a unit of area considerably. We will in that context refer to a matrix of the structure

$$\mathbf{T} = \quad \mathbf{I} + (t_{ij})\mathbf{E}_{ij} \qquad\qquad (10.4.4)$$

(i.e. the diagonal elements are all unity, and only one off-diagonal element t_{ij} is (permitted to attain) a non-zero value, as an <u>axis redirecting</u> transformation operator.

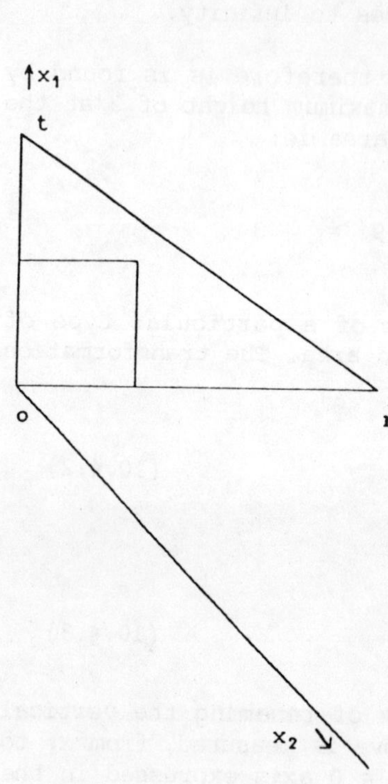

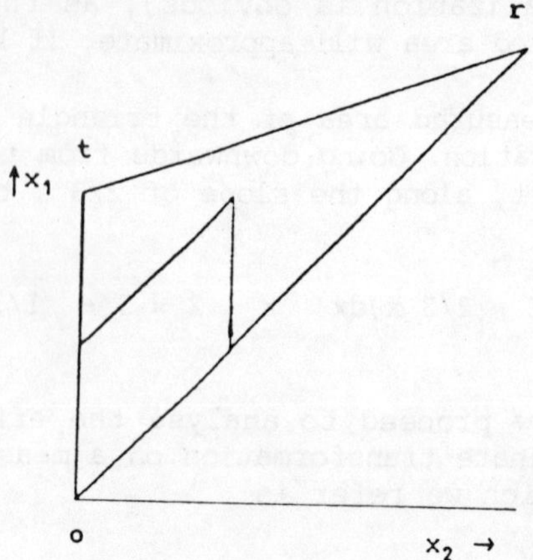

Redefining the <u>direction</u> of
one axis, while leaving its
scale unchanged, does not
affect the area spanned.

Theorem:

If **T** is an n by n invertible matrix,

then **T** can be can be expressed as:

$$\mathbf{T} = \mathbf{L}\,\hat{\mathbf{D}}\,\mathbf{R} \tag{10.4.5}$$

where:

- 1) **L**, $\hat{\mathbf{D}}$, and **R** are all real and of order n by n, and

- 2) **L** and **R** can be expressed a recursive product of axis redirecting transformation operators $(\mathbf{I} + \mathbf{E}_{ij}t_{ij})$, and permutation operators

- 3) $\hat{\mathbf{D}}$ is diagonal matrix.

Proof:

We first consider some special cases, as follows:

Case 1): $n = 1$
Then the theorem has no content.
q.e.d. for case 1

Case 2) $n = 2$, of which:

Subcase 2a) $n = 2$, $t_{11} \neq 0$

Then
$$\mathbf{T} = \begin{bmatrix} 1 & - \\ t_{21}t_{11}^{-1} & 1 \end{bmatrix} \begin{bmatrix} t_{11} & t_{12} \\ - & t_{22} - t_{21}t_{11}^{-1}t_{12} \end{bmatrix} =$$

$$\begin{bmatrix} 1 & - \\ t_{21}t_{11}^{-1} & 1 \end{bmatrix} \begin{bmatrix} t_{11} & - \\ - & t_{22} - t_{21}t_{11}^{-1}t_{12} \end{bmatrix} \begin{bmatrix} 1 & t_{11}^{-1}t_{12} \\ - & 1 \end{bmatrix}$$

q.e.d. for subcase 2a)

Subcase 2b): $n = 2$, $t_{11} = 0$

Then either pre- or postmultiplication by a permutation operator transforms $\mathbf{T}$ into a matrix to which case 2a) as already proved is applicable.

q.e.d. for subcase 2b.

q.e.d. for case 2.

Case 3): $n > 2$: of which:

Subcase 3a): $n > 2$, $\mathbf{T}$ partitions as:

$$\mathbf{T} = \left[\begin{array}{c|c} t_{11} & \\ \hline t_{21} & I \end{array} \right] \tag{10.4.6}$$

For $t_{11} = 0$, $\mathbf{T}$ is not invertible, and no proof is needed.

Otherwise, for $t_{11} \neq 0$, **T** is the recursive product of at most n-1 axis redirecting transformation operators, each of which has an element $t_{i1} \neq 0$, $i > 1$.

q.e.d. for subcase 3a.

Case 3b): n > 2, **T** partitions as

$$\mathbf{T} = \left[\begin{array}{c|c} 1 & t'_{12} \\ \hline & I \end{array}\right]$$

(10.4.7)

Proof analogous to case 3a.

q.e.d. for subcase 3b.

At this point, we introduce the provisional assumption that the theorem is true for matrices of order 3 by 3, ... n-1 by n-1, in addition to matrices of order 1 by 1 and 2 by 2, for which a proof (where needed for 2 by 2), was supplied already.

We now consider:

Subcase 3c): n > 2, $t_{11} \neq 0$

We then express **T** as follows:

$$\mathbf{T} = \left[\begin{array}{c|c} t_{11} & t'_{12} \\ \hline t_{21} & T_{22} \end{array}\right] =$$

$$\left[\begin{array}{c|c} 1 & \\ \hline t_{21}t_{11}^{-1} & I \end{array}\right] \left[\begin{array}{c|c} t_{11} & \\ \hline & T_{22}-t_{21}t_{11}^{-1}t'_{12} \end{array}\right] \left[\begin{array}{c|c} 1 & t'_{12}t_{11}^{-1} \\ \hline & I \end{array}\right]$$

(10.4.8)

The theorem is true, for the lefthand transformation operator in (10.4.8), by case 3a as proved already, and for the righthand transformation operator by case 3b, both without a non-trivial diagonal matrix.

It is held true by assumption, for $[\mathbf{T}_{22}-t_{21}t_{11}^{-1}t'_{12}]$.

By implication the theorem is then also true for the block-diagonal matrix in the middle of the righthand-side of (10.4.8), and a proof by recursive induction arises.

q.e.d. for case 3c, subject to the provisional assumption.

Since there is a **P** which makes subcase 3c) applicable to **P T**, whenever **T** is free of zero columns and n > 2 applies, subcase 3c) completes the proof.

q.e.d.

We now come to discuss the geometrical interpretation of this theorem. Of the three factors on the righthand side of (10.4.5), **L** and **R** always have a determinant of +1, or (on account of the permutation operator(s)), of -1. The absolute value of $|\mathbf{T}|$ is therefore always equal to the absolute value of $|\hat{\mathbf{D}}|$.

Multiplication by an axis-redirecting transformation-operator changes the shape of the geometrical area covered by an area spanned in the way discussed in the previous section, but not the size of the geometrical area. Multiplication by a permutation operator amounts to re-naming the variables, and may (or may not) result in a change of the determinant, but not of the area covered.

Therefore, if all the extreme points which span an area are multiplied by **T**, the geometrical area is multiplied by the absolute value of $|\mathbf{T}|$, which is the absolute value of $|\hat{\mathbf{D}}|$. Only the multiplication by $|\hat{\mathbf{D}}|$ actually changes the size of the geometrical area, by changing the units of measurement in the way discussed in the previous section, possibly (for a negative d_{ii}) combined with a reversal of the axis-direction. The rest of the transformations change the shape by moving the directions of the axes, and renaming the variables, but leave the amount of area covered as before.

10.5 **N plus one areas, parallelons and parallel matrices**

We shall refer to an area spanned by n+1 points in n-dimensional space as an n plus one area. The area contained in a generalized parallelogram (parallelepiped) is determined by n+1 extreme points. One point serves as the displaced origin, and then there are n more adjacent points. For n > 2, there are more than n+1 extreme points, but these are fully determined by the parallel requirement. This is best illustrated by the archetype of this type of area, the unit of area measurement. We illustrate this point for n=3:

The 'first' 4 extreme points of the unit of measurement of 3-dimensional space are:

$$\begin{bmatrix} - \\ - \\ - \end{bmatrix}, \quad \begin{bmatrix} 1 \\ - \\ - \end{bmatrix}, \quad \begin{bmatrix} - \\ 1 \\ - \end{bmatrix}, \text{ and } \begin{bmatrix} - \\ - \\ 1 \end{bmatrix},$$

but 3 more

$$\begin{bmatrix} - \\ 1 \\ 1 \end{bmatrix}, \quad \begin{bmatrix} 1 \\ 1 \\ - \end{bmatrix}, \text{ and } \begin{bmatrix} - \\ 1 \\ 1 \end{bmatrix}$$

are formed by entering two unity elements per vector, and finally there is the summation vector.

As each element can be either 0 or 1, there are 2^n different vectors consisting solely of the elements 0 and 1. For a combination of 2^n points to be the extreme points of a generalized parallelogram, we cannot rely on all the elements of the matrix being 0 or 1, as indeed, this is only the case for a unit of measurement.

We refer to a (linear) transformation of the unit of measurement, as a parallelon. (The two-dimensional specimen of a parallelon is a parallelogram, the three-dimensional specimen is a parallele-piped). We could then refer to a n by 2^n matrix which contains the vectors which span a parallelon, as a parallel matrix.

It is obvious that, for an n by 2^n matrix to be a parallel matrix, certain conditions must be satisfied.

A sufficient condition is that, once we displace one point to the origin, inversion of an n by n matrix of the other displaced points, and premultiplication of the other points (vectors) by its inverse, turns them into the complement of vectors with elements of 0 and 1.

Example:

$$\mathbf{A} = \begin{bmatrix} 1 & 2 & 3 & 1 & 4 & 2 & 3 & 4 \\ 2 & 5 & -2 & 7 & 1 & 10 & 3 & 6 \\ 3 & 3 & 9 & 3 & 9 & 3 & 9 & 9 \end{bmatrix}$$

It is not at a casual glance clear, whether this 3 by 8 matrix does contain the extreme points of a parallelepiped. In fact this is the case, as we proceed to show as follows:

First, we re-name the leading column as the origin, by sub-tracting it from itself and all the other columns. The matrix now becomes:

$$\mathbf{B} = \begin{bmatrix} - & 1 & 2 & - & 3 & 1 & 2 & 3 \\ - & 3 & -4 & 5 & -1 & 8 & 1 & 4 \\ - & - & 6 & - & 6 & - & 6 & 6 \end{bmatrix}$$

We now calculate the inverse:

$$\begin{bmatrix} 1 & 2 & - \\ 3 & -4 & 5 \\ - & 6 & - \end{bmatrix}^{-1} = \begin{bmatrix} 1 & - & -0.333 \\ - & - & 0.167 \\ -0.6 & 0.2 & 0.333 \end{bmatrix}$$

Pre-multiplication of the whole matrix **B** by this inverse transforms **B** into:

$$\begin{bmatrix} 1 & - & -0.333 \\ - & - & 0.167 \\ -0.6 & 0.2 & 0.333 \end{bmatrix} \begin{bmatrix} - & 1 & 2 & - & 3 & 1 & 2 & 3 \\ - & 3 & -4 & 5 & -1 & 8 & 1 & 4 \\ - & - & 6 & - & 6 & - & 6 & 6 \end{bmatrix}$$

$$
= \begin{bmatrix}
- & 1 & - & - & 1 & 1 & - & 1 \\
- & - & - & 1 & - & 1 & - & 1 & 1 \\
- & - & - & 1 & - & 1 & 1 & 1
\end{bmatrix}
$$

and we note that the columns of the transformed matrix do indeed constitute the coordinates of a unit of measurement. Therefore, those of **A** constitute the coordinates of the corners of a parallelepiped.

The content of this parallelepiped is in absolute value equal to:

$$
\begin{vmatrix}
1 & 2 & - \\
3 & -4 & 5 \\
- & 6 & -
\end{vmatrix} = -30, \quad \text{and is therefore 30.}
$$

Following the illustration above, we <u>define</u> a <u>parallelon matrix</u> as an n by n! matrix **A**, which has, for some ordering of the columns of **A**, the property that, if the leading column of **A** is subtracted from all the others, the resulting

$$
\mathbf{B} = \mathbf{A} - \mathbf{a_1} \, \mathbf{s'} \tag{10.5.1}
$$

has the property that

$\mathbf{B_2}^{-1}\mathbf{B}$ contains all the different vectors consisting of the elements 0 and 1, (where $\mathbf{B_2}$ is the block of **B** formed by columns 2 to n+1).

However, the words: for <u>some</u> ordering of the columns of **A**, are of non-trivial relevance, as may be illustrated by perturbing them.

$$
\begin{array}{cccccccc}
3 & 7 & 8 & 4 & 5 & 1 & 2 & 6
\end{array}
$$

$$
\mathbf{A} = \begin{bmatrix}
3 & 3 & 4 & 1 & 4 & 1 & 2 & 2 \\
-2 & 3 & 6 & 7 & 1 & 2 & 5 & 10 \\
9 & 9 & 9 & 3 & 9 & 3 & 3 & 3
\end{bmatrix}
$$

(The numbers at the head of the tabulation indicate the original positions of these column, and therefore serve to keep track of the relationship with the original ordering.)

We again displace the origin to the leading column.

$$B = \begin{bmatrix} - & - & 1 & -2 & 1 & -2 & -1 & -1 \\ - & 5 & 8 & 9 & 3 & 4 & 7 & 12 \\ - & - & - & -6 & - & -6 & -6 & -6 \end{bmatrix}$$

We now proceed on the assumption that we obtain B in the same way as we did above, and extract the block B_2:

$$B_2 = \begin{bmatrix} - & 1 & -2 \\ 5 & 8 & 9 \\ - & - & -6 \end{bmatrix} \qquad B_2^{-1} = \begin{bmatrix} -1.6 & 0.2 & 0.833 \\ 1 & - & -0.333 \\ - & - & 0.167 \end{bmatrix}$$

There is not actually any theorem or rule, which says that we always find the same determinant, and in any case, an attempt to re-construct a unit of measurement now fails:

$$\begin{bmatrix} -1.6 & 0.2 & 0.833 \\ 1 & - & -0.333 \\ - & - & 0.167 \end{bmatrix} \begin{bmatrix} - & - & 1 & -2 & 1 & -2 & -1 & -1 \\ - & 5 & 8 & 9 & 3 & 4 & 7 & 12 \\ - & - & - & -6 & - & -6 & -6 & -6 \end{bmatrix}$$

$$= \begin{bmatrix} - & 1 & - & - & -1 & -1 & -2 & -1 \\ - & - & 1 & - & 1 & - & 1 & 1 \\ - & - & - & 1 & - & 1 & 1 & -1 \end{bmatrix}$$

The reason why no unit of measurement is formed, is that the wrong collection of columns was picked. We need 4 columns, of which 3 are related to the fourth one as <u>adjoining</u> points.

Algebraically we first note that, once the origin has been displaced, and the transformation made, all <u>other</u> columns (than the origin) either cannot be expressed as sums of other columns in any way at all (the unit vectors which mark the coordinate directions), or they are sums of the origin, the null vector, and the simple <u>sum</u> of several others (more unity elements than one), it is <u>not</u> an axis-marker in relation to that origin.

This additive relationship is invariant under pre-multiplication by $\mathbf{B_2}$, but the subtraction of the origin from the other points needs to be considered explicitly. Thus, once we have selected an origin and n direction-vectors, then, if those n+1 points are indeed a corner-part of the parallelon, we should be able to express all <u>other</u> columns (but none of the n+1 selected ones), as the sum of the origin, plus the <u>differences</u> between one or more of the axis-markers, and the origin.

For example, we have:

$$\mathbf{a_8} = \mathbf{a_1} + [\mathbf{a_2} - \mathbf{a_1}] + [\mathbf{a_3} - \mathbf{a_1}] + [\mathbf{a_4} - \mathbf{a_1}]$$

$$\begin{bmatrix} 4 \\ 6 \\ 9 \end{bmatrix} = \begin{bmatrix} 1 \\ 2 \\ 3 \end{bmatrix} + \begin{bmatrix} 2 \\ 5 \\ 3 \end{bmatrix} - \begin{bmatrix} 1 \\ 2 \\ 3 \end{bmatrix} + \begin{bmatrix} 3 \\ -2 \\ 9 \end{bmatrix} - \begin{bmatrix} 1 \\ 2 \\ 3 \end{bmatrix} + \begin{bmatrix} 1 \\ 7 \\ 3 \end{bmatrix} - \begin{bmatrix} 1 \\ 2 \\ 3 \end{bmatrix}$$

and no similar condition can be met with column 3 as origin, and columns 7, 8, and 4 as axis-markers.

We end this section, by exploring the relation between the parallelon, and the corresponding n plus one points area. Areas spanned by one corner and n adjoining other corners always contain an area that is a fraction of 1/n! of the whole parallelon.

The graph below illustrates this point geometrically:

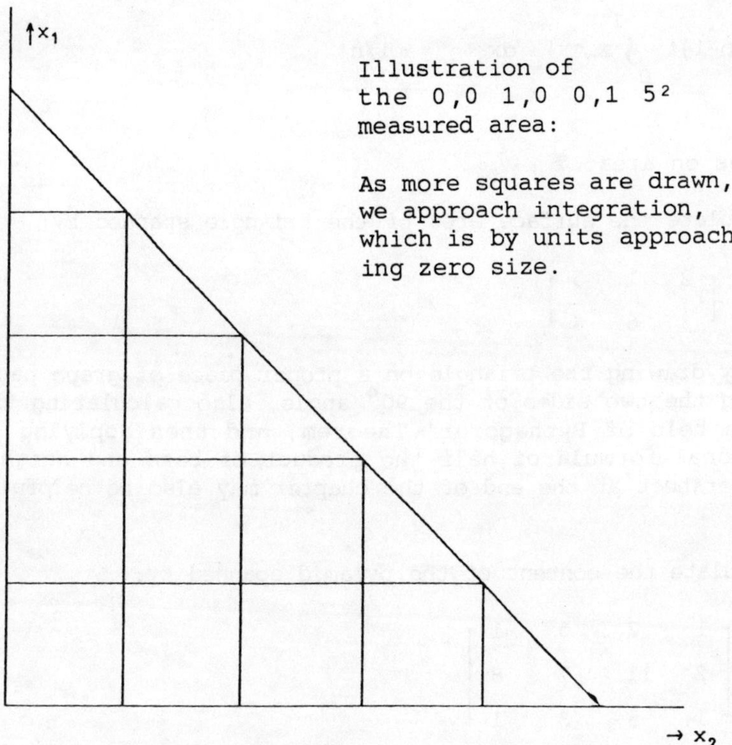

Algebraically, this 'n!' property is also most readily proved by the area spanned by the origin and n unit vectors:

$$\int_0^1 dx_n \,\ldots\ldots\ldots\, \int_0^{x_4} dx_3 \int_0^{x_3} dx_2 \int_0^{x_2} dx_1 \;=$$

$$\int_0^1 dx_n \,\ldots\ldots\ldots\, \int_0^{x_4} dx_3 \int_0^{x_3} x_2 \, dx_2 \;=$$

$$\frac{1}{2} \int_0^1 dx_n \,\ldots\ldots\ldots\, \int_0^{x_4} x_3{}^2 \, dx_3 \;=$$

$$1/(n-1)! \int_0^1 x_n^{n-1} \, dx_n = 1/n!$$

Exercises on area:

1): Calculate the surface area of the triangle spanned by:

$$A = \begin{bmatrix} 2 & 1 & 5 \\ 3 & 6 & 4 \end{bmatrix}$$

(Check by drawing the triangle on a proper piece of graph paper, measuring the two sides of the $90°$ angle, also calculating them with the help of Pythagoras' Theorem, and then applying the conventional formula of half the product of base and height.) The answersheet at the end of the chapter may also be helpful to verify.

2): Calculate the content of the pyramid spanned by:

$$B = \begin{bmatrix} 1 & 2 & 5 & 1 \\ 2 & 11 & -7 & 8 \\ 1 & -5 & 3 & 1 \end{bmatrix}$$

(There is an answersheet at the end of the chapter)

10.6 Rotation in two dimensional space

The concept of the angle between two vectors is meaningful, if we interpret them as the endpoints of lines, each of which has its starting point at the origin.

To keep track of the angles between vectors, it is convenient to assume that the distance between each point under scrutiny and the origin, is one unit.

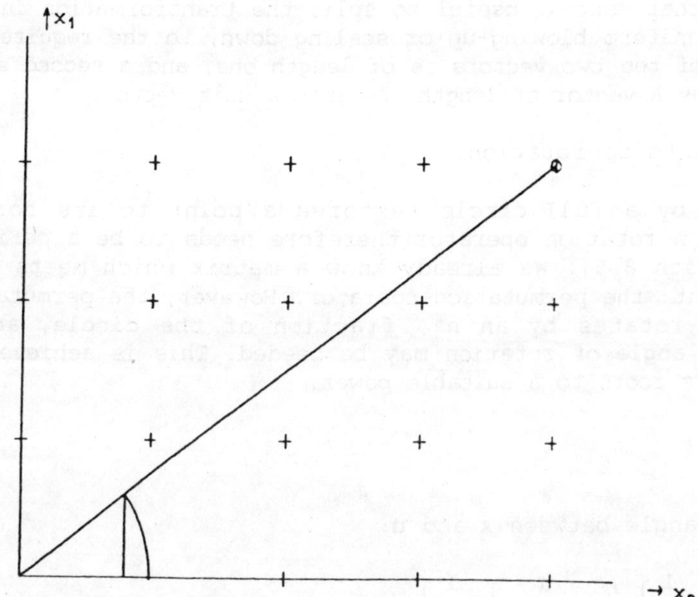

The length of the vector (3, 4), as marked in the top righthand corner of the picture is:

$$\sqrt{(3^2 + 4^2)} = \sqrt{25} = 5.$$

To be able to read the angle from the graph, we put the measurement point at a normalized distance of 1 from the origin, as marked by the circle segment. Clearly the horizontal base of the triangle in the bottom left part of the graph, is the cosine of the angle.

To make this concept operational on two arbitrary points, we need to maintain the relation with the origin, otherwise no angle is defined at all. We shall come to discuss rotation around other points than the origin later in this section. For the time being we assume rotation around the origin.

In addition to the reference to the origin, we need a transformation, which turns one of the two vectors into a unit vector, and which leaves some quantity, which can be meaningfully interpreted as the angle between the two vectors, unaffected.

It is in that context useful to split the transformation in two parts: a uniform blowing-up or scaling down, to the requirement that one of the two vectors is of length one, and a second stage which turns a vector of length one into a unit vector.

This amounts to rotation.

Rotation by a full circle restores a point to its former position. A rotation operator therefore needs to be a periodic (see section 8.5). We already know a matrix which meets that requirement: the permutation-operator. However, the permutation operator rotates by an nth fraction of the circle, and a different angle of rotation may be needed. This is achieved by raising its roots to a suitable power.

Example:

Find the angle between $\mathbf{x}$ and $\mathbf{u}$:

$$\mathbf{x} = \begin{bmatrix} 1 \\ 3 \end{bmatrix}, \qquad \mathbf{u} = \begin{bmatrix} 1 \\ 1 \end{bmatrix}$$

To transform $\mathbf{u}$ into a unit vector, we first transform the scale by a factor $\tfrac{1}{2}\sqrt{2}$ (to obtain a length of unity), and then, to transform $\tfrac{1}{2}\sqrt{2}\,\mathbf{u}$ into a unit vector, we need to rotate by 45 degrees, or a fraction of 1/8 of the full circle. The direction of rotation will be against the clock, of $\mathbf{e}_1$ is desired, and with the clock if $\mathbf{e}_2$ is desired.

In either case, the required rotation therefore needs to meet the condition:

$$\mathbf{R}^8 = \mathbf{I}$$

In the two-dimensional case, the interpretation of a complex number as a circle segment is helpful, and the two points u and v, represented by the vectors:

$$\mathbf{u}^* = \begin{bmatrix} \tfrac{1}{2}\sqrt{2} \\ \tfrac{1}{2}\sqrt{2} \end{bmatrix}, \qquad \text{and} \qquad \mathbf{x}^* = \begin{bmatrix} \tfrac{1}{2}\sqrt{2} \\ 1\tfrac{1}{2}\sqrt{2} \end{bmatrix}$$

have been mapped in the graph below.

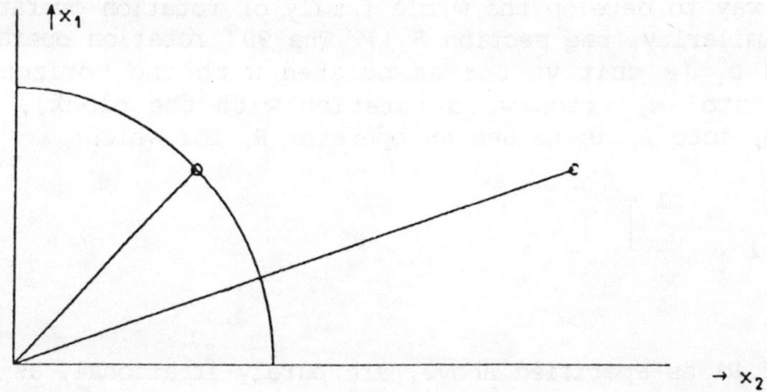

Complex numbers are conventionally associated with a point on a
circle, the absolute value of the complex number is the radius,
and the tangent of the angle between the point and one of the
axes, is the ratio between the real and the irrational part: We
map the real component in one direction, and the irrational
component in the other direction. (See, for example and Hawkins &
Hawkins [21], esp. Ch. 3.) If we put the real component in the
rightward horizontal direction, and the irrational component in
the upward vertical direction, then the 45° slope corresponds,
for a complex number on the unit circle (absolute value 1 = one),
to the point $\frac{1}{2}\sqrt{2} + \frac{1}{2}i\sqrt{2}$.

On account of the way complex numbers occur in pairs, the
irrational direction is both upward and downward, but the real
component has an unambiguous sign.

The negative root of a permutation-operator of order 2, $\lambda = -1$,
is classified as 1/4 circle rotation. However, the permutation
operator would never get a point with two positive coordinates
outside the positive quadrant. We need a matrix with a pair of
complex roots.

If x_1 is mapped on the rightward horizontal axis, then

$$T = \begin{bmatrix} \frac{1}{2}\sqrt{2} & \frac{1}{2}\sqrt{2} \\ -\frac{1}{2}\sqrt{2} & \frac{1}{2}\sqrt{2} \end{bmatrix}$$

will provide 45° rotation with the clock, and its transpose
against the clock.

The easiest way to develop the whole family of rotation operators is to use similarity (see section 8.1). The $90°$ rotation operator which turns e_1 (a unit vector associated with the horizontal direction) into $-e_2$ (rightward rotation with the clock), and conversely e_2 into e_1 is to use an operator R, for which:

$$R^4 = \begin{bmatrix} - & 1 \\ -1 & - \end{bmatrix}$$

applies.

The roots of R^4 as specified above, are purely irrational, as may be verified by:

$$\begin{bmatrix} - & 1 \\ -1 & - \end{bmatrix}\begin{bmatrix} 1 \\ i \end{bmatrix} = \begin{bmatrix} i \\ -1 \end{bmatrix} = \begin{bmatrix} 1 \\ i \end{bmatrix} \times i$$

The matrix of the vectors of R_4 inverts as:

$$\begin{bmatrix} 1 & 1 \\ -i & i \end{bmatrix}^{-1} = \tfrac{1}{2}\begin{bmatrix} 1 & i \\ 1 & -i \end{bmatrix}$$

The general 2-dimensional rotation-operator is best expressed as being similar to the $180°$ rotation operator. The vectors of the $180°$ rotation operator are the same as those of the $90°$ rotation operator outlined above, and a general formula for a 2-dimensional rotation operator is:

$$R_k = \tfrac{1}{2}\begin{bmatrix} 1 & 1 \\ -i & i \end{bmatrix}[-I]^{2/k}\begin{bmatrix} 1 & i \\ 1 & -i \end{bmatrix} \tag{10.6.1}$$

where R_k is the rotation operator, which rotates any vector a kth fraction of the full circle to the right. The kth power of that operator then is $[-I]^2 = I$.

To avoid ambiguity on account of multiple roots in (10.6.1), we need to define $[-I]^{2/k}$ $(= I^{1/k})$ more precisely.

In the interest of relating the fraction of the circle to an angle (in degrees, i.e. in 360th parts of the full circle,), we denote:

$$d = 360 / k \tag{10.6.2}$$

and hence $k = 360/d$, and $1/k = d/360$, i.e. a quarter circle turn is an angle of unity, and express the powers of I as:

$$I^{1/k} = [-I]^{2/k} = I^{d/360} =$$

$$\begin{bmatrix} \cos(d) - i\sin(d) & - \\ - & \cos(d) + i\sin(d) \end{bmatrix} \qquad (10.6.3)$$

These rotation formulae have so far been stated by analogy to the known cases of rotating by a half or a quarter circle. A more general proof is now required on the following points:

● 1) R_k is always real

Proof:

Denoting $\cos(d)$ as α, and $\sin(d)$ as β, and noting the implied restriction $\alpha^2 + \beta^2 = \cos^2(d) + \sin^2(d) = 1$, we express any R_k as:

$$R_k = \tfrac{1}{2} \begin{bmatrix} 1 & 1 \\ -i & i \end{bmatrix} \begin{bmatrix} \alpha-\beta i & - \\ - & \alpha+\beta i \end{bmatrix} \begin{bmatrix} 1 & i \\ 1 & -i \end{bmatrix}$$

$$= \tfrac{1}{2} \begin{bmatrix} \alpha - \beta i & \alpha + \beta i \\ -\alpha i - \beta & \alpha i - \beta \end{bmatrix} \begin{bmatrix} 1 & i \\ 1 & -i \end{bmatrix} =$$

$$\tfrac{1}{2} \begin{bmatrix} 2\alpha & 2\beta \\ -2\beta & 2\alpha \end{bmatrix} = \begin{bmatrix} \alpha & \beta \\ -\beta & \alpha \end{bmatrix} = \begin{bmatrix} \cos(d) & \sin(d) \\ -\sin(d) & \cos(d) \end{bmatrix} \qquad (10.6.4)$$

q.e.d. (ad 1)

● 2) Rotation, defined as multiplication by any R_k, leaves the length of a vector **x** unaffected.

Proof:

By (10.6.4), we express $R_k'R_k$ as:

$$R_k'R_k = \begin{bmatrix} \cos(d) & -\sin(d) \\ \sin(d) & \cos(d) \end{bmatrix} \begin{bmatrix} \cos(d) & \sin(d) \\ -\sin(d) & \cos(d) \end{bmatrix}$$

$$= \begin{bmatrix} \cos^2(d) + \sin^2(d) & - \\ - & \sin^2(d) + \cos^2(d) \end{bmatrix}$$

$$= \begin{bmatrix} 1 & - \\ - & 1 \end{bmatrix} \qquad\qquad (10.6.5)$$

Therefore: $\mathbf{x}'\mathbf{R_k}'\mathbf{R_k}\mathbf{x} = \mathbf{x}'\mathbf{x}$

q.e.d. (ad 2)

● 3) The distance between two points (other than the origin) is not affected.

Proof:

We compare the length of $\mathbf{x-u}$, before and after transformation by $\mathbf{R_k}$, as follows, considering (10.6.5):

$$[\mathbf{x}'\mathbf{R_k}' - \mathbf{u}'\mathbf{R_k}']\,[\mathbf{R_k}\mathbf{x} - \mathbf{R_k}\mathbf{u}] = [\mathbf{x}' - \mathbf{u}']\mathbf{R_k}'\,\mathbf{R_k}[\mathbf{x} - \mathbf{u}]$$

$$= [\mathbf{x}' - \mathbf{u}']\,\mathbf{I}\,[\mathbf{x} - \mathbf{u}]$$

q.e.d. (ad 3)

While properties 1) to 3) do indeed establish that $\mathbf{R_k}$ is a rotation operator, we also need to show that:

● 4) The degree of rotation is indeed the one stated by (10.6.4).

Proof:

Rotation of $\mathbf{e_1}$ to $\mathbf{R_k}\mathbf{e_1} = \begin{bmatrix} \cos(d) \\ -\sin(d) \end{bmatrix}$

does indeed amount to rotation by an angle of d degrees (with the clock).

q.e.d. (ad 4)

● 5) Relation (10.6.4) does indeed state the 1/k^{th} power of the identity matrix.

Proof:

For d = 90° we have cos(90°) = 0, sin(90°) = 1, $\mathbf{R}_4 = \begin{bmatrix} - & i \\ -i & - \end{bmatrix}$

for d = 180°, cos(180°) = -1, sin(180°) = 0, $\mathbf{R}_2 = \begin{bmatrix} -1 & - \\ - & -1 \end{bmatrix}$

for d = 360°, cos(360°) = 1, sin(360°) = 0, $\mathbf{R}_1 = \begin{bmatrix} 1 & - \\ - & 1 \end{bmatrix}$

For other values of k, including fractional values, the usual convention:

$$A^{p/k} = [A^{1/k}]^p \qquad (10.6.6)$$

and the equivalence of repeated rotation with multiplication of of the rotation operator by itself, provides the general proof.

q.e.d. (ad 5)

Note the convention that a positive power of the identity matrix is associated with rotation in the direction of the clock. The corresponding negative power of the identity matrix (the inverse of the rotation operator), is by (10.6.5) its transpose, and represents rotation against the clock.

We now come to discuss the issue of rotating around another point than the origin. This is best dealt with by redefining the coordinate system first, and, in the interest of having results which are relevant in the original coordinate system, restoring the old coordinate system after rotation. Hence, if we wish to rotate the point (5, 4) around by 45° in the direction of the clock around the point (1, 1), we start with rotating the point (4, 3) by 45° around the origin, and add (1, 1) to the result.

The general formula for the rotation of the two-dimensional vector **x** by d degrees in the direction of the clock around the point **v** therefore is:

$$\begin{bmatrix} x^*_1 \\ x^*_2 \end{bmatrix} = \begin{bmatrix} \cos(d) & \sin(d) \\ -\sin(d) & \cos(d) \end{bmatrix} \begin{bmatrix} x_1 - v_1 \\ x_2 - v_2 \end{bmatrix} + \begin{bmatrix} v_1 \\ v_2 \end{bmatrix}$$

$$(10.6.7)$$

It is at this point useful to clarify the relationship between rotation and De Moivres's Theorem, which is, in effect, both stated and proved in this section.

Theorem (De Moivres's Theorem):

$$\Big(\lambda(\cos(\phi) + i \sin(\phi))\Big)^n = \lambda^n\Big(\cos(n\phi) + i \sin(n\phi)\Big) \qquad (10.6.8)$$

(For a proof outside matrix algebra, see Hawkins & Hawkins [21], pp. 39-42)

The factor λ^n in (10.6.8) follows the normal rules concerning the power of a product-expression. The substantive part of the theorem is:

$$\Big((\cos(\phi) + i \sin(\phi)\Big)^n = \Big(\cos(n\phi) + i \sin(n\phi)\Big) \qquad (10.6.9)$$

and a proof for n > 0, follows from the identity:

$$R_k{}^n\underline{s} = R_k{}^n\underline{s} \qquad (10.6.10)$$

The lefthand-side of (10.6.10) is then read as as a matrix, developed conform (10.6.1) and (10.6.4) and the righthand-side as a rotation operator, rotating by a fraction n/k of the full circle, i.e. over an angle which is n times the angle specified by R_k itself.

The association of a negative power of the identity matrix, with rotation against the clock, also extends to a generalization of (10.6.6), and hence to negative fractional powers of the identity matrix.

If **x** and **u** are related by:

$$\mathbf{u} = R_k\mathbf{x}$$

i.e. **u** is obtained by rotating **x** by a kth fraction of the circle in the direction of the clock, then the rotation-operator which transforms **u** back into **x**, rotating a kth fraction of the circle against the clock, is R_k^{-1}:

$$\mathbf{x} = R_k^{-1}\mathbf{u} = R_k{}'\mathbf{u}$$

and de Moivre's theorem is seen to be true for all real n, whether integer, fractional, positive, or negative.

10.7 Rotation in three-dimensional space

We saw earlier in this chapter, how the concept of area permits a straightforward generalization in n-dimensional space. The similar generalization of the concept of rotation to the n-dimensional case runs into difficulties. One obvious point is that it operates by way of a pair of complex numbers. In fact, a comparison between the 2-dimensional case, where we rotate around a point, and the three-dimensional case, makes the problem clear.

We need a real vector, which is unaffected by the rotation operation, i.e. the axis of rotation. As in the previous section, we initially limit ourselves to the case of rotation around the origin, or, in three-dimensional space, around an axis of rotation, which goes through the origin.

In analogy with the previous section, we 'start up' with a symmetric matrix to diagonalize.

Our 'startup' matrix has (as all symmetric matrices) three real roots, and the vector of one of them is the axis of rotation. The root associated with **x**, the axis of rotation needs to be zero, and we start with rotating by a full circle. This means that the two other roots need to be unity. Their vectors, being characteristic vectors of a diagonalizable matrix, need to be orthogonal to **x**, a property which we geometrically interpret as being perpendicular to the axis of rotation.

But that still leaves a degree of freedom, which we can interpret as rotating the whole axis-system around. We resolve that ambiguity by requiring the x_1-element of the third vector to vanish, i.e. the second axis is required to be in the $x_1=0$ plane.

This does not actually affect the rotation-operator itself: a matrix with a repeated root of unity has non-unique vectors.

The following matrix conforms to these requirements, and will help us as a base to develop suitable vectors:

$$\mathbf{T} = [\mathbf{I}_3 - \mathbf{x}\,\mathbf{x}'] \qquad (10.7.1)$$

where $\mathbf{x}$ has been normalized to the requirement:

$$\mathbf{x}'\mathbf{x} = 1 \qquad (10.7.2)$$

We diagonalize $\mathbf{T}$ as:

$$\mathbf{T} = \mathbf{X}\,\mathbf{S}\,\mathbf{X}' = \mathbf{X}\,\mathbf{S}\,\mathbf{X}^{-1} \qquad (10.7.3)$$

where:

$$\mathbf{X} = \begin{bmatrix} x_1 & (x_2{}^2 + x_3{}^2)\,\beta & - \\ x_2 & -x_2\,x_1\,\beta & x_3\,\delta \\ x_3 & -x_3\,x_1\,\beta & -x_2\,\delta \end{bmatrix} \qquad (10.7.4)$$

where β and δ serve to normalize the second and third column-vectors:

$$\beta = 1 \,/\, \sqrt{(x_2{}^2 + x_3{}^2)^2 + (x_2\,x_1)^2 + (x_3\,x_1)^2} \qquad (10.7.5)$$

$$\delta = 1 \,/\, \sqrt{x_2{}^2 + x_3{}^2} \qquad (10.7.6)$$

The normalization of the two other axes (or indeed of $\mathbf{x}$), is not actually necessary (the vectors, normalized or not, and the roots, fully determine the transformation-operator itself), but it is convenient to have a matrix $\mathbf{X}$, of which the inverse is simply the transpose.

The actual 'rotation' part of the rotation operator is at this stage:

$$\mathbf{S} = \begin{bmatrix} - & - & - \\ - & 1 & - \\ - & - & 1 \end{bmatrix}$$

for the start-up matrix (rotation by a full circle)), but non-trivial rotation is then attained by:

$$
R = X B X' = X \begin{bmatrix} 1 & & \\ & \cos(d) & \sin(d) \\ & -\sin(d) & \cos(d) \end{bmatrix} X' \qquad (10.7.7)
$$

We can interpret X as a transformation operator in its own right, which transforms the axis of rotation direction into the x_1 coordinate direction, and defines two other 'horizontal' directions x_2 and x_3 as being associated with the real, and the irrational component of the pair of complex vectors of R.

Note in that context, that (10.7.7) gives the blockdiagonalization of R conform section 8.9. In that redefined coordinate system we then rotate around the x_1 axis, i.e. in the $x_1 = 0$ plane and any plane parallel to it, copying the vertical coordinate, and put the coordinate system back to what it was, by multiplying once more by $X' = X^{-1}$.

We now proceed to show that R, as defined by (10.7.7), has the properties which we geometrically associate with rotation.

● 1) A vector αx_1 (= a point which is on the intended axis of rotation), is unaffected by premultiplication by R. This is obvious: x_1 is the vector associated with the unity root of R.

● 2) The distance between two vectors v and u is not affected by premultiplication by R.

Proof:

$$[v - u]' \, R'R \, [v - u] = [v - u]' \, [X'B'X] \, [X'B\,X] \, [v - u] =$$

$$[v - u]' \, X'B' \, [X\,X'] \, B\,X \, [v - u] = [v - u]' \, X'B'B\,X \, [v - u] =$$

$$[v - u]' \, X'X \, [v - u] = [v - u]'[v - u]$$

(where X' is both the transpose and the inverse of X)

q.e.d.

● 3) The area spanned by any 4 points v_1 to v_4 is not affected by premultiplication by **R**.

Proof:

The determinant of **R** is the recursive product of the determinants of **X**, **B** and **X'** As **X** and **X'** are periodics and invertible, their determinants must be 1 in absolute value, and (being real), either +1 or -1. Since **R** is blockdiagonal, its determinant is the product of its unity top righthand element, and the determinant of its 2 by 2 bottom righthand block, which was shown to be unity in the previous section.

Therefore:

$$|R| \; = \; 1 \qquad\qquad\qquad (10.7.8)$$

The areas referred to therefore compare as follows:

$$|[v_2 - v_1 \;|\; v_3 - v_1 \;|\; v_4 - v_1 \;]| \text{ before transformation, and}$$

$$|[Rv_2 - Rv_1 \;|\; Rv_3 - Rv_1 \;|\; Rv_4 - Rv_1 \;]| =$$

$$|R[v_2 - v_1 \;|\; v_3 - v_1 \;|\; v_4 - v_1 \;]| =$$

$$|[v_2 - v_1 \;|\; v_3 - v_1 \;|\; v_4 - v_1 \;]| \text{ after transformation.}$$

q.e.d.

Note, that $|R| = 1$ is the essential property here, and that that property is not in its own sufficient to ensure that a matrix is a rotation operator. There clearly are other square matrices with unit determinants, which are not rotation operators.

Finally, we need to mention rotation around an axis which does not go through the origin. The device of displacing the origin, discussed in the previous section for the two-dimensional case, permits immediate generalization to the three-dimensional case.

The one comment is, that we need two points to determine an axis of rotation in the first place. Suppose for example we wish to rotate around an axis-line, which joins the points (2, 3, 4) and (3, 4, 5). We would then displace the origin to (2, 3, 4), rotate around the axis-line between (3, 4, 5), now re-named (1, 1, 1), the difference between (2, 3, 4) and (2, 4, 5), and after rotation we add (2, 3, 4) to the result, to get back to the original coordinate system.

Exercise:

Rotate the pyramid spanned by (2, 3, 4), (3, 4, 5), (2, 0, 4), and (2, 4, 0), by 60° anti-clockwise, when seen from (3, 4, 5) towards (2, 3, 4). Part-check the correctness of your result, by verifying the distances between the points of corners of the pyramid, before and after rotation. (There also is an answersheet at the end of the chapter).

10.8 Orthogonal projection and the Moore–Penrose inverse

The notion of a projection (into a space with fewer dimensions) permits a straightforward generalization to the n-dimensional case. We shall limit ourselves to rectangular projections.

There is a clear association between orthogonality of vectors, and 90° angles, hence the name 'orthogonal projection'.

As with rotation, it is easiest to (separately) displace the origin, and then project in a plane going through the origin.

If we want to project in a (hyper)plane, and we don't also want to change the shape of any projected objects in other respects, we need a transformation-operator, with n-1 unity roots, and an nth zero root. If n-1 unity-roots have n-1 independent vectors, we have n perpendicular axes (a full complement of n independent vectors automatically implies orthogonality), and we suppress the direction of the one associated with the zero root.

In the two-dimensional case, projection-operators which project on a line going through the origin, are relatively easily written down immediately:

$$\mathbf{T} = \begin{bmatrix} \tfrac{1}{2}\sqrt{2} \\ -\tfrac{1}{2}\sqrt{2} \end{bmatrix} \begin{bmatrix} \tfrac{1}{2}\sqrt{2} & -\tfrac{1}{2}\sqrt{2} \end{bmatrix} = \begin{bmatrix} \tfrac{1}{2} & -\tfrac{1}{2} \\ -\tfrac{1}{2} & \tfrac{1}{2} \end{bmatrix}$$

has the 45° degree line as direction of projection, i.e. it removes any distance from the the line which slopes 45° upwards through the origin, and therefore projects on the line which is orthogonal to it, the minus 45° degree line, which slopes downwards through the origin.

When projecting on a line which does not go through the origin, we need a separate calculation of the coordinates of the new origin. This is the intersection of the actual line to project on, and the line perpendicular (orthogonal) to it, going through the origin. Adding the coordinates of this point to the results of the transformation-operation ensures the correct position of the projection of the origin, and (as will be shown for the general n-dimensional case further in this section), by implication of all other projected points as well.

In the three-dimensional case, one way we could get the projection-operator, is by first developing the corresponding rotation-operator. Projection into the plane which is perpendicular to the axis of rotation is then achieved, by first writing the rotation-operator out for 0° degrees rotation, and then replacing the original unity root associated with the axis of rotation, by a zero. As in the case of rotation, there is a degree of freedom in choosing the coordinate directions, provided they are perpendicular (orthogonal) to the direction of projection ('axis of rotation').

There is, however a more elegant, n-dimensional method.

We define a transformation-operator:

$$\mathbf{T} = \mathbf{X} \mathbf{X}^- \qquad\qquad (10.8.1)$$

where $\mathbf{X}^-$ is the Moore-Penrose generalized inverse of $\mathbf{X}$, and where $\mathbf{X}$ is the matrix of independent directions away from <u>one</u> point in the plane (line), we wish to project on.

This operator copies any direction in the plane, (which is a combination of the characteristic vectors associated with its unity root), but suppresses the direction of the axis which is perpendicular to it -that is the vector associated with the zero root of $[\mathbf{X} \mathbf{X}^-]$. In the case of projection on a line in two-, or in three-dimensional space, $[\mathbf{X} \mathbf{X}^-]$ will be of rank 1. For projection into a plane in three-dimensional space, we need two direction, i.e. rank 2.

We will attend to the precise algebraic definition of the matrix
X later in this section, at this stage we proceed to illustrate
its geometrical interpretation.

If (10.8.1) is used, there is no need to normalize <u>before</u> we
start the calculations. Thus, the 45° line sloping downwards
through the origin is the line through (1, -1) and (-1, 1), and
we calculate **X**, which is in the two-dimensional case of order 2
by 1 and of rank 1, by displacing the origin towards one point
which meets the line to project on, and enter the new coordinates
of the other point as the one column of **X**:

$$\mathbf{X} = \begin{bmatrix} -2 \\ 2 \end{bmatrix}$$

i.e. we measure the position of the point which was originally at
(-1, 1) as (-2, 2), because we now take (1, -1) as origin.

$$\mathbf{X}^- = \mathbf{X}'[\mathbf{X}'\mathbf{X}]^{-1} = [\ -2 \quad 2\ ]\ [1/8] = [\ -0.25 \quad 0.25\ ]$$

$$\mathbf{X}\,\mathbf{X}^- = \begin{bmatrix} -2 \\ 2 \end{bmatrix}\ [\ -0.25 \quad 0.25\ ] = \begin{bmatrix} 0.5 & -0.5 \\ -0.5 & .5 \end{bmatrix}$$

In addition, (10.8.1) also provides the basis for an elegant
solution towards finding the appropriate displacement of the
origin, if the origin is not itself in the line (plane) to
project on.

Theorem:

If the n by k matrix **X** is of full rank k,

then the shortest distance from the n by 1 vector **y** to a combination of the $\mathbf{x}_j$ (the nearest to **y**) is the distance between **y** and

$$\mathbf{z} = \mathbf{X}\,\mathbf{X}^-\,\mathbf{y} \qquad\qquad\qquad (10.8.2)$$

Proof:

By assumption, $\mathbf{z}$ is a combination of the $\mathbf{x}_j$.

Therefore:

$$\mathbf{z} = \mathbf{X}\, \mathbf{v} \tag{10.8.3}$$

The distance δ between $\mathbf{y}$ and $\mathbf{z}$ therefore is the square root of:

$$\delta^2 = [\mathbf{y} - \mathbf{z}]'\, [\mathbf{y} - \mathbf{z}] = [\mathbf{y} - \mathbf{X}\, \mathbf{v}]'\, [\mathbf{y} - \mathbf{X}\, \mathbf{v}]$$

$$= \mathbf{y}'\mathbf{y} - 2\mathbf{y}'\mathbf{X}\, \mathbf{v} + \mathbf{v}'\mathbf{v} \tag{10.8.4}$$

The necessary first-order conditions for the minimization of δ are:

$$\partial\delta^2/\partial\mathbf{v} = -2\, \mathbf{X}'\mathbf{y} + 2\, \mathbf{X}'\mathbf{X}\, \mathbf{v} = [0] \tag{10.8.5}$$

From which:

$$\mathbf{v} = [\mathbf{X}'\mathbf{X}]^{-1}\mathbf{X}\, \mathbf{y} = \mathbf{X}^{-}\mathbf{y} \tag{10.8.6}$$

Substitution of the righthand side of (10.8.6) for $\mathbf{v}$ into (10.8.3) now confirms (10.8.2)

q.e.d.

The multiplication of of any arbitrary $\mathbf{x}$ by $\mathbf{T}$ as defined by (10.8.1) projects it on a line (more general: a hyperplane), which goes through the origin: $\mathbf{T}$ is of rank k<n, (there is nothing to project on, if n independent columns are entered in $\mathbf{X}$), and therefore permits the expression of the origin as a combination. To obtain the appropriate matrix $\mathbf{X}$ of rank n-1 (or n-2 of projecting on a line in three-dimensional space), it is still convenient to first displace the origin to one point on the line (in the projection-plane). There is no need to leave the resulting zero vector in $\mathbf{X}$.

The correction of the origin is now found as the distance of the point towards which we displaced the origin, to the projection-plane (line), which goes through the origin, using (10.8.2).

Example:

Project on the line: $3x_1 + 2x_2 = 16$.

We first of all illustrate this problem graphically:

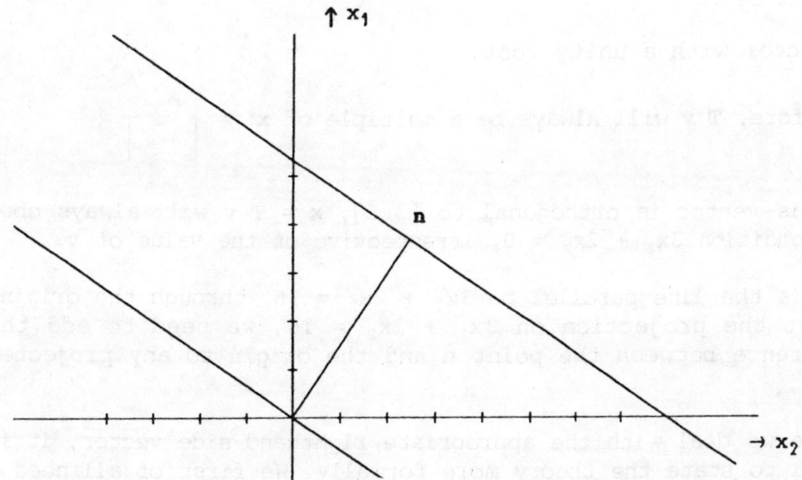

The line we wish to project on, $3x_1 + 2x_2 = 16$, crosses the x_1 axis at $\mathbf{x}_1 = (5.33 , 0)$, and the x_2 axis at $\mathbf{x}_2 = (0, 8)$. The line which is perpendicular to it and goes through the origin, $2x_1 - 3x_2 = 0$, intersects with the line $3x_1 + 2x_2 = 16$ at (the n=new origin, $\mathbf{n} = (3.69, 2.46)$.

We first of all displace the origin initially to $\mathbf{x}_1 = (5.33 , 0)$, and the matrix $\mathbf{X}$ contains the coordinates of $\mathbf{x}_2$, measured from $\mathbf{x}_1$, i.e.:

$$\mathbf{X} = \begin{bmatrix} -5.33 \\ 8 \end{bmatrix} \; ; \quad \mathbf{X'X} = [5.33^2 + 8^2] = [92.44]$$

$$[\mathbf{X'X}]^{-1} = [0.011]$$

$$\mathbf{X^-} = [\mathbf{X'X}]^{-1}\,\mathbf{X} = [0.011]\,[-5.33 \quad 8] = [-0.058 \quad 0.086]$$

$$\mathbf{T} = \mathbf{X}\,\mathbf{X^-} = \begin{bmatrix} -5.33 \\ 8 \end{bmatrix} [-0.058 \quad 0.086] = \begin{bmatrix} 0.31 & -0.46 \\ -0.46 & 0.69 \end{bmatrix}$$

This matrix is idempotent, with $\begin{bmatrix} 3 \\ 2 \end{bmatrix}$

as vector with a zero root

and $\begin{bmatrix} 2 \\ -3 \end{bmatrix}$

as vector with a unity root.

Therefore, $\mathbf{T}\,\mathbf{v}$ will always be a multiple of $\mathbf{x}^* = \begin{bmatrix} 2 \\ -3 \end{bmatrix}$

As this vector is orthogonal to $[3\ 2]$, $\mathbf{x} = \mathbf{T}\,\mathbf{v}$ will always obey the condition $3x_1 + 2x_2 = 0$, irrespective of the value of $\mathbf{v}$.

This is the line parallel to $3x_1 + 2x_2 = 16$, through the origin. To get the projection on $3x_1 + 2x_2 = 16$, we need to add the difference between the point $\mathbf{n}$ and the origin to any projected point.

Before we deal with the appropriate righthand-side vector, it is useful to state the theory more formally. We first of all need a general algebraic definition of a projection:

The (orthogonal) <u>projection</u> of the vector $\mathbf{x}$ into the set of points which satisfy the linear restrictions

$$\mathbf{A}\,\mathbf{v} = \mathbf{b} \tag{10.8.7}$$

(where $\mathbf{A}$ is of order m by n and of full rank m)

is the point $\mathbf{v} = \mathbf{v}^*$, which minimizes the length of

$$\mathbf{d} = \mathbf{x} - \mathbf{v} \tag{10.8.8}$$

We shall refer to this definition also in a slightly abbreviated form as the projection of $\mathbf{x}$ into $\mathbf{A}\mathbf{v} = \mathbf{b}$.

Theorem:

$$\mathbf{d} = \mathbf{x} - \mathbf{v}^*$$

can be expressed as a combination of those vectors $\mathbf{u}_j$, which satisfy:

$$\mathbf{A'A}\,\mathbf{u}_j = \mathbf{u}_j\,\lambda_j \neq [0] \qquad (10.8.9)$$

Proof:

Any $\mathbf{d}$ can be expressed as:

$$\mathbf{d} = \sum_{j=1}^{n} \mathbf{u}_j\, w_j \qquad (10.8.10)$$

where the $\mathbf{u}_j$ are the n independent characteristic vectors of $\mathbf{A'A}$, and are therefore all orthogonal to each other.

Therefore (assuming normalization to length one of all the $\mathbf{u}_j$):

$$\mathbf{d'd} = \sum_{j=1}^{n} w_j{}^2 \qquad (10.8.11)$$

For any j, for which $\mathbf{A'A}\,\mathbf{u}_j = [0]$, and therefore $\mathbf{Au}_j = [0]$ applies, the corresponding w_j can be set at zero (adding $\mathbf{u}_j w_j$ to $\mathbf{v}$), without disturbing (10.8.7).

q.e.d.

In geometric terms, this result was already obvious, and we were in fact applying it: $\mathbf{d}$ is the direction which is perpendicular to, in our example, the $3x_1 + 2x_2 = 16$ line.

We shall in that connection refer to the vectorial difference between $\mathbf{x}$ and $\mathbf{v}^*$ as the <u>distance vector</u>.

Theorem:

$$[\mathbf{I} - \mathbf{A}^-\mathbf{A}] \; \mathbf{x}$$

is the projection of $\mathbf{x}$ into the set of points $\mathbf{v}$ satisfying

$$\mathbf{A} \; \mathbf{v} = [0] \tag{10.8.12}$$

Proof:

By the previous theorem $\mathbf{d} = \mathbf{x} - \mathbf{v}^*$ is orthogonal to an $\mathbf{u}_j$, only if $\mathbf{A} \; \mathbf{u}_j = [0]$ applies.

Therefore, on pre-multiplication of (10.8.10) by $[\mathbf{A}^-\mathbf{A}]$:

$$[\mathbf{A}^-\mathbf{A}] \; \mathbf{d} = [\mathbf{A}^-\mathbf{A}] \; \mathbf{x} - [\mathbf{A}^-\mathbf{A}] \; \mathbf{v}^* =$$

$$\sum_{j=1}^{r} [\mathbf{A}^-\mathbf{A}] \; \mathbf{u}_j \; w_j \; + \; \sum_{j=r+1}^{n} [\mathbf{A}^-\mathbf{A}] \; \mathbf{u}_j \; wj \; =$$

$$\sum_{j=1}^{r} [\mathbf{A}^-\mathbf{A}] \; \mathbf{u}_j \; w_j \tag{10.8.13}$$

Since $\mathbf{v}^*$ also needs to satisfy $\mathbf{A} \; \mathbf{v}^* = [0]$ -(10.8.12)-, relation (10.8.13) is equivalent to:

$$[\mathbf{A}^-\mathbf{A}] \; \mathbf{d} = [\mathbf{A}^-\mathbf{A}] \; \mathbf{x} = \sum_{j=1}^{r} [\mathbf{A}^-\mathbf{A}] \; \mathbf{u}_j \; w_j \tag{10.8.14}$$

As the $\mathbf{u}_j$ are also vectors of $[\mathbf{A}^-\mathbf{A}]$, with the roots of the first r vectors being unity, we infer from (10.8.14):

$$[\mathbf{A}^-\mathbf{A}] \; \mathbf{d} = [\mathbf{A}^-\mathbf{A}] \; \mathbf{x} = \sum_{j=1}^{r} [\mathbf{A}^-\mathbf{A}] \; \mathbf{u}_j \; w_j \; =$$

$$\sum_{j=1}^{r} \mathbf{u}_j \; w_j \; = \; \mathbf{d} \; = \; \mathbf{x} - \mathbf{v}^* \tag{10.8.15}$$

From which, on re-ordering:

$$\mathbf{v^*} = [I - A^-A]\ \mathbf{x} \tag{10.8.16}$$

q.e.d.

We now come back to the matrix $\mathbf{X}$, as referred to in (10.8.1) and succeeding formulae, which we did not so far specify fully.

In fact, any matrix $\mathbf{X}$, of order n by n-m, which satisfies:

$$\mathbf{A}\,\mathbf{X} = [0] \tag{10.8.17}$$

and is of full rank n-m (where m=r is the rank of the m by n matrix $\mathbf{A}$, defining the side-restrictions (10.8.7), will do.

Our geometrical interpretation of the columns of $\mathbf{X}$ is that they are directions into which a solution to (10.8.7) can be moved without breaching that requirement:

$$\mathbf{A}\,[\mathbf{x_k} - \mathbf{x_1}] = \mathbf{Ax_k} - \mathbf{Ax_1} = \mathbf{b} - \mathbf{b} = [0] \tag{10.8.18}$$

hence the requirement (10.8.17).

In normalized form, the columns of $\mathbf{X}$ are the vectors $\mathbf{p_j}$ which satisfy:

$$\mathbf{A}\,\mathbf{p_j} = [0], \text{ hence } [\mathbf{A^-A}]\mathbf{p_j} = [0], \text{ and therefore } [I - \mathbf{A^-A}]\mathbf{p_j} = \mathbf{p_j}.$$

as well as (not really essential), the normalization requirement:

$$\mathbf{p_j}'\mathbf{p_j} = 1\ (j = m+1,\ m+2,\ \ldots\ n),$$

These are the vectors to which we referred in section 9.8 as the dummy row-components of $\mathbf{A}$, and we express $\mathbf{T}$ as:

$$\mathbf{T} = \mathbf{X}\,\mathbf{X^-} = [I - \mathbf{A^-A}] = \sum_{j=1}^{r} \mathbf{p_j}\,\mathbf{p_j}' \tag{10.8.19}$$

We now come to discuss the appropriate righthand-side of a projection-relation.

Theorem:

For any **x**, which satisfies **A x** = **b**,

the vectorial distance to its projection in **A x** = [0] is:

$$\mathbf{d} = \mathbf{A}^- \mathbf{b} \qquad\qquad (10.8.20)$$

Proof:

Pre-multiplication of the assumed **A x** = **b** by **A**$^-$ results in:

$$\mathbf{A}^-\mathbf{A}\,\mathbf{x} = \mathbf{A}^-\mathbf{b} \qquad\qquad (10.8.21)$$

Therefore the projection is:

$$[\mathbf{I} - \mathbf{A}^-\mathbf{A}]\,\mathbf{x} = \mathbf{v}^* = \mathbf{x} - \mathbf{A}^-\mathbf{b} \qquad\qquad (10.8.22)$$

Therefore, on re-ordering:

$$\mathbf{d} = \mathbf{x} - \mathbf{v}^* = \mathbf{A}^-\mathbf{b} \qquad\qquad (10.8.23)$$

q.e.d.

In two-dimensional graphics, an obvious way to get that vector,
is to calculate the projection of some other point on the line
through the origin, and then take the difference between the
point itself and the projection.

Any point on the line $3x_1 + 2x_2 = 16$ will in fact suffice (the
distance between the two parallel lines is the same everywhere),
but having displaced the origin initially to $\mathbf{x}_1$, it is natural to
use the same vector again.

$$\mathbf{T}\,\mathbf{x}_1 = \begin{bmatrix} 0.31 & -0.46 \\ -0.46 & 0.69 \end{bmatrix} \begin{bmatrix} 5.33 \\ - \end{bmatrix} = \begin{bmatrix} 1.64 \\ -2.46 \end{bmatrix}$$

is the projection of $\mathbf{x}_1$ on the $3x_1 + 2x_2 = 0$ line.

Exercises:

Write the projection-operators, which will project on the line:

$$x_1 + 3x_2 = 5 \text{ (in two-dimensional space)},$$

and on the plane:

$$x_1 + 2x_2 + 3x_3 = 10$$

(There is an answersheet/worked example-sheet at the end of the chapter)

10.9 Some comments on n-dimensional geometry

There are a number of criteria, by which one can judge the meaningfulness of generalizing, or attempting to generalize, a geometrical concept, which has an algebraic counterpart in the familiar realms of two-dimensional and three-dimensional geometry to any number of dimensions.

The first requirement is that there must <u>be</u> straightforward algebraic generalization. The second requirement is that there must be credible association with a concept which is familiar in 'ordinary' geometry, and thirdly, it would be desirable if there were some practical or operational use for such an n-dimensional generalization.

We have seen in this chapter, that the notions of distance and content pass on the first and the second count. It is also worth mentioning, that the concept of an n-dimensional volume also passes on the third count. It has relevance to the problem of simultaneous estimation methods in econometrics.

The simultaneous least squares fit problem requires to estimate certain elements of the matrix **A**, which specifies a system of stochastic economic relations as:

$$\mathbf{A}\, \mathbf{y}_t \;=\; \mathbf{B}\, \mathbf{x}_t \;+\; \mathbf{u}_t \tag{10.9.1}$$

where **A** and **B** are coefficients matrices, some of whose elements need to be fitted to data, which are observed values of the vectors $\mathbf{y}_t$ and $\mathbf{x}_t$, t being the observation index.

If the system is estimated per separate relation, then, depending on how good the statistical estimate of A and B is, the area within which the bulk of the implied values of the elements of the $\mathbf{u}_t$ will be smaller for a good fit, and bigger for a poor fit. The corresponding uncertainty area in the coordinate system in which $\mathbf{y}_t$ is mapped, is $|A^{-1}|$ times the similar area in the coordinate system in which the $\mathbf{u}_t$ are mapped, or (depending on the sign of $|A|$, $-|A^{-1}|$ times that area in the u space.

Consequently, the determinant $|A^{-1}|$ figures prominently in simultaneous estimation methods for econometric models, and ill-conditioned systems, for which the uncertainty area in the $\mathbf{y}_t$ space could be large, are avoided.

The concept of volume, or n-dimensional area, passes on all three counts. The same applies to the concept of being 'inside' an area, a 'convex combination', a notion which figures large in mathematical programming.

But an attempt to generalize the notion of rotation to n dimensions, runs into immediate trouble, even on the first count. In two-dimensional space we rotate around a point, and in three-dimensional space we rotate around a line. A three-dimensional rotation operator will therefore need to have one real root of $\lambda=0$, with the axis of rotation as vector, and a pair of complex roots.

It is therefore unclear, how we could generalize the notion of rotation to four or more dimensions. Should there be two zero roots, and a pair of complex roots, rotating around a plane ? Or should there be two pairs of complex roots, i.e. we rotate again around the origin, but somehow twist at the same time ? Rotation cannot meaningfully generalized to the n-dimensional case.

Then again, projection appears not to create any problems, and certainly passes on the first and the second count.

Answersheet on surface and content

1): Calculate the surface area of the triangle spanned by:

$$\mathbf{A} = \begin{bmatrix} 2 & 1 & 5 \\ 3 & 6 & 4 \end{bmatrix}$$

We displace the origin to (2, 3), and the transformed coordinates matrix is then:

$$\begin{array}{ccc} \mathbf{o} & \mathbf{p} & \mathbf{q} \end{array}$$

$$\mathbf{C} = \begin{bmatrix} - & -1 & 3 \\ - & 3 & 1 \end{bmatrix}$$

Accordingly the surface of the parallelogram opq is (the absolute value of) the determinant:

$$\begin{vmatrix} -1 & 3 \\ 3 & 1 \end{vmatrix} = -1 \times 1 - (3 \times 3) = -1 -9 = -10.$$

Reference to the visual illustration of the triangle opq in the graph below indicates that the triangle is a rectangular one, and the 'parallelogram' therefore a rectangle, and in fact a square. We calculate the length of the two sides adjoining the right angle by Pythagoras' Theorem as:

$$\sqrt{1^2 + 3^2} = \sqrt{10}$$

confirming that the surface of the full square of which the triangle is half, is indeed 10. The surface of the triangle is therefore 10/2 = 5.

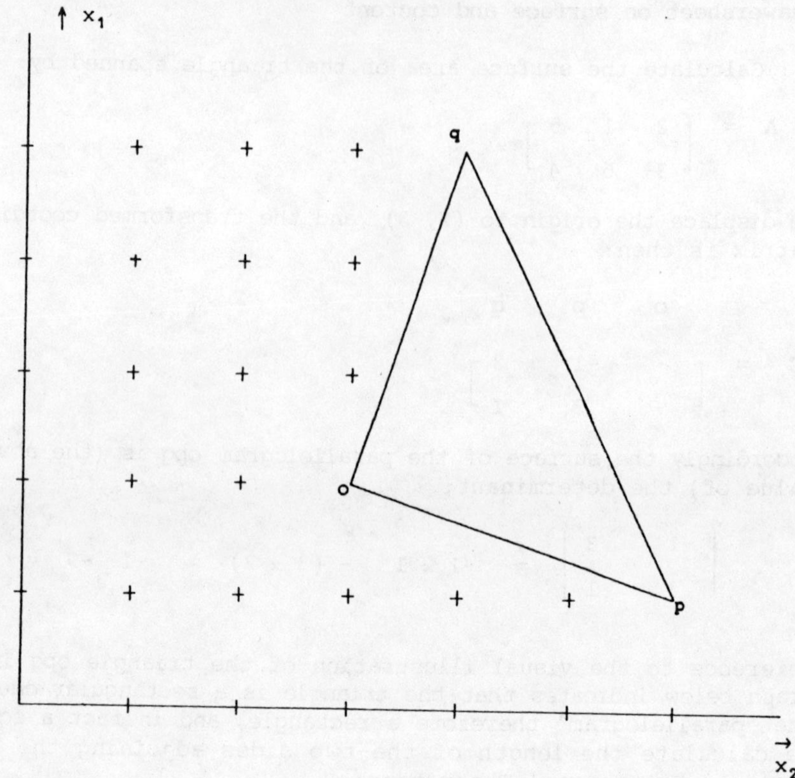

2): Calculate the content of the pyramid spanned by:

$$
B = \begin{array}{cccc}
\mathbf{o} & \mathbf{p} & \mathbf{p} & \mathbf{r}
\end{array}
\begin{bmatrix}
1 & 2 & 5 & 1 \\
2 & 11 & -7 & 8 \\
1 & -5 & 3 & 1
\end{bmatrix}
$$

Displacement of the origin to **o** results in a matrix of secondary coordinates which is:

$$
D = \begin{array}{cccc}
\mathbf{o} & \mathbf{p} & \mathbf{p} & \mathbf{r}
\end{array}
\begin{bmatrix}
- & 1 & 4 & - \\
- & 9 & -9 & 6 \\
- & -6 & 2 & -
\end{bmatrix}
$$

and we develop the determinant

$$
\begin{vmatrix}
1 & 4 & - \\
9 & -9 & 6 \\
-6 & 2 & -
\end{vmatrix}
$$

by its third column as

$$
6 \times \begin{vmatrix} 1 & 4 \\ -6 & 2 \end{vmatrix} = 6 \times (2 + 24)
$$

$$
6 \times 26 = 156.
$$

This is the content of the parallelepiped of which **o** is one corner, and **p**, **q**, and **r** are adjoining corners. The content of the corresponding pyramid is therefore 156/6 = 26.

Answersheet on rotation

We first of all displace the origin to one of the two points around which we wish to rotate, (2, 3, 4) is chosen here. The result of this transformation is summarized below:

$$
\begin{bmatrix} 2 & 3 & 2 & 2 \\ 3 & 4 & - & 4 \\ 4 & 5 & 4 & - \end{bmatrix}
\rightarrow
\begin{bmatrix} - & 1 & - & - \\ - & 1 & -3 & 1 \\ - & 1 & - & -4 \end{bmatrix}
$$

We now need to rotate around the (normalized) axis between the displaced point (3, 4, 5), which became (1, 1, 1), and the displaced point (2, 3, 4), which became the origin. In first instance that is a summation vector, but after normalization we have:

$$
\mathbf{x} = \begin{bmatrix} \sqrt{1/3} \\ \sqrt{1/3} \\ \sqrt{1/3} \end{bmatrix} = \begin{bmatrix} 1/3\ \sqrt{3} \\ 1/3\ \sqrt{3} \\ 1/3\ \sqrt{3} \end{bmatrix}
$$

We now need to calculate the other elements of **X**.

We start with their unscaled form of the second column:

$x_1^2 + x_3^2 = 1/3 + 1/3 = 2/3$ of which the square is: 4/9

$x_1\ x_3 = 1/3$ of which the square is: 1/9

$x_1\ x_2 = 1/3$ of which the square is: 1/9

adds to: 6/9

The length of this vector $1/3\ \sqrt{6}$, and the reciprocal of this length is $\tfrac{1}{2}\sqrt{6} = 1.225$.

Therefore:

$$x_{12} = 2/3 \times \tfrac{1}{2}\sqrt{6} = 0.816 \text{ of which the square is: } 0.666$$

$$x_{22} = -1/3 \times \tfrac{1}{2}\sqrt{6} = -0.408 \text{ of which the square is: } 0.167$$

$$x_{32} = -1/3 \times \tfrac{1}{2}\sqrt{6} = -0.408 \text{ of which the square is: } 0.167$$

$$\text{adds to: } 1.000$$

Now the unscaled form of the third column:

$$x_{38} = \sqrt{1/3} = 0.577 \qquad \text{of which the square is: } 1/3$$

$$-x_2 = -\sqrt{1/3} = -0.577 \qquad \text{of which the square is: } 1/3$$

$$\text{adds to: } 2/3$$

The length of this vector is therefore: $\sqrt{2/3} = 0.816$

The reciprocal of the length therefore is 1.225

Therefore:

$$(x_{13} = 0)$$

$$x_{23} = \sqrt{1/3} \times 1.225 = 0.707 \text{ the square of which is: } 0.5$$

$$x_{33} = \sqrt{1/3} \times 1.225 = 0.707 \text{ the square of which is: } 0.5$$

The matrix **X** is now assembled as:

$$\mathbf{X} = \begin{bmatrix} 0.577 & 0.816 & - \\ 0.577 & -0.408 & 0.707 \\ 0.577 & -0.408 & -0.707 \end{bmatrix}$$

The full rotation-operator is now assembled as:

R = **X B X'** =

$$\begin{bmatrix} 0.577 & 0.816 & - \\ 0.577 & -0.408 & 0.707 \\ 0.577 & -0.408 & -0.707 \end{bmatrix} \begin{bmatrix} 1 & - & - \\ - & 0.500 & -0.866 \\ - & 0.866 & 0.500 \end{bmatrix} \begin{bmatrix} 0.577 & 0.577 & 0.577 \\ 0.816 & -0.408 & -0.408 \\ - & 0.707 & -0.707 \end{bmatrix}$$

where $0.500 = \cos(60°)$, and $0.866 = \sin(60°)$

Or, working out the matrix-multiplication:

$$[\mathbf{X\ B}]\ \mathbf{X'} = \begin{bmatrix} 0.577 & 0.408 & -0.707 \\ 0.577 & 0.408 & 0.707 \\ 0.577 & -0.816 & - \end{bmatrix} \begin{bmatrix} 0.577 & 0.577 & 0.577 \\ 0.817 & -0.408 & -0.408 \\ - & 0.707 & -0.707 \end{bmatrix}$$

$$= \begin{bmatrix} 0.667 & -0.333 & 0.667 \\ 0.667 & 0.667 & -0.333 \\ -0.333 & 0.667 & 0.667 \end{bmatrix}$$

Note, that this matrix has the summation-vector both as column-vector, and as row-vector, as it should have.

We now perform the actual rotation:

$$\begin{bmatrix} 0.667 & -0.333 & 0.667 \\ 0.667 & 0.667 & -0.333 \\ -0.333 & 0.667 & 0.667 \end{bmatrix} \begin{bmatrix} - & 1 & - & - \\ - & 1 & -3 & 1 \\ - & 1 & - & -4 \end{bmatrix} = \begin{bmatrix} - & 1 & 1 & -3 \\ - & 1 & -2 & 2 \\ - & 1 & -2 & -2 \end{bmatrix}$$

We now need to put the origin again at its original position:

$$\begin{bmatrix} - & 1 & 1 & -3 \\ - & 1 & -2 & 2 \\ - & 1 & -2 & -2 \end{bmatrix} \rightarrow \begin{bmatrix} 2 & 3 & 3 & -1 \\ 3 & 4 & 1 & 5 \\ 4 & 5 & 2 & 2 \end{bmatrix}$$

We now check on unchanged distances:

The distance between (2, 3, 4) and (3, 4, 5) does not need checking, as these points are themselves unchanged. But we need to check on in total five distances, as follows:

Distances from (2, 3, 4):

$$\begin{bmatrix} 2 \\ - \\ 4 \end{bmatrix} - \begin{bmatrix} 2 \\ 3 \\ 4 \end{bmatrix} = \begin{bmatrix} - \\ -3 \\ - \end{bmatrix} \quad \text{became} \quad \begin{bmatrix} 3 \\ 1 \\ 2 \end{bmatrix} - \begin{bmatrix} 2 \\ 3 \\ 4 \end{bmatrix} = \begin{bmatrix} 1 \\ -2 \\ -2 \end{bmatrix}$$

length $\sqrt{1^2 + 2^2 + 2^2} = \sqrt{9}$ unchanged

$$\begin{bmatrix} 2 \\ 4 \\ - \end{bmatrix} - \begin{bmatrix} 2 \\ 3 \\ 4 \end{bmatrix} = \begin{bmatrix} - \\ 1 \\ -4 \end{bmatrix} \quad \text{became} \quad \begin{bmatrix} -1 \\ 5 \\ 2 \end{bmatrix} - \begin{bmatrix} 2 \\ 3 \\ 4 \end{bmatrix} = \begin{bmatrix} -3 \\ 2 \\ -2 \end{bmatrix}$$

length was $\sqrt{1^2 + 4^2} = \sqrt{17}$

became $\sqrt{3^2 + 2^2 + 2^2} = \sqrt{17}$ unchanged

Distances from (3, 4, 5):

$$\begin{bmatrix} 2 \\ - \\ 4 \end{bmatrix} - \begin{bmatrix} 3 \\ 4 \\ 5 \end{bmatrix} = \begin{bmatrix} -1 \\ -4 \\ -1 \end{bmatrix} \quad \text{became} \quad \begin{bmatrix} 3 \\ 1 \\ 2 \end{bmatrix} - \begin{bmatrix} 3 \\ 4 \\ 5 \end{bmatrix} = \begin{bmatrix} - \\ -3 \\ -3 \end{bmatrix}$$

length was $\sqrt{1^2 + 4^2 + 1^2} = \sqrt{18}$

became $\sqrt{3^2 + 3^2} = \sqrt{18}$ unchanged.

$$\begin{bmatrix} 2 \\ 4 \\ - \end{bmatrix} - \begin{bmatrix} 3 \\ 4 \\ 5 \end{bmatrix} = \begin{bmatrix} -1 \\ - \\ -5 \end{bmatrix} \quad \text{became} \quad \begin{bmatrix} -1 \\ 5 \\ 2 \end{bmatrix} - \begin{bmatrix} 3 \\ 4 \\ 5 \end{bmatrix} = \begin{bmatrix} -4 \\ 1 \\ -3 \end{bmatrix}$$

distance was $\sqrt{1^2 + 5^2} = \sqrt{26}$

became $\sqrt{4^2 + 1^2 + 3^2} = \sqrt{26}$

The distance between the two rotated points:

$$\begin{bmatrix} 2 \\ - \\ 4 \end{bmatrix} - \begin{bmatrix} 3 \\ 4 \\ 5 \end{bmatrix} = \begin{bmatrix} -1 \\ -4 \\ -1 \end{bmatrix} \quad \text{became} \quad \begin{bmatrix} 3 \\ 1 \\ 2 \end{bmatrix} - \begin{bmatrix} 3 \\ 4 \\ 5 \end{bmatrix} = \begin{bmatrix} - \\ -3 \\ -3 \end{bmatrix}$$

length was $\sqrt{1^2 + 4^2 + 1^2} = \sqrt{18}$

became $\sqrt{3^2 + 3^2} = \sqrt{18}$ unchanged.

Answersheet on projection

We first deal with the relatively simple two-dimensional case.

We are asked to project on the line:

$$x_1 + 3x_2 = 5,$$

and we initially deal with projecting on the line:

$$x_1 + 3x_2 = 0.$$

The perpendicular (orthogonal) direction to this line is:

$$3x_1 - x_2 = 0.$$

We normalize both vectors, and the matrix of the two normalized characteristic vectors is:

$$\begin{bmatrix} 0.1\sqrt{10} & 0.3\sqrt{10} \\ 0.3\sqrt{10} & -0.1\sqrt{10} \end{bmatrix}$$

The corresponding transformation-operator is:

$$\mathbf{T} = \begin{bmatrix} 0.3\sqrt{10} \\ -0.1\sqrt{10} \end{bmatrix} \begin{bmatrix} 0.3\sqrt{10} & -0.1\sqrt{10} \end{bmatrix} = \begin{bmatrix} 0.9 & -0.3 \\ -0.3 & 0.1 \end{bmatrix}$$

which is an idempotent matrix:

$$\begin{bmatrix} 0.9 & -0.3 \\ -0.3 & 0.1 \end{bmatrix} \begin{bmatrix} 0.9 & -0.3 \\ -0.3 & 0.1 \end{bmatrix} = \begin{bmatrix} 0.9 & -0.3 \\ -0.3 & 0.1 \end{bmatrix}$$

The new origin, expressed in the old coordinate system is the intersection of

$$x_1 + 3x_2 = 5$$

with

$$3x_1 - x_2 = 0 \quad \rightarrow \quad x_2 = 3x_1$$

Therefore:

$$x_1 + 3x_2 = 5 \quad \rightarrow \quad x_1 + 9x_1 = 10x_1 = 5 \quad \rightarrow \quad x_1 = 0.5$$

Therefore:

$$3x_2 = 4.5 \rightarrow x_2 = 1.5$$

The projection of the origin on the $x_1 + 3x_2 = 5$ line therefore needs to be at $(0.5, 1.5)$, and to bring it there we simply add this vector to the transformation operator, which therefore is:

$$\begin{bmatrix} x^*_1 \\ x^*_2 \end{bmatrix} = \begin{bmatrix} 0.9 & -0.3 \\ -0.3 & 0.1 \end{bmatrix} \begin{bmatrix} x_1 \\ x_1 \end{bmatrix} + \begin{bmatrix} 0.5 \\ 1.5 \end{bmatrix}$$

The following graph confirms these results, and in particular the projection of the origin to $(0.5, 1.5)$:

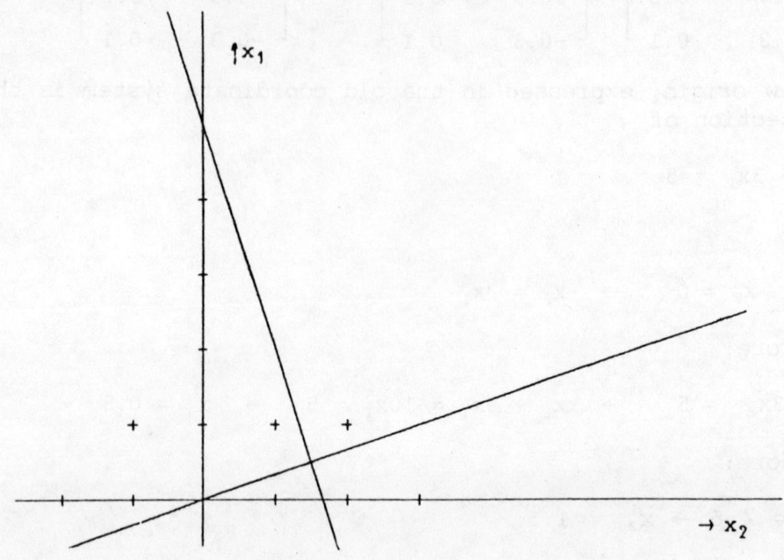

Now using the Moore-Penrose inverse ($T = \mathbf{X}\,\mathbf{X}^-$):

The line to project on goes through the points (5, 0) at the intersection with the vertical axis, and and (0, 1.67) at the intersection with the horizontal axis.

On displacement of the origin to (5, 0), the coordinates of the other point become (-5, 1.67).

For a 2 by 1 matrix of full rank 1, the generalized inverse is best calculated as:

$$\mathbf{X}^- = [\mathbf{X}'\mathbf{X}]^{-1}\mathbf{X}' = \left[[-5 \quad 1\,2/3] \begin{bmatrix} -5 \\ 1\,2/3 \end{bmatrix} \right]^{-1} [-5 \quad 1\,2/3] =$$

$$(27\,7/9)^{-1} \times [-5 \quad 1\,2/3] = [-9/50 \quad 3/50]$$

Therefore:

$$\mathbf{X}\,\mathbf{X}^- = \begin{bmatrix} -5 \\ 1.67 \end{bmatrix} \begin{bmatrix} -9/50 & 3/50 \end{bmatrix} = \begin{bmatrix} 0.9 & -0.3 \\ -0.3 & 0.1 \end{bmatrix}$$

Having used $\mathbf{T} = \mathbf{X}\,\mathbf{X}^-$ rather than $\mathbf{T} = [\mathbf{I} - \mathbf{A}^-\mathbf{A}]$, calculation of the origin-displacement as $\mathbf{d} = \mathbf{A}^-\mathbf{b}$ is not the most practical method, and we shall reserve that for the three-dimensional case.

For the three-dimensional projection-operator, it is impractical to use any other method than by way of the Moore-Penrose inverse, and we shall use $\mathbf{T} = \mathbf{A}^-\mathbf{A}$.

The one side-restriction:

$$x_1 + 2x_2 + 3x_3 = 10 :$$

gives rise to $\mathbf{A} = [1 \; 2 \; 3]$

Therefore $\mathbf{A}^-$ is $\mathbf{A}'$ divided by the square of the length of the one row, which is the root of $\mathbf{A}\mathbf{A}' = [14]$. (The one vector is $\mathbf{u}_1 = [1]$):

$$\mathbf{A}^- = \begin{bmatrix} 1/14 \\ 1/7 \\ 3/14 \end{bmatrix} ; \qquad \mathbf{A}^-\mathbf{A} = \begin{bmatrix} 1/14 & 1/7 & 3/14 \\ 1/7 & 2/7 & 3/7 \\ 3/14 & 3/7 & 9/14 \end{bmatrix}$$

Therefore $\mathbf{T} = [\mathbf{I} - \mathbf{A}^-\mathbf{A}] = \begin{bmatrix} 13/14 & -1/7 & -3/14 \\ -1/7 & 5/7 & -3/7 \\ -3/14 & -3/7 & 5/14 \end{bmatrix}$

$$\mathbf{A}^-\mathbf{b} = \begin{bmatrix} 1/14 \\ 1/7 \\ 3/14 \end{bmatrix} \times 10 = \begin{bmatrix} 5/7 \\ 1\,3/7 \\ 2\,1/7 \end{bmatrix}$$

The transformation we are asked to define therefore is:

$$\mathbf{x}^* = \begin{bmatrix} x^*_1 \\ x^*_2 \\ x^*_3 \end{bmatrix} = \begin{bmatrix} 13/14 & -1/7 & -3/14 \\ -1/7 & 5/7 & -3/7 \\ -3/14 & -3/7 & 5/14 \end{bmatrix} \begin{bmatrix} x_1 \\ x_2 \\ x_3 \end{bmatrix} + \begin{bmatrix} 5/7 \\ 1\,3/7 \\ 2\,1/7 \end{bmatrix}$$

CHAPTER XI

NON-NEGATIVE SQUARE MATRICES

11.1 The properties of all-positive square matrices

Theorems concerning non-negative square matrices come in stronger and weaker forms, and not surprising, the strongest theorems apply to matrices which are positive non-zero in each element.

The matrix algebra of nonnegative square matrices has been pioneered by the German mathematician Frobenius, who we cite in this section primaraly on account of [12].

We shall use the notation $\gg$ (two $>$ signs), to indicate that a vector, or a matrix is greater than another vector (or matrix), in every element. The meaning of the symbol $\ll$ is analogous. The single '$>$' symbol, when applied to a vector or a matrix, then means $\geq$ in all elements, $\neq$ in at least one element.

If $\mathbf{A}^p \gg [0]$ applies for some p, we refer to $\mathbf{A}$ as <u>semipositive</u>.

The further theorem that if $\mathbf{A}$ is not semi-positive, then some power $\mathbf{A}^p$ of $\mathbf{A}$ is block-diagonal, is also due to Frobenius [13].

The result of section 11.5, identifying the structure of a matrix which is non-negative and indecomposable, but not semi-positive, is due to Wielandt [48]. The results of both Wielandt and Frobenius, have been brought into the English language, by Debreu and Herstein [9].

One point of notation is useful to mention at this stage: Unless the opposite is stated, the symbol λ (without index) is reserved, throughout this chapter, for the biggest real root of $\mathbf{A}$.

We now need to lay the groundwork, and establish the properties of matrices consisting solely of positive non-zero elements.

Theorem:

It the n by n matrix **A** obeys the condition **A** $\gg$ [0],

then:

- 1) There is a root λ of **A**, which is:

  1a) real and $\lambda > 0$, and

  1b) greater than any other real root of **A**, and

  1c) greater than the absolute value of any negative root of **A**,

  1d) greater than the absolute value of any complex root of **A**, and for which:

  1e) $\partial|I\gamma - A|/\partial\gamma > 0$ applies for $\gamma \geq \lambda$, and

  1f) $|I\gamma - A| > 0$ applies for $\gamma > \lambda$, and

  1g) If γ increases towards $+\infty$, $|I\gamma-A|$ also goes to $+\infty$.

  1h) For $\gamma > \lambda$, $[I\gamma-A]^{-1} \gg [0]$ and $[I-\gamma^{-1}A]^{-1} \gg [0]$ exist, and if γ approaches λ from the $\gamma > \lambda$ side, each element of $[I\gamma-A]^{-1}$, and of $[I-\gamma^{-1}A]^{-1}$, will go to $+\infty$.

  1i) $|I\gamma - A_{ii}| > 0$ applies for $\gamma \geq \lambda$, for all principal minors.

  1j) $\partial|I\gamma - A_{ii}|/\partial\gamma > 0$ applies for $\gamma \geq \lambda$, for all principal minors.

  1k) $\partial[I-\gamma^{-1}A]^{-1}/\partial\gamma \ll [0]$ applies for $\gamma > \lambda$

  1l) $\partial[I-\gamma^{-1}A_{ii}]^{-1}/\partial\gamma \ll [0]$ applies for $\gamma > \lambda$, for all principal minor matrices.

- 2) λ is a distinct (non-repeated) root of the characteristic equation of **A**

- 3) The corresponding vector **x** can be written as **x** $\gg$ [0].

- 4) $\partial\lambda/\partial a_{ij} > 0$ (i = 1, 2 ..... n, j = 1, 2, ... n)

Proof:

We introduce the provisional assumption, that the theorem is true
for matrices of orders 1 by 1 to n-1 by n-1.

We now partition $[I\delta - A]$, as follows:

$$[I\delta - A] = \left[\begin{array}{c|c} I\delta - A_{11} & -a_{1,n} \\ \hline -a'_{n,1} & \delta-a_{n,n} \end{array}\right] \tag{11.1.1}$$

By the theorem, assumed true for A_{11}, there is a number $\phi > 0$,
for which we can express, for $\delta > \phi$, $|I\delta-A|$ as:

$$|I\delta-A| = |I\delta-A_{11}|(\delta -a_{nn} -a'_{n,1}[I\delta-A_{11}]^{-1}a_{1n}) \tag{11.1.2}$$

By 1h), assumed true for A_{11}, $|I\delta-A|$, as expressed by (11.1.2),
goes to $+\infty$, if δ does so.

Furthermore, if δ approaches ϕ, the similar root of A_{11} from the
$\delta > \phi$ side, then by 1h), assumed true for A_{11}, the expression
within brackets ( ) on the righthand side of (11.1.2) will
approach $-\infty$, while the factor $|I\delta-A_{11}|$ outside the ( ) will
stay positive (and approach $+\infty$). Since $|I\delta-A|$ is a continuous
function of δ, a real root of $|I\delta-A| = 0$, must lie between $\delta = \phi$,
and $\delta = +\infty$.

q.e.d. ad 1a) and 1b), subject to the provisional assumption.

If, contrary to what the theorem states, λ were not greater than
the absolute value of some negative real root of A, then 1a) as
proved, subject to the p.a., would be contradicted for A^2.

q.e.d. ad 1c), subject to the provisional assumption.

Now suppose that, contrary to what the theorem states:

$$A v + ui = [v + ui] (\mu + \epsilon i) \tag{11.1.3}$$

were true, for $\delta = \mu + \epsilon i$, $|\mu + \epsilon| \geq \lambda$ $(\epsilon \neq 0)$

Then $[I\delta - A_{11}]^{-1}$ permits development by the power series:

$$[I\delta - \mathbf{A}_{11}]^{-1} =$$

$$= I\delta^{-1} + \delta^{-2}\mathbf{A}_{11} + \delta^{-3}\mathbf{A}_{11}{}^2 + \delta^{-4}\mathbf{A}_{11}{}^3 + \ldots \ldots \qquad (11.1.4)$$

Furthermore, $|I\delta - \mathbf{A}|$ could be expressed as:

$$|I\delta - \mathbf{A}| =$$

$$|I\delta - \mathbf{A}_{11}| \left(\delta - a_{nn} - \mathbf{a'}_{n1}[I\delta - \mathbf{A}_{11}]^{-1}\mathbf{a}_{1n} \right) \qquad (11.1.5)$$

And, if δ were indeed a root, the righthand side of (11.1.5) would need to vanish.

Note, that (11.1.4) and (11.1.5) also need to hold for $\delta = \lambda$.

Therefore:

$$a_{nn} = \lambda - \mathbf{a'}_{n1}[I\lambda - \mathbf{A}_{11}]^{-1}\mathbf{a}_{1n} \qquad (11.1.6)$$

and the similar relationship must hold for other roots.

Therefore, if δ were a root:

$$a_{nn} + \mathbf{a'}_{n1}[I\delta - \mathbf{A}_{11}]^{-1}\mathbf{a}_{1n} =$$

$$a_{nn} + \mathbf{a'}_{n1}[I\lambda - \mathbf{A}_{11}]^{-1}\mathbf{a}_{1n} = \lambda \qquad (11.1.7)$$

However, (11.1.4) and (11.1.6) imply, on account of the arithmetic of absolute values and complex numbers, for $|\delta| \geq \lambda$:

$$a_{nn} + \mathbf{a'}_{n1}[I\delta\mathbf{A}_{11}]^{-1}\mathbf{a}_{1n}\delta^{-1} < \lambda \qquad (11.1.8)$$

except for $\delta = \lambda$, when (11.1.7) will apply.

(See also section 10.6 -de Moivre's Theorem-), where it may be seen that the real part of some terms in (11.1.4) will be negative for a complex number δ, (or a negative real δ).

Therefore:

$$|\delta| < \lambda, \text{ except for } \delta = \lambda$$

q.e.d. ad 1d), subject to the provisional assumption.

Differentiation of (11.1.2) with respect to δ leads to:

$$\partial\delta/\partial|I\delta-A| = |I\delta-A_{11}| \, (1 - a'_{n1}[\partial[I\delta-A_{11}]^{-1}/\partial\delta]a_{1n})$$

$$+ \partial|I\delta-A_{11}|/\partial\delta \, (\delta - a_{nn} - a'_{n1}[I\delta-A_{11}]^{-1}a_{1n}) \qquad (11.1.9)$$

and 1j) and 1l) are seen to imply 1e).

q.e.d. ad 1e), subject to the provisional assumption.

Since 1e) and the definition of λ as a root imply 1f), we now conclude:

q.e.d. ad 1f), subject to the provisional assumption.

The first part of 1h) is stated by the expansion theorem, and the presence of δ itself in (11.1.2), and 1e) now implies 1h) q.e.d. for 1h).

Property 1k) now is implied by (11.1.4).
q.e.d. ad 1k), subject to the p.a.

The remaining properties listed under 1) restate properties for which a proof was supplied for **A**, for its principal minors. Since all of 1) reduces to the true statement $a_{11} > 1$ for n=1, the general proof of all of 1) now follows by recursive induction.

q.e.d. ad 1) (independent of the remaining properties)

2) is implied by 1i), given that a repeated root requires the sum of the principal minors of $[I\lambda-A]$ to vanish.

q.e.d. ad 2)

Concerning 3), we express the relation between λ, **A** and **x** in partitioned form, as far as the leading block-row is concerned, as follows:

$$[I\lambda - A_{11}]x_1 \quad = \quad a_{1n}x_n \qquad\qquad (11.1.10)$$

Therefore, for $x_n = 1$, considering 1h) as proved:

$$x_1 = [I\lambda - A_{11}]^{-1}a_{1n} \gg [0] \qquad (11.1.11)$$

q.e.d. ad 3)

Concerning 4) :

Recall the differentiation of a root with respect to the elements of the matrix (section 8.10), and consider 1) and 3) as proved, and the generalization of 3) to A', and therefore $w' \gg [0]'$.

If the vector $x \gg [0]$ and $w' \gg [0]'$ are normalized to the requirement $w'x = 1$, the partial derivative of the root relative to any element of the matrix is expressed as:

$$\partial\lambda/\partial a_{ij} = w_i \; \lambda \; x_j > 0 \qquad (11.1.12)$$

q.e.d. ad 4)

q.e.d.

We are now also in a position to say something about $[I\delta-A]^{-1}$, for $\delta < \lambda$. We partition $[I\delta-A]$, as follows:

$$[I\delta-A] = \left[\begin{array}{c|c} I\delta-A_{11} & -a_{1n} \\ \hline -a'_{n1} & \delta-a_{nn} \end{array} \right] \qquad (11.1.13)$$

If both $[I\delta-A_{11}]^{-1}$ and $[I\delta-A]^{-1}$ exist, we develop $[I\delta-A]^{-1}$ by blockinversion as:

$$[I\delta-A]^{-1} = \left[\begin{array}{c|c} [[I\delta-A]^{-1}]_{11} & [I\delta-A_{11}]^{-1}a_{1n}\phi^{-1} \\ \hline \phi^{-1}a'_{n1}[I\delta-A_{11}]^{-1} & \phi^{-1} \end{array} \right] \qquad (11.1.14)$$

where:

$$\phi = \delta - a_{nn} - a'_{n1}[I\delta-A_{11}]^{-1}a_{1n} \qquad (11.1.15)$$

For $\lambda_1 < \delta < \lambda$, the above theorem, applied to A_{11}, states:

$$[I\delta-A_{11}]^{-1} \gg [0] \qquad (11.1.16)$$

Since for $\emptyset=0$, $\gamma=\lambda$ would apply, and for $\emptyset>0$, $\gamma>\lambda$ is implied, we infer $\emptyset < 0$, and hence, if $\lambda_1<\gamma$ applies for all simultaneous reorderings:

$$[I\gamma-\mathbf{A}]^{-1} << [0] \tag{11.1.17}$$

11.2 Of semi-positive and non-negative matrices

Frobenius' second paper, on non-negative square and indecomposable matrices [12], contains the phrase that this class of matrices shares 'almost all' the properties of positive matrices.

That is a fair comment, but identification of the types of matrices for which some of the properties of all-positive matrices do not hold, is useful, both in the interest of replicating the proof from the previous section to other types of matrices, and for its own sake, i.e. to show that for a broad class of non-negative and indecomposable matrices 'almost all' is even more 'almost all'.

We shall be using the term <u>semipositive</u>, for a non-negative (and square) matrix **A**, for which for some k, $\mathbf{A}^k >> [0]$ is true.

An example of a matrix, which is non-negative and indecomposable, but not semi-positive, is the <u>permutoid</u>:

Recall (from section 5.4), the definition of a permutoid:

The n by n matrix **A** is a permutoid, if, (for some simultaneous reordering of its rows and columns):

$$a_{1n} \neq 0, \text{ and } a_{i-1,i} \neq 0 \ (i = 2, 3, \ldots\ldots n) \text{ is true,}$$

is possible, while all other elements are zeros.

Our interest in this section is in matrices where there are not just n non-zero elements, but more specifically n positive elements, and we shall speak of a positive permutoid. It should also be clear, that we are concerned with <u>indecomposable</u> matrices. The term 'positive permutoid' should therefore be understood as implying indecomposability. For a positive permutoid, we develop the characteristic equation by recursive substitution, in association with a vector, for which $x_n = 1$ is required. As this is an invertible matrix, we can simplify the algebra, by writing the relation between the vector and the root as:

$$[I - \lambda^{-1}A] \; x = [0] \tag{11.2.1}$$

Otherwise, our analysis is a straight generalization of section 8.6, where we discussed the roots and vectors of the indecomposable permutation operator.

We begin with requiring:

$$x_n = 1$$

and recursively develop the consistent values for the other elements of the vector:

$$x_1 = a_{1n}\lambda^{-1}x_n = a_{1n}\lambda^{-1}$$

$$x_2 = a_{21}\lambda^{-1}x_1 = a_{21}a_{1n}\lambda^{-2} \; x_n = a_{21}a_{1n}\lambda^{-2}$$

$$\ldots\ldots\ldots$$

$$x_n = a_{n,n-1}\lambda^{-1}x_{n-1,n-2} = a_{1n} \; a_{21} \; a_{32} \; \ldots\ldots \; a_{n,n-1} \; \lambda^n \; x_n$$

This recursive development of the vector identifies both the vector and the characteristic equation:

$$x_n = 1$$

can be set by arbitrary requirement, and we can solve the other elements of the vector, conform:

$$x_k = \lambda^k \; a_{1,n} \prod_{j=1}^{k} a_{j+1,j} \tag{11.2.2}$$

The characteristic equation of this matrix therefore is:

$$\lambda^n = |-A| \qquad (11.2.3)$$

where $|-A|$ is the recursive product of the non-zero elements of A and is always positive non-zero.

We now find A to be similar to λP.

This is shown by transforming the vector into a summation-vector

From

$$A \, x = x \lambda \qquad (11.2.4)$$

we derive

$$[\hat{x}]^{-1} A \, x = [\hat{x}]^{-1} A \, [\hat{x}] \, s = s \, \lambda \qquad (11.2.5)$$

The transformation according to (11.2.5) leaves the distinction between zero and non-zero elements unaffected, and a matrix which has only one non-zero element in each row and a summation-vector as column vector, must have equal non-zero elements:

$$[\hat{x}]^{-1} A \, [\hat{x}] = \lambda \, P \qquad (11.2.6)$$

where P is an indecomposable permutation operator, ordered in lower semi-diagonal form.

Therefore:

$$A = \lambda \, [\hat{x}] \, P \, [\hat{x}]^{-1} \qquad (11.2.7)$$

Because of the way in which a permutation operator of the type we are dealing with (indecomposable, ordered in lower semi-diagonal form), moves the columns of any matrix to the left, except for the leading one, which becomes the nth, P has the following property:

$$\sum_{k=1}^{n} P^k = \sum_{k=0}^{n-1} P^k = E \qquad (11.2.8)$$

where E is the matrix of n^2 unity elements.

It follows that, although a positive permutoid is not semi-positive, it nevertheless shares with the all-positive matrix the property, that for $\gamma > \lambda$:

$$[I - \gamma^{-1}A]^{-1} \gg [0] \quad \text{and} \quad [I\gamma - A]^{-1} \gg [0]$$

is true (for a positive permutoid).

Theorem

If the n by n matrix A has a real dominant root $\lambda > 0$, and corresponding vectors $x \gg [0]$ and $w' \gg [0]$,

then A is semipositive.

Proof:

We now express A as:

$$A = R + x \lambda w' \qquad (11.2.9)$$

where x and w' are, respectively, the column- and row-vector, associated with λ as root, and have been made to fit the requirement:

$$w'x = 1 \qquad (11.2.10)$$

The rows of $R = [A - x \lambda w']$ are orthogonal to x, as may be seen by evaluating:

$$A x = \lambda x = R x + x \lambda w' x = R x + x\lambda \qquad (11.2.11)$$

For similar reasons, the columns of R are orthogonal to w'

Therefore:

$$A^2 = [R + x \lambda w']^2 =$$

$$R^2 + [R x \lambda w'] + [x \lambda w' R] + [x \lambda w']^2 =$$

$$R^2 + [x \lambda w']^2 = R^2 + [x \lambda w'] \qquad (11.2.12)$$

and by recursion:

$$A^k = R^k + [\mathbf{x}\,\lambda\,\mathbf{w'}] \tag{11.2.13}$$

Furthermore, for any other root of **A** we have, on postmultiplication of (11.2.13), by the corresponding vector $\mathbf{x_j}$, (regardless of whether or not there are n distinct vectors):

$$\mathbf{A}\,\mathbf{x_j} = \mathbf{x_j}\lambda_j = R\,\mathbf{x_j} + [\mathbf{x}\,\lambda\,\mathbf{w'}]\,\mathbf{x_j} \tag{11.2.14}$$

Since, by assumption $\lambda_j < \lambda$, $\mathbf{x_j}$ is orthogonal to $\mathbf{w'}$, and (11.2.14) reduces to:

$$\mathbf{A}\,\mathbf{x_j} = \mathbf{x_j}\lambda = R\,\mathbf{x_j} \tag{11.2.15}$$

and R is seen to have the same vectors as **A**, except for the reduction of one root to zero.

We therefore infer from (11.2.13), that, for increasing k,

A^k will converge to $[\mathbf{x}\,\lambda^k\,\mathbf{w'}]$

q.e.d.

Note, that this theorem is not specifically restricted to non-negative and/or indecomposable matrices.

We also state the following:

Corollary:

If **A** is semi-positive, then **A** has a distinct real root $\lambda > 0$, which dominates all other roots.

(If $A^k \gg [0]$ is true, then A^k has a root λ^k, which has those characteristics, but $|\lambda_i| = \lambda$ would imply $|\lambda_i{}^p| = \lambda^p$.)

Theorem

If the k by k diagonal block $\mathbf{A_{11}}$ of the n by n non-negative and indecomposable matrix **A** is semipositive, then **A** is semipositive.

Proof:

We <u>denote</u>:

an arbitrary power of $\mathbf{A}$, as:

$\mathbf{A}^b = \mathbf{B}$, and its next power as $\mathbf{A}^{b+1} = \mathbf{AB} = \mathbf{BA} = \mathbf{C}$, and:

for $\mathbf{A}_{11}{}^p \gg [0]$, $\mathbf{A}^p = \mathbf{P}$, and $\mathbf{A}^{p+1} = \mathbf{A}^q = \mathbf{Q}$,

$\mathbf{A}^{p+2} = \mathbf{QA} = \mathbf{AQ} = \mathbf{R}$, and $\mathbf{A}^{p+3} = \mathbf{RA} = \mathbf{AR} = \mathbf{U}$.

Since $\mathbf{C}_{11} = \mathbf{A'}_1\mathbf{B}_1 = \mathbf{B'}_1\mathbf{A}_1 = \mathbf{A}_{11}\mathbf{B}_{11} + \mathbf{A}_{12}\mathbf{B}_{22} =$

$$\mathbf{B}_{11}\mathbf{A}_{11} + \mathbf{B}_{12}\mathbf{A}_{22}, \text{ we infer:}$$

$\mathbf{C}_{11} \geq \mathbf{A}_{11}\mathbf{B}_{11}$, and hence by recursive induction:

$\mathbf{P}_{11} \geq \mathbf{A}_{11}{}^p \gg [0]$.

Therefore $\mathbf{A}_{11}{}^p \gg [0]$ implies $\mathbf{P}_{11} \geq \mathbf{A}_{11} \gg [0]$.

We now consider the following cases:

Case 1): $k=n-1$.
Then $a_{1n} \neq [0]$, and $a'_{n1} \neq [0]$, as otherwise $\mathbf{A}$ would be decomposable, and therefore:

$q_{1n} = \mathbf{P}_{11}a_{1n} + p_{1n} \geq \mathbf{P}_{11}a_{1n} \gg [0]$

as well as:

$q'_{n1} = a'_{n1}\mathbf{P}_{11} + a_{nn}p'_{n1} \geq a'_{n1}\mathbf{P}_{11} \gg [0]$

and therefore also:

$r_{nn} = q'a_{1n} + q_{nn}a_{nn} \geq q'a_{1n} > 0$.

q.e.d. for case 1.

Case 2): $k=n-2$

$$
A = \begin{bmatrix}
A_{11} & a_{1,n-1} & a_{1,n} \\
a'_{n-1,1} & a_{n-1,n-1} & a_{n-1,n} \\
a'_{n,1} & a_{n,n-1} & a_{n,n}
\end{bmatrix}
$$

Since case 1 does not depend on the decomposability of A as such, but on $a_{1n} \neq [0]$, $a'_{n1} \neq [0]$, that situation, if present in case 2, would imply a proof by recursive induction, and the similar conclusion applies to $a_{1,n-1}$ and $a'_{n-1,1}$.

The remaining substantive part of case 2 therefore is:

$$
A = \begin{bmatrix}
A_{11} & a_{1,n-1} & \\
 & a_{n-1,n-1} & a_{n-1,n} \\
a'_{n,1} & a_{n,n-1} & a_{n,n}
\end{bmatrix}
$$

where $a_{1,n-1} > [0]'$ needs to be assumed, on account of the indecomposability of A.

Therefore, for $Q = PA$:

$$q_{1,n-1} = P_{11}\, a_{1,n-1} + p_{1,n-1}\, a_{n-1,n-1} + p_{1,n}\, a_{n,n-1} \geq$$

$$P_{11}\, a_{1,n-1} \;\gg\; [0]$$

and by implication, using $Q_{11} \gg [0]$, $r_{1,n-1} \gg [0]$.

Similarly, from $R = QA$:

$$r_{1,n} = q_{1,n-1}a_{n-1} + q_{1,n}a_{n,n} \geq q_{1,n-1}a_{n-1} \gg [0]$$

The similar results for $r'_{n-1,1} \gg [0]'$ and $r'_n \gg [0]'$ are analogous, using $Q=AP$ and $R=AQ$.

The diagonal elements $u_{n-1,n-1} > 0$ and $u_{n,n} > 0$ now follow from either $U = RA$ or $U = AR$.

q.e.d. for case 2.

Case 3): $k < n-2$, $a_{1,i} > [0]$.
Since case 1 is not dependent on the indecomposability of **A** as such, only on $a_{1,n} \neq [0]$, $a'_{n,1} \neq [0]$, it generalizes to a k+1 by k+1 block under case 3, providing a proof by recursive induction.

q.e.d. for case 3.

That leaves: Case 4)

$$
\mathbf{A} \;=\; \begin{bmatrix} \mathbf{A}_{11} & \mathbf{A}_{12} & \\ \hline & \mathbf{A}_{22} & \mathbf{A}_{23} \\ \hline \mathbf{A}_{31} & \mathbf{A}_{32} & \mathbf{A}_{33} \end{bmatrix}
$$

where $\mathbf{A}_{22}$ and $\mathbf{A}_{33}$ are square.

Since $\mathbf{A}_{31} = [0]$, and/or $\mathbf{A}_{23} = [0]$ would imply the decomposability of **A**, we have for some i and j:

$$
\mathbf{A^*}_{22} \;=\; \begin{bmatrix} \mathbf{A}_{1,1} & a_{1,j} & \\ \hline & a_{j,j} & a_{j,i} \\ \hline a'_{i,1} & a_{i,j} & a_{i,i} \end{bmatrix} \qquad (a'_{i,1} \neq [0], \quad a_{1,j} \neq [0])
$$

Since case 2 does not depend on the indecomposability of **A** as such, but on $a_{1,n-1} \neq [0]$ and $a'_{n,1} \neq [0]$, case 2 is applicable to $\mathbf{A^*}_{22}$, and a proof by recursive induction follows.

q.e.d. for case 4

q.e.d.

11.3 Of input-output matrices

We shall refer to a non-negative square matrix of which all the roots are $|\lambda_i| < 1$, as an input-output matrix. This term obviously relates to the uses of such matrices in inter-industry analysis, but it should be made clear, that we are here using this term as a generic mathematical term, and that any theory developed with the help of its use, is valid in other fields of application.

There is some degree of analogy between input-output matrices, and positive definite matrices. Note, however, that input-output matrices do not have to be symmetric.

If **A** is an input-output matrix, then [I-A] can be inverted by row-operations, taking positive pivots along the main diagonal only.

Example (from Chapter I):

$$
\begin{array}{cccccccc}
x_1 & x_2 & x_3 & = & y_1 & y_2 & y_3 & \Sigma \\
\end{array}
$$

$$
\begin{bmatrix}
|0.867| & -0.067 & - \\
-0.150 & 0.850 & - \\
-0.167 & -0.100 & 1
\end{bmatrix}
\begin{bmatrix}
1 & - & - \\
- & 1 & - \\
- & - & 1
\end{bmatrix}
\begin{array}{c}
1.800 \\
1.700 \\
1.733
\end{array}
$$

$$
\begin{array}{cccccccc}
x_1 & x_2 & x_3 & = & y_1 & y_2 & y_3 & \\
\end{array}
$$

$$
\begin{bmatrix}
1 & -0.077 & - \\
- & |0.838| & - \\
- & -0.113 & 1
\end{bmatrix}
\begin{bmatrix}
1.153 & - & - \\
0.173 & 1 & - \\
0.193 & - & 1
\end{bmatrix}
\begin{array}{c}
2.076 \\
2.011 \\
2.080
\end{array}
$$

etc.

Input-output analysis is an obvious field of application of the theory of non-negative square matrices. However, actual input-output matrices generally contain large numbers of zeros, and usually are decomposable, by way of containing zero rows and/or zero columns. We need to develop the theory of non-negative square matrices, whether positive or indecomposable or not, more generally.

Theorem:

If **A** is an n by n non-negative matrix,

then:

● 1) There is a root λ of **A**, which is:

1a) real and $\lambda \geq 0$, and

1b) not smaller than the any other root of **A**, and

1c) not smaller than the absolute value of any negative root of **A**, and

1d) not smaller than the absolute value of any complex root of **A**, and for which:

1e) $\partial|I\delta - A|/\partial\delta \geq 0$ applies for $\delta \geq \lambda$, and
$\partial|I\delta - A|/\partial\delta > 0$ applies for $\delta > \lambda$

1f) $|I\delta - A| > 0$ applies for $\delta > \lambda$, and

1g) If δ increases towards $+\infty$, $|I\delta-A|$ also goes to $+\infty$.

1h) For $\delta > \lambda$, $[I\delta-A]^{-1} > [0]$ and $[I-\delta^{-1}A]^{-1} > [0]$ exist.

1i) $|I\delta - A_{ii}| > 0$ applies for $\delta > \lambda$, and
$|I\delta - A_{ii}| \geq 0$ applies for $\delta \geq \lambda$, for all principal minors.

1j) $\partial|I\delta - A_{ii}|/\partial\delta > 0$ applies for $\delta > \lambda$, for all principal minors.

1k) $\partial[I-\delta^{-1}A]^{-1}/\partial\delta < [0]$ applies for $\delta > \lambda$

1l) $\partial[I-\delta^{-1}A_{ii}]^{-1}/\partial\delta < [0]$ applies for $\delta > \lambda$, for all principal minor matrices.

● 2) ******** There is no equivalent of 2) as stated for **A** $\gg$ [0]

● 3) The corresponding vector can be written as **x** > [0].

● 4) ******** We do not state a weaker equivalent of 4) as stated for **A** $\gg$ [0] at this stage.

Summary of the proof:

Concerning 1):

As in section 11.1, we express $|I\delta - A|$ as:

$$|I\delta-A| \; = \; |I\delta-A_{11}| \; (\delta - a_{nn} - a'_{n,1}[I\delta-A_{11}]^{-1}a_{1n}) \qquad (11.1.2)$$

Development of $[I\delta - A_{11}]^{-1}$ by the power series implies:

$$a'_{n,1}[I\delta-A_{11}]^{-i}a_{1n} \quad \geq \quad 0$$

for $\delta > \lambda_1$, where λ_1 is the biggest root of A_{11}.

Due to the possibility that A, as well as the two vectors, contain zeros, we cannot exclude the possibility that $a'_{n,1}A_{11}{}^{p}a_{1,n}$ will apply for all p, and $a'_{n,1}[I\delta-A_{11}]^{-1}a_{1n} = 0$ will hold in that case.

Then (11.1.2) needs to be written as:

$$|I\delta-A| \; = \; |I\delta-A_{11}| \; (\delta - a_{nn}) \qquad (11.3.1)$$

and the roots of A are those of A_{11}, and a_{nn}.

(As $|I\delta-A|$, $|I\delta-A_{11}|$, and $(\delta-a_{nn})$ are all continuous functions of λ, (11.3.1) then also applies for $\delta=\lambda$.)

If a particular ordering which places the index i in position n conforms to (11.3.1), proofs of 1a) to 1g), and of 1j), then immediately follow for that index i.

If, for a particular ordering $a'_{n,1}[I\delta^{-1}-A_{11}]a_{1n} > 0$ applies, then the proofs from section 11.1 replicate for the index i which that ordering puts in position n (with some of the stated properties applying in the stronger forms stated in section 11.1).

Proofs of 1h), 1k), and 1l) follow on developing $[I-\delta^{-1}A]$ by the power series, whether (11.3.1) applies, or (11.1.2) with a non-trivial term $a'_{n,1}[I\delta-A_{11}]a_{1n} > 0$.

q.e.d. ad 1)

Concerning 3):

If $\lambda > \lambda_1$ applies for at least one ordering and a corresponding principal minor block A_{11}, the proof from section 11.1 for the $A \gg [0]$ case $x_n = 1$, $x_1 = [I\lambda - A_{11}]^{-1} a_{1n} > [0]$, replicates, but a separate proof is needed if $|I\lambda - A_{11}| = 0$ applies for all i.

In the latter case we may assume, for at least one ordering $w_1 \neq 0$, as well as $|I\lambda - A_{11}| = 0$. ($w' \neq [0]'$ is part of the definition of a characteristic vector).

Therefore $[I\lambda - A_{11}]v_1 = [0]$ implies $-a'_{n1}v_1 = 0$, and $v_1 > [0]$, which is by the theorem, assumed true for A_{11} a vector of A_{11}, can be completed with a zero in position n, to become a complete vector of A, and a proof by recursive induction follows.

q.e.d. ad 3)

q.e.d.

Theorem

A is an input-output matrix ($|\lambda_i| < 1$, i = 1, 2, ... n), if and only if $[I - A]$ can be inverted by means of positive pivots along the main diagonal.

Proof:

Property 1f) above states $|I-A| > 0$, and 1i) states $|I-A_{ii}| > 0$ for all the principal minors of an input-output matrix.

These properties both ensure the presence of positive pivots, and would be contradicted by their absence.

q.e.d.

Theorem:

If, for the n by n matrix $\mathbf{A} \geq [0]$,

$$\mathbf{f} = \mathbf{t} - \mathbf{A}\,\mathbf{t} \gg [0] \qquad\qquad (11.3.2)$$

applies, for $\mathbf{t} \geq [0]$

then $\mathbf{A}$ is an input-output-matrix.

Proof:

On pre-multiplication of (11.3.2) by $\mathbf{w}' > [0]'$, we obtain:

$$\mathbf{w}'\mathbf{f} = \mathbf{w}'\mathbf{t} - \mathbf{w}'\mathbf{A}\,\mathbf{t} = (1-\lambda)\mathbf{w}'\mathbf{t} > 0 \qquad\qquad (11.3.3)$$

(Input-output-analysis interpretation: $\mathbf{f}$ is the vector of final output, $\mathbf{t}$ is the vector of total output per industry, which needs to be non-negative, and $\mathbf{f} \gg [0]$ is sufficient to ensure that $\mathbf{A}$ meets the Hawkins-Simon [19] conditions. See also myself [22], chapter III.)

The similar theorem for the value added per unit of output vector:

$$\mathbf{v}' = \mathbf{p}' - \mathbf{p}'\mathbf{A} \gg [0]', \quad \mathbf{p}' \geq [0]'$$

where $\mathbf{v}'$ is the value added per unit of output, and $\mathbf{p}'$ is the price-vector, will be obvious.

11.4 The positive vector property

If $\mathbf{A}$ is non-negative and indecomposable, we have the following

Theorem (positive vector theorem)

If the n by n non-negative matrix is indecomposable,

then:

● 1) There is a root λ of **A** which is:

1a) real and $\lambda > 0$

1i) $|I\delta - A_{ii}| > 0$ applies for all principal minor matrices, for $\delta \geq \lambda$ (and not just $\delta > \lambda$ as with all non-negative matrices), and λ exceeds the absolute value of any root of A_{ii}.

● 2) λ is a distinct (non-repeated) root of **A**

● 3) The corresponding vector can be written as **x** $\gg$ [0]

● 4) $\partial\lambda/\partial a_{ij} > 0$ (i = 1, 2, ... n, j = 1, 2, ... n)

(And obviously the other properties listed under 1) for all non-negative matrices apply).

Proof:

Concerning 1a) and 3):

By the weaker form of 3) as proved for all non-negative matrices, we may require **x** > [0].

Now suppose by contra-assumption to 1a) and/or 3):

$$A = \left[\begin{array}{c|c} A_{11} & A_{12} \\ \hline A_{21} & A_{22} \end{array}\right] \left[\frac{\quad}{x_2}\right] = \left[\frac{\quad}{x_2}\right] \lambda \qquad (11.4.1)$$

where A_{11} is square and has been chosen to match all the zero elements of **x**, therefore $x_2 \gg [0]$.

Since an indecomposable matrix does not contain any zero columns, we must immediately exclude $\lambda = 0$

q.e.d. ad 1a)

For $\lambda > 0$, the leading block-row of (11.4.1) implies $A_{12} = [0]$, and therefore the decomposability of **A**. Therefore no non-trivial partitioning conform (11.4.1) may be assumed to exist.

q.e.d. ad 3).

Concerning 1i):

We introduce the provisional assumption that the theorem is true for smaller matrices, and therefore for all A_{ii}, if A has any indecomposable diagonal blocks A_{ii}.

If A has no indecomposable diagonal blocks, all diagonal blocks of A have an n_i-fold repeated zero root. (The roots of a decomposable matrix, whether non-negative or not, are those of its diagonal blocks.) No further proof is needed in that case.

Otherwise, it will be sufficient to provide a proof for the largest real root of any diagonal block of A, which is then also the largest real root of an n-1 by n-1 block A_{11}. (The combination of the present theorem with the more general but weaker theorem which applies to all square and non-negative matrices then implies dominance over any other roots of diagonal blocks of A.)

Now suppose that, contrary to what the theorem states, $\lambda_1 = \lambda$ were a root, of A_{11} as well as of A.

Since

$$[I\lambda - A_{11}]\ x = a_{1n}x_n = [0]$$

is known to be resolvable, z'_1, the row-vector of A_{11} associated with $\lambda_1 = \lambda$ as root, can be completed with a zero, to form a full row-vector of A.

This contradicts either 3) as proved, or else 2).

Now suppose that both 2) and 1i) were false. By the weaker form of the theorem as proved for all square and non-negative matrices, we can require $z'_1 > [0]'$. This would show the existence of two independent row-vectors of A:

$$z' = [z'_1,\ 0] > [0]' \quad \text{and} \quad w' \gg [0]'.$$

One of these would need to be orthogonal to $x \gg [0]$.

This is impossible, therefore $\lambda_i < \lambda$ needs to be assumed instead.

q.e.d. ad 1i)

Concerning 2):
Given that by 1i) as proved, $|I\lambda - A_{11}| > 0$ applies to all principal minors of $\mathbf{A}$, λ is a distinct root.
q.e.d. ad 2)

Concerning 4):
Given 3) $\mathbf{x} \gg [0]$, and by implication $\mathbf{w}' \gg [0]'$, the proof of section 11.1, using 3), is applicable.
q.e.d. ad 4)

q.e.d.

We are now in a position to state some limits on λ.

Similarity transformation of $\mathbf{A}$ into:

$$\mathbf{A}^* = [\hat{\mathbf{x}}]^{-1} \mathbf{A} [\hat{\mathbf{x}}] \tag{11.4.2}$$

and therefore:

$$\mathbf{A}^* \mathbf{s} = \mathbf{s} \lambda \tag{11.4.3}$$

now reveals:

$$\lambda = \sum_{j=1}^{n} a^*_{ij} = \sum_{j=1}^{n} a_{ij} (x_j / x_i) \qquad (i = 1, 2, \dots n)$$

$$\tag{11.4.4}$$

There will be at least one i, for which

$$x_j \geq x_i$$

will apply for all j

and therefore:

$$\lambda = \sum_{j=1}^{n} a_{ij} (x_j / x_i) \geq \sum_{j=1}^{n} a_{ij} \qquad \text{(some i)} \tag{11.4.5}$$

and similarly, for an i for which $x_j \leq x_i$ applies for all j:

$$\lambda = \sum_{j=1}^{n} a_{ij} (x_j / x_i) \leq \sum_{j=1}^{n} a_{ij} \quad (\text{some } i) \qquad (11.4.6)$$

We can actually require (11.4.5) and (11.4.6) to hold in the stronger '>' and '<' forms, except in the limiting case that all x_i are equal. This is immediately obvious for the all-positive matrix, where at least one term in (11.4.5), resp. (11.4.6) will display that stronger relationship, but for an indecomposable matrix which contains zero elements a more formal analysis is needed.

This relates to the case where

$$A = \left[\begin{array}{c|c} A_{11} & A_{12} \\ \hline A_{21} & A_{22} \end{array}\right] \left[\frac{x_1}{s}\right] = \left[\frac{x_1}{s}\right] \lambda \qquad (11.4.7)$$

holds, where we would require either, in analogy to (11.4.5), $x_j > 1$ or, in analogy to (11.4.6), $x_j < 1$, $(j = 1, 2, \ldots n_1)$. In both cases, $A_{21} x_1 = [0]$ would imply the decomposability of A.

We therefore conclude that λ is a number between the biggest and the smallest row-total of the rows of A, and that, with the exception of all rowtotals being the same, λ will not actually be equal to either extreme.

Exercise:

For $A = \begin{bmatrix} 1 & 2 & 3 \\ 3 & 2 & 1 \\ 4 & 2 & 4 \end{bmatrix}$

find upper and lower limits for the Frobenius root. There is an answersheet at the end of the chapter.)

We can use the positive vector theorem to show that non-negative square and indecomposable matrices share a further common characteristic with all-positive matrices:

Theorem

If the n by n non-negative matrix **A** is indecomposable, then the biggest positive root λ of **A**, is in excess of any similar root of any minor-matrix $\mathbf{A}_{ii}$ of **A**.

Proof

Suppose by contra-assumption:

$$\mathbf{A} = \left[\begin{array}{c|c} \mathbf{A}_{11} & \mathbf{A}_{12} \\ \hline \mathbf{A}_{21} & \mathbf{A}_{22} \end{array}\right] \qquad \mathbf{Ax} = \mathbf{x}\lambda \qquad \mathbf{A}_{11}\mathbf{v}_1 = \mathbf{v}_1\lambda$$

If $\mathbf{A}_{11}$ was decomposable, and had no indecomposable diagonal blocks, then $\mathbf{A}_{11}$ would be lower-triangular with only zeros on the main diagonal, and therefore have an n_1-fold repeated zero root, and no root equal to λ. We may therefore assume, without loss of generality, that $\mathbf{A}_{11}$ is indecomposable. Therefore λ is a distinct root of $\mathbf{A}_{11}$ and $\mathbf{v} \gg [0]$ may be required.

11.5 Indecomposable matrices which are not semipositive

We shall refer to an n by n matrix, which partitions (or which permits simultaneous reordering) as:

$$\mathbf{A} = \left[\begin{array}{ccccccc} & & & \cdots\cdots & & & \mathbf{A}_{1q} \\ \mathbf{A}_{21} & & & & & & \\ & \mathbf{A}_{32} & & & & & \\ & & \cdots\cdots\cdots\cdots & & & & \\ & & & & & \mathbf{A}_{q,q-1} & \end{array}\right]$$

where all the diagonal blocks are square, with the non-zero blocks in block-positions $1,q$ and $j+1,j$ ($j = 1, 2, \ldots q-1$), as a <u>blockpermutoid</u> (of block-order q).

It should be made clear that this term is specific to this book, in the absence of a recognized conventional term. There are other textbooks, which discuss the same type of matrix, but use a

different term for it. Minc [33] refers to a matrix which is here
called a permutoid as a generalized permutation matrix
(operator), but does not use any special term, even while (p.51)
the substantial result of this section is covered. Graham [17]
speaks of a cyclic matrix. It was felt that the term block-
permutoid conveys the definition rather more clearly.

The characteristic equation of a blockpermutoid is best developed
by block-recursion, as follows:

$$x_2 \, \lambda = A_{21} \, x_1$$

$$x_3 \, \lambda = A_{32} \, x_2 \qquad \rightarrow \qquad x_3 \, \lambda^2 = A_{32} \, x_2 \, \lambda = A_{32} \, A_{21} \, x_1$$

$$\cdots\cdots\cdots\cdots\cdots\cdots\cdots$$

$$x_q \, \lambda^{q-1} = \prod_{r=q-1}^{1} A_{r+1,r} \, x_1$$

$$x_1 \, \lambda = A_{1q} \, x_q \qquad \rightarrow \qquad x_1 \, \lambda^q = A_{1q} \prod_{r=q-1}^{1} A_{r+1,r} \, x_1$$

The similar recursive substitution can be started for any x_j, and
leads to the general formula:

$$x_j \, \lambda^q = \prod_{r=j-1}^{1} A_{r+1,r} \, A_{q,1} \prod_{r=q-1}^{j} A_{r+1,r} \, x_j \qquad (11.5.1)$$

Note the <u>falling</u> order of the indices with the 'Π' signs. For
ordinary multiplications, these would have been reversed, but
this is not valid for matrix multiplication. For j=1, the
lefthand of the two 'Π' recursive product expressions disappears
(as may be seen above), and for j=n the righthand of the two
recursive product expressions disappears.

Note that

$$Q_{jj} = \prod_{r=j-1}^{1} A_{r+1,r} \, A_{q,1} \prod_{r=q-1}^{j} A_{r+1,r} \, x_j \qquad (11.5.2)$$

are also the diagonal blocks of:

$$Q = A^q \qquad (11.5.3)$$

Theorem (excess elements theorem)

If the n by n matrix **A** has more than n non-zero elements,

then

either a): **A** has at least one non-vanishing principal minor.

or: b): **A** permits simultaneous re-ordering, to a form where

$$a_{ij} = 0 \text{ applies for all } j \geq i$$

(**A** is lower-triangular, with the non-zero elements only strictly <u>below</u> the diagonal.)

Proof:

For n=1 the theorem has no content.

For n=2, three non-zero element do indeed imply that at least one diagonal element is non-zero.

q.e.d. for n=2.

In the non-trivial n > 2 case, we begin by partitioning **A** as:

$$A = \begin{bmatrix} & & a'_{13} \\ a_{21} & A_{22} & A_{23} \\ & A_{32} & A_{33} \end{bmatrix} \qquad (11.5.4)$$

where $a_{11} = 0$ has been assumed, as otherwise a) applies by assumption.

If **A** is decomposable, (and hence permits re-ordering into lower block-triangular form), an assumption that the theorem is true for two separate n_j by n_j diagonal blocks, leads to a proof by recursive induction.

(Should one or more diagonal blocks of **A** be indecomposable, then n_j or more non-zero elements are contained in it. It is then either itself invertible as a permutoid, or it contains non-vanishing principal minors.)

If either the leading row or the leading column was a complete null vector, **A** is decomposable, and no further proof would be needed, but applicability of the theorem to an indecomposable matrix **A** requires (for some ordering of indices 2 to n:

$$a_{21} \neq [0], \quad a'_{13} \neq [0]'.$$

We will indicate the composite bottom righthand diagonal block as:

$$\mathbf{A^*}_{22} = \left[\begin{array}{c|c} \mathbf{A}_{22} & \mathbf{A}_{23} \\ \hline \mathbf{A}_{32} & \mathbf{A}_{33} \end{array} \right]$$

$$(11.5.5)$$

If $\mathbf{A^*}_{22}$ is an invertible permutoid a) applies by assumption, and no further proof is needed.

Otherwise, an assumption that the theorem is true for $\mathbf{A^*}_{22}$ implies that that we may assume $\mathbf{A^*}_{22}$ to contain at least one zero row or column or else no further proof would be needed (**A** contains either exactly n-1 elements and is an indecomposable permutoid, or less than n-1, and contains a zero row as well as a zero column, or more than n-1, and then the theorem, assumed true for $\mathbf{A^*}_{22}$, implies that $\mathbf{A^*}_{22}$, and hence **A**, has at least one non-vanishing principal minor, or else itself permits reordering into the lower-triangular form in which a zero row and column arise).

The following cases now arise:

Case 1): $\mathbf{A^*}_{22}$ contains a null row in the bottom block-row.
Then **A** is decomposable, and a proof by recursive induction arises.
q.e.d. for case 1.

Case 2) $\mathbf{A^*}_{22}$ contains a zero column in its leading block-column.
As case 1.
q.e.d. for case 2.

Case 3) $\mathbf{A^*}_{22}$ contains a zero row in its leading block-row.
Then we may, after assigning the index 1 to this row, and n to
the vector with the old index 1, assume that $\mathbf{a'}_{13}$ contains just
one non-zero element (at the end).

After such re-ordering, case 3 can only be assumed to persist as:

$$
\mathbf{A} = \left[\begin{array}{c|c|c} & & a_{1n} \\ \hline a_{21} & \mathbf{A}_{22} & \\ \hline & a'_{n2} & \end{array} \right]
\tag{11.5.6}
$$

q.e.d. for case 3 for n=3.

An assumption, that the theorem is true for $\mathbf{A}_{22}$, now implies the
lower triangularity of $\mathbf{A}_{22}$, with zeros on the diagonal.
(otherwise, $\mathbf{A}_{22}$ would either itself be a minor-matrix with a non-
vanishing determinant, or it would have non-vanishing principal
minors, which are also non-vanishing principal minors of $\mathbf{A}$.)

The following subcases of case 3 now arise:

Subcase 3a) $a_{ij} = 0$, for some $i = j+1$, but $a_{hj} \neq 0$, for $h > j+1$.

We can put the old h^{th} column and row in position i, and increase
indices i to h-1 all by 1. This procedure leaves indices 1 and n
in place, and respects the lower-triangular form of the composite
block which consists of the intersection of rows 2 to n with
columns 1 to n-1. Therefore, unless $\mathbf{A}$ contains null vectors (in
which case no further proof is needed), its recursive use ends
with:

Subcase 3b)

$$
\left. \begin{array}{l} a_{i+1,j} \neq 0, \ (i = 1, 2, \ \ldots \ n-1), \\[4pt] \text{and } a_{ij} = 0 \text{ for } j > i, \text{ except } a_{1n} \neq 0. \end{array} \right\}
\tag{11.5.7}
$$

If $a_{ij} = 0$ then also applies for all $i > j+1$, then $\mathbf{A}$ is an indecomposable permutoid (with exactly n elements), and no further

proof is needed.

Otherwise we have the following

Lemma:

$a_{hq} \neq 0$ (h > q+1) -more than exactly n non-zero elements-

and (11.5.7) together imply a non-vanishing principal minor.

Proof (of the lemma):

On removal of indices q+1 to h-1, inclusive, an invertible rump-block, with non-zero elements in the positions:

$a_{1,n-h+q+1} \neq 0$ (formerly $a_{1n} \neq 0$)

and

$a_{i+1,j} \neq 0$ (i = 1, 2, ... q-1), (positions unchanged)

            (i=q+1)   (formerly $a_{hq} \neq 0$)

and   (i = q+2, ... n-h+q)   (formerly i = h-1 to n)

(See the example below.)

(That block is invertible, irrespective of recursive induction, as may be seen by developing its determinant by the leading row, and recursive induction implies that, if it contains more than just n-h+1 elements, it also has non-vanishing principal minors.)

q.e.d. for the lemma.

q.e.d. for case 3.

Since A^*_{22} must be assumed to contain a zero row as well as a zero column, this exhausts all possible cases.

q.e.d.

Example of the re-ordering invoked by the lemma stated in the the above proof:

$$
A =
\begin{array}{c}
\begin{array}{ccccccc} 1 & 2 & 3 & 4 & 5 & 6 & 7 \end{array} \\
\left[
\begin{array}{ccccccc}
- & - & - & - & - & - & 1 \\
2 & - & - & - & - & - & - \\
- & 3 & - & - & - & - & - \\
- & - & 4 & - & - & - & - \\
- & 8 & - & 5 & - & - & - \\
- & - & - & - & 6 & - & - \\
- & - & - & - & - & 7 & -
\end{array}
\right]
\begin{array}{c} 1 \\ 2 \\ 3 \\ 4 \\ 5 \\ 6 \\ 7 \end{array}
\end{array}
$$

Here the various indices are: n=7, h=5, q=2. The indices to be removed therefore are from q+1=3, to h-1=4, i.e. 3 and 4, and the resulting minor-block is:

$$
A =
\begin{array}{c}
\begin{array}{ccccc} 1 & 2 & 5 & 6 & 7 \end{array} \\
\left[
\begin{array}{ccccc}
- & - & - & - & 1 \\
2 & - & - & - & - \\
- & 8 & - & - & - \\
- & - & 6 & - & - \\
- & - & - & 7 & -
\end{array}
\right]
\begin{array}{c} 1 \\ 2 \\ 5 \\ 6 \\ 7 \end{array}
\end{array}
$$

Because the line of non-zero elements is parallel to rather than <u>on</u> the diagonal, this operation systematically removes not just h-1-q = 2 non-zero elements, the number of removed vectors, but full h-q = 3 non-zero elements disappear:

Elements $a_{i,i-1}$ (i = q+1, ... h-1) go because they belong to the ith row, and elements $a_{i+1,i}$ (i = q+1, ... h-1) go because they belong th column i. This coincides with the requirement that rows q = 2 and h = 5 become adjoining, thus bringing the old element a_{hq} = 8 one row below the old element $a_{q,q-1}$ = 2, and that columns q = 2 and h=5 become adjoining, thus bringing the old element $a_{h+1,h}$ = 6 to be immediately to the right of the old element a_{hq} = 8.

The above theorem does not specifically refer to non-negative
matrices, but it was included in this chapter, mainly because of
the following application:

If the n by n non-negative and indecomposable matrix has no non-
vanishing principal minors,

then **A** is a positive permutoid.

(If there were less than n elements, **A** could not be indecomposable, with just n positive non-zero elements **A**, if free of null
vectors, is indeed a positive permutoid, and if there were more
than n non-zero elements, than **A** has at least one principal
minor).

Now note, that the proof of the above theorem was supplied,
strictly in terms of ordering.

This consideration gives rise to the following generalization:

Theorem (excess blocks theorem):

If **A** partitions into k block-rows and k block-columns, with
square diagonal blocks

then

either: a) The block-indices permit reordering in such a way that
            a diagonal (multi-) block appears, which does not
            permit internal reordering of its block-indices to a
            block-triangular multiblock.

or: b) The block-indices permit re-ordering in such a way
            that **A** is seen to be lower block-triangular, with null
            blocks on the diagonal.

(To generalise the proof, read '**A**$_{ij}$ > [0]' for 'a$_{ij}$ $\neq$ 0', and for
'non-vanishing minor', read 'square multi-block, not seen to be

decomposable by the specified block-structure'.)

Theorem (Consistent semipositive block theorem)

If any diagonal block Q_{jj} of a qth power $Q = A^q$ of the block-permutoid A is semi-positive, $[Q_{jj}]^p \gg [0]$,

then all the Q_{jj} are semi-positive.

Proof:

No Q_{jj} is semi-positive, unless all the blocks of A which figure in (11.5.2), are free of null vectors.

We may therefore complete any expansion of $[Q_{jj}]^p$, by adding (in total q) appropriate factors to the left and/or the right, to form any $[Q_{ii}]^{p+1} \gg [0]$.

q.e.d.

Recall also (from section 9.5), that if $A\,B$ and $B\,A$ both exist, and $A\,B\,y = y\,\lambda \neq [0]$ applies, then $[B\,A]$ has the same non-zero root, i.e. $B\,A\,v = B\,A\,[B\,y] = B\,[A\,B]\,y = y\,\lambda$ is implied, and hence $B\,A\,v = v\,\lambda$.

Application of this property to the recursive matrix-product given by (11.5.2), where we note that we obtain a different Q_{jj}, simply by taking a factor of the left of the recursive product and put it at the right instead, leads to the conclusion that if A is an indecomposable blockpermutoid of block-order k, than all the diagonal blocks of A^k have the same non-zero roots.

The non-zero roots of A are therefore the qth power roots of any Q_{jj}, which are the same for all Q_{jj}. Obviously, the larger diagonal blocks of Q, have one or more zero roots, which may not be roots of the smaller blocks, even while they are roots of A.

The rank of a blockpermutoid A, which is of block-order q, is q times the lowest of the ranks of any of its non-zero blocks.

The definition of a blockpermutoid matrix does not require

indecomposability: it merely states that the non-zero elements of **A** only occur in certain blocks. However, the previous theorem allows us to draw a clear distinction between blockpermutoids, for which the diagonal blocks of **A**^{q} are all semi-positive, and those were none of these blocks are semi-positive. If the diagonal blocks of **A**^{q} are semi-positive, then we will say that the blockpermutoid is block-semipositive.

The structure of a blockpermutoid severely limits the number of non-vanishing minors, and non-vanishing principal minors in particular.

The diagonal blocks are all null blocks. Therefore, to begin with, all principal minor blocks of order 1 by 1 (= single diagonal elements), are null blocks, and (for a block-permutoid of block-order q, any combination of n_j rows and n_j corresponding columns, needs to contain, for every row drawn from the jth block-row, a column drawn from a different block-column, because the jth block-row does not contain any non-zero elements in the jth block-column.

The blocks of which the determinants are the principal minors, then have a block-structure which is similar to that of a block-permutoid, except that we still have to clarify the requirement that the diagonal blocks must be square.

In fact, as far as minor-blocks which give rise to non-vanishing minors is concerned, all non-zero blocks of the minor-matrix must also be square, as otherwise, there is at least one block-column which has its number of non-zero rows as rank, but its number of non-zero columns as the smallest order-parameter. That is a condition, which implies square diagonal blocks as well.

If $n_2 > n_1$ applies for the two leading square diagonal blocks, then the 1,2 non-zero block of of a blockpermutoid contains n_2 rows, and fewer, viz. only n_1 non-zero columns, and since that are all the non-zero columns in the entire second block-row, the minor-block is non-singular, only if it does not just have a block-structure which is similar to a block-permutoid, but more specifically is one in the full sense of the the definition, inclusive the requirement that the diagonal blocks are square.

The orders of the non-vanishing principal minors are therefore all multiples of the block-order (or just equal to the block-order, i.e. permutoids).

Therefore, the characteristic equation of a blockpermutoid of block-order q is:

$$\sum_{j=0}^{r} c_j \lambda^{n-jr} = 0 \qquad\qquad (11.5.8)$$

i.e. contains only r+1 non-vanishing terms, including $c_0 = 1$ for λ^n (only r terms to be calculated), where r is the smallest rank of any minor-block, and $r \times q \leq n$ is the rank of the parent-matrix.

The separate terms in (11.5.8) relate to minor-blocks, which are of orders q by q, 2q by 2q, ... rq by rq.

Example:

$$
\mathbf{A} = \left[
\begin{array}{ccc|ccc|cc}
 &   &   &   &   &   & 2 & 3 \\
 &   &   &   &   &   & 4 & - \\
 &   &   &   &   &   & - & 9 \\
\hline
1 & 2 & 3 &   &   &   &   &   \\
4 & - & 5 &   &   &   &   &   \\
\hline
 &   &   & 6 & 7 &   &   &   \\
 &   &   & 8 & - &   &   &   \\
\end{array}
\right]
$$

If we begin with extracting the leading column, we note that the intersection of the leading column with the leading row, and indeed, with any row not drawn from the second block-row, is a zero element. We therefore also extract, let us say the leading row of the second block-row and the leading column of the second block-column. The resulting 2 by 2 minor-block is:

$$
\begin{array}{c}
 \\
\text{row 1} \\
\text{row 4}
\end{array}
\begin{array}{cc}
x_1 & x_4 \\
\left[\begin{array}{cc} - & - \\ 1 & - \end{array}\right]
\end{array}
$$

Only if we extract a row and a column from the third blockrow and the third block-column as well, an invertible minor-block, e.g.

$$
\begin{array}{c}
\text{row 1} \\
\text{row 4} \\
\text{row 6}
\end{array}
\begin{bmatrix}
\overset{x_1}{-} & \overset{x_4}{-} & \overset{x_6}{2} \\
1 & - & - \\
- & 6 & -
\end{bmatrix}
$$

could possibly arise.

And adding just <u>one</u> column and one row, again results in a singular block, as may be illustrated, by adding the second row and the second column, to obtain:

$$
\begin{array}{c}
\text{row 1} \\
\text{row 2} \\
\text{row 4} \\
\text{row 6}
\end{array}
\begin{bmatrix}
\overset{x_1}{-} & \overset{x_2}{-} & \overset{x_4}{-} & \overset{x_6}{2} \\
- & - & - & 4 \\
1 & 2 & - & - \\
- & - & 6 & -
\end{bmatrix}
$$

This block has a leading block-column of order 4 by 2, which is of rank 1.

The rank of the parent matrix is $3 \times 2 = 6$, and there will be only 2 groups of non-vanishing principal minors, which are either determinants of blocks of order 3 by 3, or of blocks of order 6 by 6.

Note also that the usual term for the determinant is entirely absent for $r \times q < n$. (In this example: $2 \times 3 = 6 < 7$.)

Blockpermutoids which contain non-square blocks have no non-vanishing determinant.

Theorem (blockpermutoid theorem)

The n by n non-negative and indecomposable matrix **A** is not also semi-positive,

if and only if:

● 1) **A** is a block-semipositive blockpermutoid (of block-order k),

and if so:

● 2) **A**^{k} blockdiagonalizes fully into k semi-positive diagonal blocks.

● 3) If k factorizes as k = q × r (q and r both integer),

> then **A**^{q} blockdiagonalizes into q diagonal blocks of block-order r, which are indecomposable and block-semipositive, if and only if r is a prime number (and are themselves blockdiagonal if r is not a prime number).

● 4) If k is a prime number, **A**^{t} is an indecomposable block-permutoid, whenever t is not equal to, or a multiple of k.

● 5) $[\mathbf{A}^t]_{jj} = [0]$ applies, whenever t is not an integer multiple of k.

Proof

Assume the theorem to be true for matrices of order 1 by 1, 2 by 2, ... n-1 by n-1.

We now discuss the following cases:

Case a) **A** has no non-vanishing principal minors.

Then **A** is an indecomposable permutoid (Corollary of the excess elements theorem).
q.e.d. ad 1) and 2) for case a), with k=n.

The characteristic equation of the indecomposable (non-negative) permutoid:

$$\lambda^n = |-P| > 0$$

has its roots on a complex circle, with the reciprocal of the order as the angle between separate roots, and, by de Moivre's Theorem, a qth power of the rth root becomes real and positive, if and only if q is an integer multiple of the angle r/k.

Therefore $\mathbf{A}^q$ has a repeated real root $\lambda^q = \lambda_r{}^q > 0$, and is decomposable, if and only if $q \times r$ is a multiple of, or equals k.
q.e.d. ad 4), for case a.

Application of the theorem itself, to any decomposable blocks of $\mathbf{A}^q$ and $\mathbf{A}^r$ confirms 3) by recursive induction, and since 5) is indeed true for t=1, by implication 5).

q.e.d. for case a.

Case b) $\mathbf{A}$ has at least one non-vanishing principal minor.

If that minor is semi-positive, then $\mathbf{A}$ is seen to be semi-positive, and no further proof is needed.

Otherwise, an assumption that the theorem is true for matrices of orders n-1 by n-1 and less, and therefore for the minors of $\mathbf{A}$, implies that that minor-matrix must be assumed to be a block-semipositive blockpermutoid, and, by the theorem, assumed to be true for the minor, a power of that minor blockdiagonalizes into semi-positive diagonal blocks. Since $[\mathbf{A}^p]_{jj} \geq [\mathbf{A}_{jj}]^p$ applies for all non-negative matrices, $[\mathbf{A}^p]_{jj} \gg [0]$ is implied.

If $\mathbf{A}^p$ is indecomposable, then $\mathbf{A}^p$, and by implication $\mathbf{A}$, is seen to be semi-positive, and no further proof is needed.

A power of a non-negative and indecomposable matrix is never blocktriangular ($\mathbf{A}^p\,\mathbf{x} = \mathbf{x}\,\lambda \gg [0]$ would be contradicted), therefore, if $\mathbf{A}^p$ is not semi-positive, then $\mathbf{A}$ is block-diagonal.

An assumption that the theorem is true for the indecomposable diagonal blocks of $\mathbf{A}^p$ therefore implies for some (not necessarily the same p, and we do not have to assume p=k), the block-diagonalization of $\mathbf{A}^p$ into diagonal blocks $[\mathbf{A}^p]_{jj} > [0]$.

The product of an an all-positive matrix and a non-negative vector is all-positive except when the vector is a null-vector. As this applies both for pre-multiplication and for post-multiplication, $\mathbf{A}_{ij} > [0]$ implies $[\mathbf{A}^{p+1}]_{ij} \gg [0]$.

By implication, $[\mathbf{A}^t]_{ij} > [0]$ implies $[\mathbf{A}^{p+t}]_{ij} \gg [0]$, and also: $[\mathbf{A}_t]_{ij} > [0]$, $[\mathbf{A}^r]_{jk} \gg [0]$ implies $[\mathbf{A}_t]_{ij}\,[\mathbf{A}^r]_{jk} \gg [0]$, in the same way as $a_{ij} > 0$, $a_{jk} > 0$ implies $a_{ij} \times a_{jk} > 0$.

This consideration allows us to generalise case a) as proved for the permutoid, to the block-permutoid.

The permutoid is the only indecomposable matrix for which a power A^p is diagonal (Corollary of the excess elements theorem).

By implication, the blockpermutoid is the only non-negative and indecomposable matrix, of which a power A^p is block-diagonal with square diagonal blocks and all the diagonal blocks meeting the condition $[A^p]_{jj} \gg [0]$, and 2) to 5) then also generalise from case a) to case b).

q.e.d.

Since semi-positive matrices have a strictly dominant Frobenius root $\lambda > 0$, we have the following

Corollary:

If the n by n non-negative and indecomposable matrix A (with a real root and vector $Ax = x\lambda$, $x \gg [0]$), has a root $\lambda_i \neq \lambda$, but $|\lambda_i| = \lambda$ (and therefore not semipositive), then A is a (block-semipositive) blockpermutoid matrix (of block-order q), and A^q block-diagonalizes into q diagonal semipositive blocks.

It obviously follows, that all powers p, for which A^p is block-diagonal with semi-positive diagonal blocks, are multiples of q, the block-order of the block-permutoid matrix A, including p=q.

Now that we have identified the one type of non-negative square matrix, which is indecomposable but not semi-positive, we can generalise one more property of the all-positive matrix to the non-negative and indecomposable matrix:

1h): For $\delta > \lambda$, $[I\delta - A]^{-1} \gg [0]$ and $[I - A\delta^{-1}] \gg [0]$ exist. (as long as A is non-negative and indecomposable, even in the $|\lambda_i| = \lambda$ case, where A is a blockpermutoid). (A proof by the power-expansion, arises immediately if A is semi-positive, and for the sum of q successive powers A^p, A^{p+1}, ... A^{p+q-1}, of an indecomposable block-permutoid, which add to a $\gg [0]$ expression, if A is a block-permutoid of block-order q.)

11.6 The vectors of decomposable non-negative matrices

We are now in a position to fill in some more detail on the properties of non-negative and decomposable matrices.

In general a decomposable matrix $\mathbf{A}$ partitions as:

$$\begin{bmatrix} \mathbf{A}_{11} & & \\ \mathbf{A}_{21} & \mathbf{A}_{22} & \\ \mathbf{A}_{31} & \mathbf{A}_{32} & \mathbf{A}_{33} \end{bmatrix} \qquad\qquad (11.6.1)$$

where the biggest real and positive root λ of $\mathbf{A}$ is also a root of of the indecomposable block $\mathbf{A}_{22}$, and where $\mathbf{A}_{11}$ and $\mathbf{A}_{33}$ can be decomposable, indecomposable, or of zero order, as the case may be.

Now let us first assume, that λ is a dominant root of $\mathbf{A}$, i.e. $\mathbf{A}_{11}$ and $\mathbf{A}_{33}$, if of non-zero order, only have roots, of which the absolute value is less than λ.

Then $\mathbf{A}$ has in any case the following non-negative vectors:

$$\mathbf{w'}_2 \mathbf{A}_{22} = \lambda \mathbf{w'}_2 \gg [0]'$$

completing conform

$$\mathbf{w'}_1 \mathbf{A}_{11} + \mathbf{w'}_2 \mathbf{A}_{21} = \lambda \mathbf{w'}_1 \qquad (\mathbf{w'}_3 = [0]') \qquad\qquad (11.6.2)$$

therefore:

$$\mathbf{w'}_1 = \mathbf{w'}_2 \mathbf{A}_{21} [\mathbf{I}\lambda - \mathbf{A}_{11}]^{-1} \geq [0] , \qquad \mathbf{w'}_3 = [0] \qquad\qquad (11.6.3)$$

If $\mathbf{A}$ is blocktriangular rather than blockdiagonal then, if $\mathbf{A}_{11}$ and therefore $\mathbf{A}_{21}$ is indeed of non-zero order, (and not a null matrix), then the $\geq$ in (11.6.3) needs to be replaced by the stronger $>$ (i.e. $\geq$, not $=$) form:

given $\mathbf{w'}_2 \gg [0]'$, and $[\mathbf{I}\lambda - \mathbf{A}_{11}]^{-1} \gg [0]$, $\mathbf{w'}_2 \mathbf{A}_{21} [\mathbf{I}\lambda - \mathbf{A}_{11}] = [0]'$ can apply, only for $\mathbf{A}_{21} = [0]$.

Similarly, for the column-vector $\mathbf{x}$:

$$\mathbf{A}_{22} \mathbf{x}_2 = \mathbf{x}_2 \lambda \gg [0]$$

completes conform:

$$\mathbf{A}_{32}\mathbf{x}_2 + \mathbf{A}_{33}\mathbf{x}_3 = \mathbf{x}_3\lambda \gg [0] \qquad (\mathbf{x}_1 = [0]) \qquad (11.6.4)$$

therefore:

$$\mathbf{x}_3 = [\mathbf{I}\lambda - \mathbf{A}_{33}]^{-1}\mathbf{A}_{32}\mathbf{x}_2 \geq [0] \qquad (\mathbf{x}_1 = [0]) \qquad (11.6.5)$$

Again, if $\mathbf{A}_{33}$ and therefore $\mathbf{A}_{32}$ is indeed of non-zero order, (and not a null matrix), then the $\geq$ in (11.6.5) needs to be replaced by a '>' sign.

Thus, there is, as in the case of an indecomposable matrix, a non-negative column-vector $\mathbf{x}$ and a corresponding non-negative row-vector $\mathbf{w}'$, which are not orthogonal to each other. As far as vectors associated with λ as root is concerned, we need look no further, given the assumption that λ is a distinct dominant root.

These vectors $\mathbf{x}$ and $\mathbf{w}'$ are not, however, the only non-negative vectors of a decomposable matrix $\mathbf{A}$. However, for other non-negative vectors, associated with the dominant positive roots of other blocks of $\mathbf{A}$, there isn't a pair of non-negative vectors, unless $\mathbf{A}$ is blockdiagonal.

If $\lambda_3 < \lambda$ is the biggest real and positive root of $\mathbf{A}_{33}$, then we will have:

$$\mathbf{A}_{33}\mathbf{u}_3 = \mathbf{u}_3\lambda_3 > [0] \quad (\mathbf{u}_1 = [0], \ \mathbf{u}_2 = [0])$$

(In the absence of an assumption of indecomposability of $\mathbf{A}_{33}$, $\mathbf{u}_3 \gg [0]$ may not be true.)

The corresponding row-vector needs to meet the requirements:

$$\mathbf{z}'_3\mathbf{A}_{33} = \lambda_3\mathbf{z}' > [0] \qquad (11.6.6)$$

and

$$\mathbf{z}'_2\mathbf{A}_{22} + \mathbf{z}'_3\mathbf{A}_{32} = \lambda_3\mathbf{z}'_2 \qquad (11.6.7)$$

Given $\mathbf{z}'_3 > [0]$, $\mathbf{z}'$ can only be orthogonal to $\mathbf{x}$, if $\mathbf{z}'_2$ contains negative non-zero elements.

If $\mathbf{A}_{33}$ is of zero order, the similar conclusion still applies to $\lambda_1 < \lambda$, $\mathbf{y}'_1\mathbf{A}_{11} = \lambda_1\mathbf{y}'_1 \gg [0]'$ and its relation to the

corresponding column vector, and the orthogonality of the
corresponding column vector to $\mathbf{w}'$.

We now come to discuss the case of co-equally dominant roots:

$$\lambda_1 = \lambda_2 = \lambda.$$

We immediately identify a row-vector and a column-vector:

$$\mathbf{w}'_1\mathbf{A}_{11} = \lambda\mathbf{w}'_1 > [0]' \qquad (\mathbf{w}'_2 = [0]', \quad \mathbf{w}'_3 = [0]') \qquad (11.6.8)$$

and

$$\left[\begin{array}{c|c} \mathbf{A}_{22} & \\ \hline \mathbf{A}_{32} & \mathbf{A}_{33} \end{array}\right] \left[\begin{array}{c} \mathbf{x}_2 \\ \mathbf{x}_3 \end{array}\right] = \left[\begin{array}{c} \mathbf{x}_2 \\ \mathbf{x}_3 \end{array}\right] \lambda \qquad (11.6.9)$$

These two vectors, which are not orthogonal to each other, are
also the only vectors associated with a twice-repeated root λ,
under those circumstances.

Theorem:

If the n by n non-negative matrix $\mathbf{A}$ is

1) lower blocktriangular, and

2) a root $\lambda_i=\lambda$ of the indecomposable diagonal block $\mathbf{A}_{ii}$ of $\mathbf{A}$ is
not smaller than the absolute value of another real root $\lambda_j = \lambda$
of $\mathbf{A}$, and

3) Neither $\mathbf{A}$, $\mathbf{A}'$, nor a diagonal block of $\mathbf{A}$ partitions as:

$$\mathbf{A} = \left[\begin{array}{c|c|c} \mathbf{A}_{11} & & \\ \hline \mathbf{A}_{21} & \mathbf{A}_{22} & \\ \hline \mathbf{A}_{31} & & \mathbf{A}_{33} \end{array}\right]$$

with λ being a root of two diagonal blocks $\mathbf{A}_{22}$ and $\mathbf{A}_{33}$ of $\mathbf{A}$,

then:

a) $\mathbf{Ax} = \mathbf{x}\lambda$ implies $\mathbf{x} > [0]$, $\mathbf{x}_j = [0]$, for $j < i$,
and $\mathbf{A}_{jj}\mathbf{x}_j = \mathbf{x}_j\lambda = \mathbf{A}_{jj}\mathbf{v}_j = \mathbf{v}_j\lambda \gg [0]$,
for the highest j for which $\mathbf{A}_{jj}\mathbf{v}_j = \mathbf{v}_j\lambda \gg [0]$
is true.

b) $\mathbf{Aw'} = \lambda\mathbf{w'}$ implies $\mathbf{w'} > [0]'$, $\mathbf{w'}_j = [0]'$, for $j > i$,
and $\mathbf{w'}_j\mathbf{A}_{jj} = \lambda\mathbf{w'}_j = \mathbf{z'}_j\mathbf{A}_{jj} = \lambda\mathbf{z'} \gg [0]'$,
for the lowest j for which $\mathbf{z'}_j\mathbf{A}_{jj} = \lambda\mathbf{z'} \gg [0]'$
is true.

c) $[\mathbf{I}\lambda-\mathbf{A}]$ is of rank n-1, regardless of multiplicity of λ as
root.

Proof:

We consider the following cases:

Case 1): λ is a distinct root of $\mathbf{A}$.
In that case, we may assume that $\mathbf{A}$ partitions conform
(11.6.1), with $\mathbf{A}_{22}$ being the one and only indecomposable diagonal
block of which λ is root.

In the special case that the block $\mathbf{A}_{11}$ of $\mathbf{A}$ partitioned conform
(11.6.1) is of zero order (therefore for $j < i$, all elements of
$\mathbf{x}_j$ are zeros on account of the lack of any of them), or if $\mathbf{A}_{33}$ is
of zero order, the same applies to $\mathbf{w'}_j$, for $j > i$. No further
proof is needed in these special cases.

Otherwise, if the partitioning is non-trivial, $\mathbf{Ax} = \mathbf{x}\lambda$ implies
$\mathbf{A}_{11}\mathbf{x}_1 = \mathbf{x}_1\lambda$.

Since, by assumption λ is not a root of $\mathbf{A}_{11}$, $\mathbf{x}_1 = [0]$ must be
assumed, and the vector $\mathbf{x}$ is: $\mathbf{x}_2 \gg [0]$, $\mathbf{x}_3 = [\mathbf{I}-\mathbf{A}_{33}]^{-1}\mathbf{x}_2 > [0]$.
q.e.d. at a), for case 1.

Similarly, $\mathbf{w'A} = \lambda\mathbf{w'}$ implies $\mathbf{w'}_3\mathbf{A}_{33} = \lambda\mathbf{w'}_3$, and therefore, given
that λ is not a root of $\mathbf{A}_{33}$, $\mathbf{w'}_3 = [0]'$, and the vector $\mathbf{w'} > [0]'$
is also determined conform the formulae earlier in this section.
q.e.d. ad b), for case 1.

Since c) has no content for case 1), we now conclude:
q.e.d. for case 1

Case 2): λ is a multiple root of **A**.

We now partition **A** as:

$$\mathbf{A} = \begin{bmatrix} \mathbf{A}_{11} & & & & \\ \mathbf{A}_{21} & \mathbf{A}_{22} & & & \\ \mathbf{A}_{31} & \mathbf{A}_{32} & \mathbf{A}_{33} & & \\ \mathbf{A}_{41} & \mathbf{A}_{42} & \mathbf{A}_{43} & \mathbf{A}_{44} & \\ \mathbf{A}_{51} & \mathbf{A}_{52} & \mathbf{A}_{53} & \mathbf{A}_{54} & \mathbf{A}_{55} \end{bmatrix} \qquad (11.6.10)$$

where λ is a root of the indecomposable diagonal blocks $\mathbf{A}_{22}$ and $\mathbf{A}_{44}$.

Proofs of $\mathbf{x}_1 = [0]$ and $\mathbf{w'}_5 = [0]'$ replicate from case 1, but they do not amount to a) and b), as a) also states $\mathbf{x}_2 = [0]$, $\mathbf{x}_3 = [0]$, and b) also states $\mathbf{w'}_3 = [0]'$ and $\mathbf{w'}_4 = [0]'$.

We now first of all introduce the provisional assumption, that the theorem is valid for matrices with 1, 2, ... k-1 roots $\lambda_i = \lambda$, where k is the number of repeats of λ as a root of **A**.

Therefore, by assumption, the theorem applies to the top lefthand and bottom righthand blocks

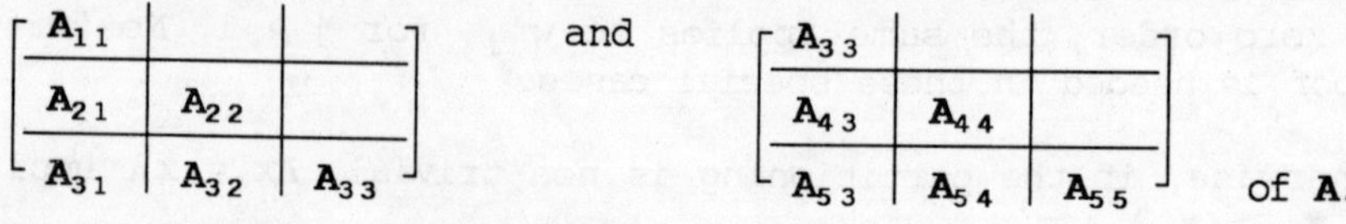

$$\begin{bmatrix} \mathbf{A}_{11} & & \\ \mathbf{A}_{21} & \mathbf{A}_{22} & \\ \mathbf{A}_{31} & \mathbf{A}_{32} & \mathbf{A}_{33} \end{bmatrix} \quad \text{and} \quad \begin{bmatrix} \mathbf{A}_{33} & & \\ \mathbf{A}_{43} & \mathbf{A}_{44} & \\ \mathbf{A}_{53} & \mathbf{A}_{54} & \mathbf{A}_{55} \end{bmatrix} \quad \text{of } \mathbf{A}.$$

Therefore, by assumption:

by property a), j=3 < i=4, $\mathbf{x}_3 = [0]$.
by property a), j=4, $\mathbf{x}_4 \gg [0]$.
by property b), j=3 > i=2, $\mathbf{w'}_3 = [0]'$.
by property b), j=2, $\mathbf{w'}_2 \gg [0]'$.

Furthermore, by b), applied to the bottom righthand block, $\mathbf{w'}_4 \neq [0]'$ implies $\mathbf{w'}_4 \gg [0]'$, and by a), applied to the top lefthand block, $\mathbf{x}_2 \neq [0]$ implies $\mathbf{x}_2 \gg [0]$.

Now suppose that $\mathbf{x}_2 \gg [0]$ were indeed true, instead of $\mathbf{x}_2 = [0]$ as the theorem states.

We consider the following requirement, extracted from $\mathbf{Ax} = \mathbf{x}\lambda$ and considering $\mathbf{x_1} = [0]$ and $\mathbf{x_3} = [0]$ as already proved:

$$\mathbf{A_{42}x_2} + \mathbf{A_{44}x_4} = \mathbf{x_4}\lambda \tag{11.6.11}$$

Pre-multiplication of (11.6.11) by $\mathbf{z'_4} \gg]0]'$, the rowvector of $\mathbf{A_{44}}$, associated with λ as the Frobenius-root of $\mathbf{A_{44}}$, results in:

$$\mathbf{z'_4}\,\mathbf{A_{42}x_2} + \mathbf{z'_4}\,\mathbf{A_{44}x_4} =$$

$$\mathbf{z'_4}\,\mathbf{A_{42}x_2} + \lambda\,\mathbf{z'_4}\,\mathbf{x_4} = \mathbf{z'4}\,\mathbf{x_4}\lambda \tag{11.6.12}$$

This requirement can only be met for $\mathbf{A_{42}} = [0]$, which was excluded by assumption under 3).

The proof for $\mathbf{w_4}' = [0]'$ is analogous, using $\mathbf{w'A} = \lambda\mathbf{w'}$ instead of $\mathbf{A\,x} = \mathbf{x}\lambda$, and $\mathbf{v_4} \gg [0]$.

Since a proof for k=1 was already supplied under case 1, the general proof now follows by recursive induction

q.e.d.

Example:

$$A = \begin{bmatrix} 1 & & & & & \\ - & 1 & 2 & & & \\ 1 & 2 & 1 & & & \\ - & - & - & - & 3 & \\ - & 5 & - & 3 & - & \\ 6 & - & - & 8 & - & 2 \end{bmatrix}$$

$\lambda=3$ (twice) $\quad \mathbf{x_3} = \begin{bmatrix} 1 \\ 1 \end{bmatrix} \quad \mathbf{x_4} = [8]$

$$\mathbf{x_1} = [0], \quad \mathbf{x_2} = [0].$$

$$\mathbf{w'} = [\ 1\ |\ 1\quad 1\ |\ -\quad -\ |\ -\ ]$$

and $\mathbf{A}$ has no other vectors associated with $\lambda=3$, than those which are proportional to the ones shown.

$\mathbf{w'_2} \neq [0]'$ or $\mathbf{w'_4} \neq [0]'$ cannot be true, because post-multiplication of:

$$\mathbf{w'}_2 \begin{bmatrix} 1 & 2 \\ 2 & 1 \end{bmatrix} + \mathbf{w'}_3 \begin{bmatrix} - & - \\ 5 & - \end{bmatrix} + w_4 \begin{bmatrix} - & - \end{bmatrix} = 3 \times \mathbf{w'}_2$$

by $\mathbf{v}_2 = \begin{bmatrix} 1 \\ 1 \end{bmatrix}$

Note, however, that the theorem does not apply to all block-triangular matrices, nor is the absence of a block-diagonal structure sufficient to be able to apply the theorem. The following contra-example illustrates this point:

$$\mathbf{A} = \begin{bmatrix} 1 & - & - & - & - \\ 1 & 2 & - & - & - \\ 1 & - & 1 & - & - \\ 1 & - & 2 & 2 & - \\ - & - & 3 & - & 1 \end{bmatrix}$$

This matrix is not block-diagonal, but its below-diagonal part is sufficiently sparsely filled, to permit two independent non-negative vectors to be associated with the repeated dominant root $\lambda = 2$: $\mathbf{x} = \mathbf{e}_2$ and $\mathbf{x} = \mathbf{e}_4$.

The third condition stated for the applicability of the theorem, is not fulfilled, as may be seen by partitioning $\mathbf{A}$ as:

$$\mathbf{A} = \left[\begin{array}{c|c|ccc} 1 & - & - & - & - \\ \hline 1 & 2 & - & - & - \\ \hline 1 & - & 1 & - & - \\ 1 & - & 2 & 2 & - \\ - & - & 3 & - & 1 \end{array} \right]$$

We finish this section (and the main body of the book), with rounding off the weaker form of the main theorem of this chapter, as it relates to decomposable matrices. In section 11.3 we stated: 'We do not state a weaker form of 4) at this stage', but the time has come to do that.

If there is a distinct real root λ, with corresponding vectors $\mathbf{x} > [0]$ and $\mathbf{w'} > [0]'$, i.e. there is an indecomposable diagonal block $\mathbf{A}_{rr}$, of which the biggest real root λ_r is greater than the similar roots of other blocks, then we will have:

$$\partial\lambda/\partial\mathbf{A}_{rr} \gg [0]$$

for the one indecomposable diagonal block of which λ is the root.

For other indecomposable diagonal blocks, we will have:

$$\partial\lambda_j/\partial\mathbf{A}_{jj} \gg [0],$$

but for $j\neq r$, λ_j is not the dominant positive root of $\mathbf{A}$ as a whole.

For the dominant positive root of the whole matrix,

$$\partial\lambda/\partial\mathbf{A}_{ij} = [0]$$

will apply for diagonal and off-diagonal blocks alike, except for the diagonal block of which λ is the root.

This follows both from section 8.10 on the differentiation of roots, bearing in mind what has been said above about the vectors, and indeed, from the simple fact that λ _is_ the root of one diagonal block.

A more complicated situation arises, if λ is the equal root of several diagonal blocks. In that case the meaning of a partial differential ratio of the dominant positive root with respect to any element of the matrix is ambiguous. An upward change in an element of any of the $\mathbf{A}_{ii}$ of which λ is a root, causes $\lambda_i=\lambda$ to become the distinct dominant positive root, but a downward change leaves the value of the biggest real root unaffected, as the same root $\lambda=\lambda_j$ is still a root of some other block $\mathbf{A}_{jj}$.

Answersheet on the limits of the Frobenius root

The rows of $\mathbf{A} = \begin{bmatrix} 1 & 2 & 5 \\ 2 & 2 & 1 \\ 3 & 2 & 4 \end{bmatrix}$

add to 8, 5, and 9.

This is an indecomposable (indeed an all-positive) matrix.

Therefore: $5 < \lambda < 9$.

The similar argument applied to $\mathbf{A}'$, and therefore to the column-sums of $\mathbf{A}$, which are 6, 6, and 10, leads us to infer:

 $6 < \lambda < 10$

Therefore, combining both requirements: $6 < \lambda < 9$.

REFERENCES

1   Aitken, A.C.:
On the Evaluation of Determinants, the Formation of their
Adjugates, and the Practical Solution of Simultaneous
Linear Equations. Proceedings of the Edinburgh
Mathematical Society (2), 3 (1932), pp. 207-219.

2   Aitken, Alexander C.:
Determinants and Matrices.
London/Edinburgh, Oliver & Boyd. (A 1958 reprint of the
1956 9th edition was consulted, first edition 1939)

3   Aitken, A.C., and Turnbull, H.W.:
An introduction to the theory of canonical matrices.
London, Blackie & Son, 1932

4   Albert, A.:
Regression and the Moore-Penrose Pseudo-inverse. Academic
Press, New York, 1972.

5   Ben Israel, A:
Generalized inverses, theory and applications. New York,
Wiley-Interscience, 1974.

6   Bodewig, E:
Matrix Calculus. Amsterdam, North Holland, 1959.

7   Clasen, B.J.:
Sur une nouvelle méthode de résolution des équations
linéaires et sur l'application de cette méthode au calcul
des déterminants.
Extrait des Annales de la Societé scientifique de
Bruxelles, 12e année, 1887-1888, Paris 1889, Gauthier-
Villars et fils.

8   Dantzig, G.B.:
Linear Programming and Extensions.
Princeton University Press, U.S.A., 1963.

9   Debreu, G., and Herstein, I.N.:
Non-Negative Square Matrices.
Econometrica, Vol 21, 1953, pp. 597-607.

10   El Hawary, M.E.:
Further comments on 'A note on the inversion of complex matrices. <u>IEEE Transactions on Automatic Control</u>, April 1975.

11   Forsythe, George, and Moler, Cleve B:
Computer Solution of Linear Algebraic Systems. Prentice Hall, Englewood Cliffs, N.J., U.S.A., 1967.

12   Frobenius, G:
Über Matrizen aus positiven Elementen.
Sitzungsberichte der königlich preuszische Akademie für Wissenschaften, Berlin, 1908, pp. 471-467.

13   Frobenius, G:
Über Matrizen aus nich negativen Elementen.
Sitzungsberichte der königlich preuszische Akademie für Wissenschaften, Berlin, 1912, pp. 456-477.

14   Gantmacher, F.R.:
The Theory of Matrices. New York, Chelsea Publishing Cy., 1959. (Russian original: Teoria Matriz, Moscow, 1954. -date from the German translation: Matritzenrechnung, Berlin, East Germany, 1958-)

15   Gauss, Carl Friedrich: Werke (Collected works).
Göttingen-Berlin, 1870-1929, 12 Vols. (The original dates of the two source-papers are given as 1810 and 1813)

16   Graham, Alexander:
Matrix Theory and Applications for Engineers and Mathematicians. Chichester, Ellis Horwood, 1979.

17   Graham, Alexander:
Nonnegative matrices and applicable topics in linear algebra. Chichester, Ellis Horwood, 1987.

18   Haavelmo, Trygve:
The probability approach in econometrics. <u>Econometrica</u>, supplement to Vol 12, July 1944.

19   Hawkes, H.E.:
The reduction of families of bilinear forms. <u>Amer. Journal of Mathematics</u> ,<u>32</u>, 1910, pp. 101-114

20   Hawkins, D, and Simon, H.A.:
      Some conditions of Macro-Economic Stability. <u>Econometrica</u>,
      <u>17</u> pp. 245-248.

21   Hawkins, F.M., and Hawkins, J.Q:
      Complex numbers and elementary complex functions.
      London, Mc.Donald & Co., 1968.

22   Heesterman, A.R.G:
      Special Simplex Algorithm for Multi-sector problems.
      <u>Numerische Mathematik</u> <u>12</u> (1968), pp 288-306.

23   Heesterman, A.R.G:
      Forecasting Models for National Economic Planning.
      Reidel, Dordrecht, Netherlands, 1970, 1972.

24   Heesterman, A.R.G.:
      The dynaminc (in)stability of the dynamic Input-Output
      model. <u>Economic systems Research</u>, <u>forthcoming</u>.

25   Hohn, Franz E.:
      Elementary Matrix Algebra. Collier Macmillan, 1973.

26   Householder, Alston S.:
      The Theory of Matrices in Numerical Analysis.
      Ginn/Blaisdell, Boston, Mass. U.S.A., 1974.

27   Jordan, Camille:
      Sur les transformations d'une forme quadratique en elle
      même. (J. de Math., (4) t. V, 1879, pp. 345-379), as
      reproduced in Jordans collected works (Paris, Gauthier-
      Villars, 1962)

28   Klein, L.R, and Goldberger, A.S.:
      Impact multipliers and dynamic properties of the Klein-
      Goldberger model. Amsterdam, North Holland Publishing Co,
      1959.

29   Koecher, Max:
      Lineare Algebra und analytische Geometrie. Springer,
      Berlin/Heidelberg, 1983

30　　Maurer, S:
Pivotal Theory of Determinants.
<u>in</u>: Balinski, M.L., <u>editor</u>: Pivoting and Extensions: In
Honour of A.W. Tucker. Amsterdam, North Holland, 1974.

31　　Mehmke, R:
Praktische Lösung der Grundaufgaben über Determinanten,
Matrizen und lineare Transformationen.
<u>Mathematische Annalen</u> <u>103</u> (1930) pp. 300-318.

32　　Mehmke, R:
Über die zweckmäszigste Art, lineare Gleichungen durch
Elimination aufzulösen. <u>Zeitschrift für angewandte
Mathematik und Mechanik</u>, <u>10</u>, (1930), pp. 508-514.

33　　Minc, Henryk:
Nonnegative Matrices, Wiley 1988

34　　Moore, E.H.:
On the reciprocal of the general algebraic matrix. <u>Bull of
the American Mathematical Society</u>, <u>26</u> (1920), pp. 394-395.

35　　Muir, Thomas:
The theory of determinants in the historical order of
development (Second edition consulted). London, Macmillan &
co. 1906.

36　　Penrose R.:
A generalized inverse for matrices <u>Proceedings of the
Cambridge Philosophical Society</u> <u>51</u> (1955), pp 406-413.

37　　Pullman, N.J.:
Matrix Theory and its applications. New York/Basel, Marcel
Dekker, 1976

38　　Rao, C.R.:
Generalized inverses for matrices and its application in
mathematical statistics. <u>in</u>: Research papers <u>in</u> Statistics:
Festschrift for J.Neyman, New York, Wiley, 1966.

39　　Rao, C.R.:
Calculus of generalized inverse of matrices, <u>Sankhya</u>,
Ser A, <u>29</u> (1967), pp 317-342.

40   Rao, C.R.:
     Generalized Inverse of Matrices and its Applications. New
     York, John Wiley, 1971.

41   Schreier, O, and Sperner: E.:
     Modern algebra an matrix theory. New York, Chelsey
     Publishing Cy., 1951, 1955, 1959

42   Spiegel, Murray R:
     Complex variables. Mc Graw Hill, 1974.

43   Stewart, G.W.:
     Introduction to Matrix Computations. Academic Press, 1973.

44   Stone, R.:
     Linear Expenditure Systems and Demand Analysis: An
     application to the pattern of British demand. Economic
     Journal 64, (Sept 1954), pp. 511-527.

45   Strang, Gilbert:
     Linear Algebra and its applications. New York, 1976

46   Theil, Henri:
     Principles of Econometrics. John Wiley, 1971

47   Vandermonde (No initials given by Muir [35], VOL I, p. 17)
     Mémoire sur l'élimination. in: Histoire de l'Academie
     Royale des Sciences, Ann. 1772, Vol 2, pp. 516-532)

48   Wielandt, H.:
     Unzerlegbare, nicht negative Matrizen.
     Mathematische Zeitschrift, Vol 52, March 1950, pp. 642-648.

49   Wilde, Carroll:
     Linear Algebra. New York, Addison Wesley, 1987

INDEXES

The following details of the software used, and their implications are mentioned here.

There is a rounding inaccuracy between the pagination of the
physical printing, and the indexing branches of the program,
which does not interpret graphics characters. This has been
consistently corrected by after-checking, only insofar as the
difference is at least two full lines.

In addition, to catch word-parts separated by '-' at the end of a
line indexing is over combinations of two lines. It is therefore
possible for a word to be reported on page x, but in fact to be
on the bottom line, or even one line before the bottom of the
previous page x-1.

As computer-indexing is strictly by character-strings, not
meaning, one will sometimes find the same idea indicated by
several different character strings.

The '&' means the listed words occurring in any order in two
successive lines, whereas the '*' means any ending.

INDEX OF AUTHORS CITED

SUBJECT INDEX,
INCLUDING AUTHORS WITHOUT SOURCE-REFERENCE